Springer Proceedings in Physics 81

Springer
Tokyo
Berlin
Heidelberg
New York
Barcelona
Budapest
Hong Kong
London
Milan
Paris
Santa Clara
Singapore

Springer Proceedings in Physics

Managing Editor: H. K. V. Lotsch

46 *Cellular Automata and Modeling of Complex Physical Systems*
Editors: P. Manneville, N. Boccara, G. Y. Vichniac, and R. Bidaux

47 *Number Theory and Physics*
Editors: J.-M. Luck, P. Moussa, and M. Waldschmidt

48 *Many-Atom Interactions in Solids*
Editors: R. M. Nieminen, M. J. Puska, and M. J. Manninen

49 *Ultrafast Phenomena in Spectroscopy*
Editors: E. Klose and B. Wilhelmi

50 *Magnetic Properties of Low-Dimensional Systems II: New Developments*
Editors: L. M. Falicov, F. Mejía-Lira, and J. L. Morán-López

51 *The Physics and Chemistry of Organic Superconductors*
Editors: G. Saito and S. Kagoshima

52 *Dynamics and Patterns in Complex Fluids: New Aspects of the Physics–Chemistry Interface*
Editors: A. Onuki and K. Kawasaki

53 *Computer Simulation Studies in Condensed-Matter Physics III*
Editors: D. P. Landau, K. K. Mon, and H.-B. Schüttler

54 *Polycrystalline Semiconductors II*
Editors: J. H. Werner and H. P. Strunk

55 *Nonlinear Dynamics and Quantum Phenomena in Optical Systems*
Editors: R. Vilaseca and R. Corbalán

56 *Amorphous and Crystalline Silicon Carbide III, and Other Group IV-IV Materials*
Editors: G. L. Harris, M. G. Spencer, and C. Y. Yang

57 *Evolutionary Trends in the Physical Sciences*
Editors: M. Suzuki and R. Kubo

58 *New Trends in Nuclear Collective Dynamics*
Editors: Y. Abe, H. Horiuchi, and K. Matsuyanagi

59 *Exotic Atoms in Condensed Matter*
Editors: G. Benedek and H. Schneuwly

60 *The Physics and Chemistry of Oxide Superconductors*
Editors: Y. Iye and H. Yasuoka

61 *Surface X-Ray and Neutron Scattering*
Editors: H. Zabel and I. K. Robinson

62 *Surface Science: Lectures on Basic Concepts and Applications*
Editors: F. A. Ponce and M. Cardona

63 *Coherent Raman Spectroscopy: Recent Advances*
Editors: G. Marowsky and V. V. Smirnov

64 *Superconducting Devices and Their Applications*
Editors: H. Koch and H. Lübbig

65 *Present and Future of High-Energy Physics*
Editors. K.-I. Aoki and M. Kobayashi

66 *The Structure and Conformation of Amphiphilic Membranes*
Editors: R. Lipowsky, D. Richter, and K. Kremer

67 *Nonlinearity with Disorder*
Editors: F. Abdullaev, A. R. Bishop, and S. Pnevmatikos

68 *Time-Resolved Vibrational Spectroscopy V*
Editor: H. Takahashi

69 *Evolution of Dynamical Structures in Complex Systems*
Editors: R. Friedrich and A. Wunderlin

70 *Computational Approaches in Condensed-Matter Physics*
Editors: S. Miyashita, M. Imada, and H. Takayama

71 *Amorphous and Crystalline Silicon Carbide IV*
Editors: C. Y. Yang, M. M. Rahman, and G. L. Harris

72 *Computer Simulation Studies in Condensed-Matter Physics IV*
Editors: D. P. Landau, K. K. Mon, and H.-B. Schüttler

73 *Surface Science: Principles and Applications*
Editors: R. F. Howe, R. N. Lamb, and K. Wandelt

74 *Time-Resolved Vibrational Spectroscopy VI*
Editors: A. Lau, F. Siebert, and W. Werncke

75 *Computer Simulation Studies in Condensed-Matter Physics V*
Editors: D. P. Landau, K. K. Mon, and H.-B. Schüttler

76 *Computer Simulation Studies in Condensed-Matter Physics VI*
Editors: D. P. Landau, K. K. Mon, and H.-B. Schüttler

77 *Quantum Optics VI*
Editors: D. F. Walls and J. D. Harvey

78 *Computer Simulation Studies in Condensed-Matter Physics VII*
Editors: D. P. Landau, K. K. Mon, and H.-B. Schüttler

79 *Nonlinear Dynamics and Pattern Formation in Semiconductors and Devices*
Editor: F.-J. Niedernostheide

80 *Computer Simulation Studies in Condensed-Matter Physics VIII*
Editors: D. P. Landau, K. K. Mon, and H.-B. Schüttler

81 *Materials and Measurements in Molecular Electronics*
Editors: K. Kajimura and S. Kuroda

Volumes 1–45 are listed at the end of the book

K. Kajimura S. Kuroda (Eds.)

Materials and Measurements in Molecular Electronics

Proceedings of the International Symposium on Materials and Measurements in Molecular Electronics
Tsukuba, Japan, February 6–8, 1996

With 192 Figures

Springer

Professor, Dr. Koji Kajimura
Electrotechnical Laboratory, and University of Tsukuba
Tsukuba, Ibaraki, 305 Japan

Dr. Shin-ichi Kuroda
Electrotechnical Laboratory
Tsukuba, Ibaraki, 305 Japan

ISBN 4-431-70185-0 Springer-Verlag Tokyo Berlin Heidelberg New York

Library of Congress Cataloging-in-Publication Data. International Symposium on Materials and Measurements in Molecular Electronics (1996 : Tsukuba-shi, Japan). Materials and measurements in molecular electronics : proceedings of the International Symposium on Materials and Measurements in Molecular Electronics, Tsukuba, Japan, February 6–8, 1996 / K. Kajimura, S. Kuroda, (eds.). p. cm. — (Springer proceedings in physics; v. 81) Includes bibliographical references. ISBN 4-431-70185-0. 1. Molecular electronics—Materials—Congresses. I. Kajimura, K. (Koji), 1943– . II. Kuroda, S. (Shin-ichi), 1950– III. Title. IV. Series. TK7874.8.I56 1996 620.1'1299—dc20 96-18854 CIP

Printed in Hong Kong

Typesetting: Camera ready copy from the authors/editors
SPIN 10539946 Printed on acid-free paper

Preface

We present here the proceedings of the International Symposium on Materials and Measurements in Molecular Electronics (M^3E '96) held February 6–8, 1996, in Tsukuba, Japan, under the cosponsorship of the Electrotechnical Laboratory and the Foundation Advanced Technology Institute. More than 110 participants, including eight invited speakers, from seven oversea countries, attended the symposium. The aim of the symposium was to provide an opportunity for the discussion of research frontiers in molecular electronics, with emphasis on the development of new molecular materials and measuring techniques based on modern spectroscopy.

Molecular electronics is regarded as one of the key technologies that may bring innovation to electronics in the 21st century. Many new molecular materials such as conjugated polymers and carbon clusters have been synthesized or discovered in the past two decades, and some, including organic electroluminescent devices, are now approaching the stage of commercial application. In the course of the development of molecular materials, detailed knowledge of the structures and electronic states of molecular aggregates is indispensable. It is also important to understand how to control the aggregation of molecules. With this perspective, the symposium was organized to promote the exchange of ideas among scientists involved in both the development of materials or devices and their characterizations. The lectures consisted of invited talks by 21 members of the international research community who are actively involved in such relevant fields as modern spectroscopy, Langmuir–Blodgett films, cluster materials, organic conductors, and conjugated electroluminescent polymers.

The chapters contained in this volume cover a rather wide range of subjects, but we believe that they represent vital aspects of research frontiers in materials science. We acknowledge all the authors for their contributions.

We are grateful to the members of the organizing committee, Prof. J. Isoya, Prof. M. Izumi, Dr. T. Ohnishi, Dr. K. Tanigaki, and Mr. T. Uchiyama for their collaboration. Support from the Physical Society of Japan is gratefully acknowledged. Finally, we thank the members of the secretariat of the symposium, Dr. K. Murata, Mr. T. Seya, and Ms. N. Takahashi, for their excellent work.

Tsukuba
March 1996

Koji Kajimura
Shin-ichi Kuroda

Contents

Part III Langmuir–Blodgett Films

Part IV Cluster Materials

Part V Organic Conductors

Part VI Conjugated Electroluminescent Polymers

Part I

Presymposium Lecture

Wide-gap semiconducting materials superior to silicon

K.Kajimura, S.Hara, K.Hayashi*, and H.Okushi
Electrotechnical Laboratory, 1-1-4 Umezono, Tsukuba, Ibaraki 305, Japan

Abstract

Fundamental material research on super-silicon materials with wide band gaps is described in this report. Silicon carbide and diamond are important wide band-gap materials because of their high performances of endurance under high temperature and low conductivities with high avalanche breakdown voltages. In such new materials, crystal hardness results in low crystallinity and difficult incorporation of impurity doping. To overcome the main issue, utilization of advanced silicon material technology is essential. We have grown homo-epitaxial diamond with atomically flat surfaces by clean epitaxy with controlled plasma. Also, using ideal techniques to terminate a surface by hydrogen, established on Si(111) surface, we have reduced interfacial electronic states density at metals/6H-SiC(0001) interfaces enough to generate a flat band, causing a pinning-free interface, by which we are able to control Schottky barrier heights and to achieve ideal Ohmic contacts with zero barrier heights.

1. Introduction

Silicon-based electronics has been progressing in the last 4 decades mainly supported by the technology of device integration, which is based on continuous developments of crystal growth, device process, and device design. The frameworks in the technology, however, have already been established, giving rise to complex and precise techniques. This technological accumulation leads to present predictable steady developments with high costs but no breakthrough for higher integration and performances in silicon based devices. In such a saturating stage, the silicon technology has rather great advantages when it is applied to other semiconducting materials which has better physical properties than silicon. In particular, it is valuable to apply it to semiconducting materials with wide band gaps because they now need material improvements and forth-coming device techniques specific to them.

Using the wide gap materials, we will enlarge the field of electronics out of the Si based one because they involve superior performances for high power and high frequency devices owing to their high insulating performances and mechanical hardness. In figure 1, a perspective in the early stage of the next century using the materials is shown. There will occur three new application fields of the electronics, *i.e.*, energy electronics, information electronics, and ultimate environmental electronics when the technology using the materials grows up enough. In the energy electronics, new electronic devices with high power, high frequency, and low loss innovate power transmission system,

*On leave of absence from Kobe Steel, LTD.

Springer Proceedings in Physics, Vol. 81
Materials and Measurements in Molecular Electronics
Editors: K. Kajimura · S. Kuroda

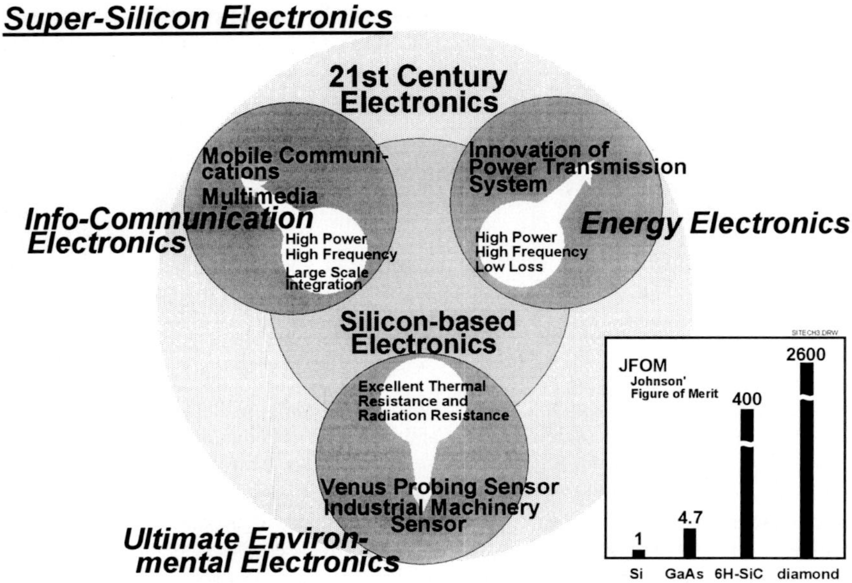

Figure 1. Super-silicon electronics and advanced semiconductors required for it.

which saves a large amount of energy loss occurred in the present system. In the info-communication electronics, the high power and high frequency operation in addition to the large scale integration promote mobile communications and multimedia. In the ultimate environmental electronics, the excellent thermal and radiation resistance of the wide gap materials is applicable to various sensors like industrial machinery sensor and Venus probing sensor and surrounding drive devices.

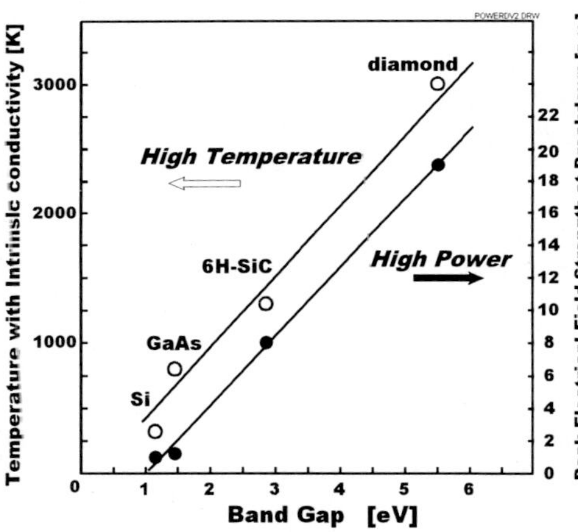

Figure 2. Performances for high temperature and high power in Si and wide band gap materials

Johnson' figure of merit (JFOM) to indicate a material performance in terms of high power and high frequency is also shown in Fig.1. JFOM is defined as

$$\mathrm{JFOM} = \frac{E_M^{\,2} \upsilon_s^{\,2}}{4\pi^2}, \qquad (1)$$

where E_M is the peak electric field strength at breakdown, which relates to the band gap width E_g, and υ_s is the scattering limited saturated velocity of carriers in the semiconductor material. JFOM relates to the frequency and power product of a semiconductor transistor[1]. When JFOM is normalized by Si, those for GaAs,

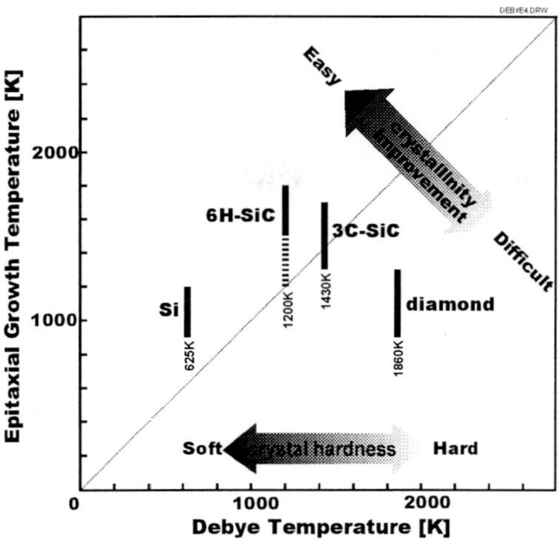

Figure 3 Debye and epitaxial growth temperatures

6H-SiC, and diamond are 4.7, 400, and 2600, respectively, indicating an extraordinary superiority of SiC and diamond over Si. Figure 2 shows a more undamental material superiority. The absolute temperature above which the semiconductor becomes intrinsic in the figure for Si, GaAs, 6H-SiC, and diamond are 600, 900, 1300, and 3000K, respectively. None of functions in Si devices acts at high temperatures over 600K since all *p*- and *n*-type regions change into intrinsic while 6H-SiC and diamond work as each type still at the temperature range. Also, the peak electric field strength at breakdown E_M itself has larger values for 6H-SiC and diamond than that of Si. E_M is defined as

$$N_B = \frac{\varepsilon_S E_M{}^2}{2qV_B}, \quad (2)$$

where N_B is the doping density in drift region and V_B is the avalanche breakdown voltage[2]. To obtain a larger N_B, which results in a lower conductivity, larger E_M is required. For high power devices, large E_M for 6H-SiC and diamond are essential.

To realize devices using SiC and diamond, systematic research to overcome a common issue among the wide semiconducting gap materials is required. Figure 3 shows a relation between Debye temperature which indicates crystal hardness specific to the material and a typical temperature range of epitaxial growth. In the regime that the epitaxial temperature is higher than the Debye temperature, which is in the upper left half area of Fig.3, surface atoms migrate intensively by searching more stable sites during the epitaxy, leading to a better crystallinity. In the other regime, surface atoms tend to stick to each adsorption site on the surface, resulting in a lower crystallinity. This requires more difficult techniques to improve the crystallinity. It is predictable that 3C-SiC and diamond needs advanced techniques to fabricate their better crystals because of the lower epitaxial temperatures. Actually, diamond, which is easily translated to graphite phase in the former regime, needs assistance of plasma which would give energy to epitaxial surface species in the latter regime.

The relation between the Debye temperature and the surface migration length dominating the atomic surface morphology is shown in Fig.4. On Si(001), wide terraces with an average length of around 50nm are observed by the scanning tunneling spectroscopy(STM). On 3C-SiC(001) surface whose crystals are heteroepitaxially grown

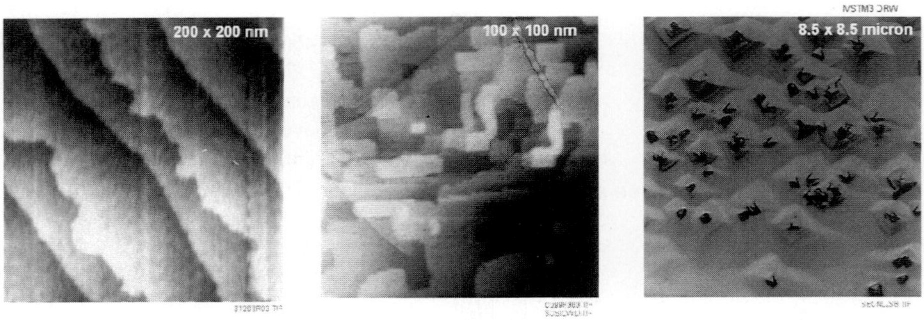

(a) Si(001) **(b) 3C-SiC(001)** **(c) diamond(001)**

Figure 4. Surface morphologies of (a) Si(001), (b)3C-SiC(001), and (c)diamond(001) surfaces.

on the same Si(001) substrate as that shown in the former STM image, intricate and dense steps are observed by STM [Fig.4(b)]. This is derived from the short surface migration length of the 3C-SiC surface. Further, on diamond(001) surface, misoriented nuclei are observed in the optical microscope image [Fig.4(c)]. Such nuclei are generated by the short surface migration length. The trend of surface migration length observed from the three images in the figure are consistent with the crystallinities predicted from Fig.3. Since the wide band-gap materials have high Debye temperatures, improvements of the crystallinities by the formation of flat epitaxial surfaces with wide terraces is the common issue.

In general, fundamental researches of a new semiconducting material for device fabrication start from bulk growth, followed by epitaxial growth in cooperation with surface investigations, and further reach interfacial investigations with metals and oxides. Diamond is in the research stage of epitaxial growth and surface study because of the low epitaxial crystallinity. An actual main subject is to enhance the surface atom migration. As mentioned earlier, the introduction of the Si material technology to the new material research has advantages to make them efficient. We introduce concepts of clean epitaxial reactions and controlled plasma reaction, which are those of the Si material technology. For diagnostics of grown films, we analyze surfaces of the films by reflection high energy electron diffraction (RHEED) , scanning electron microscopy, and atomic force microscopy (AFM).

In SiC research, interface investigation is a central issue because bulk 6H single crystal is commercially available and many investigations on epitaxy and the surface analysis have already been reported[3-8]. Since a wide band-gap semiconductor has a stronger insulating property, the conductivity of an interface between SiC and a metal as well as the bulk conductivity is lower than that of Si because of the formation of a higher Schottky barrier due to their wide band gap. So, a conductive electrode, always referred to as an Ohmic contact, is difficult to achieve. In addition to the difficulty, no unified and systematical way of formations for Schottky and Ohmic interfaces for all semiconductors have been provided historically. We introduce up-to-date techniques established in Si surface research to flatten SiC surfaces atomically to unify the phenomena of the Schottky and Ohmic interfaces.

2. Diamond

2.1 Diamond as a semiconductor

Diamond is an attractive semiconducting material which has superior physical and electrical properties, such as wide band gap, high mobilities of electrons and holes, small dielectric constant, high breakdown voltage, high thermal conductivity, and excellent radiation hardness. Hence, diamond has been recognized as a suitable material for electronic devices which can operate at high temperatures and/or in chemically harsh environments as well as for high-frequency and high-power electronic devices. Recently, another application field of diamond arises from its unique surface properties: negative electron affinity, which appears on hydrogen terminated diamond surfaces, making these materials candidates for flat panel displays. Successful operation of active electronic devices at high temperatures [9] indicates the potential of this material; however, electronic properties of the diamond are not sufficient (Hall mobility of over 1000 $cm^2/V{\cdot}s$ has been recently reported [10], but it is still lower than that of world best natural semiconducting diamond) and not well understood at present. Further, many technological issues such as heteroepitaxial growth [11] and doping control still remain to be established. In particular, control of *n*-type conduction is essential for the development of electronic devices and must be resolved based on the fundamental understanding of diamond in the material aspect.

Achievement of a device-quality single-crystal diamond is the key factor to realized future electronic devices based on diamond. The following conditions are required for this purpose: (1) the applicability to various device structures, which makes it possible to obtain advanced device structures; (2) atomically flat surfaces, which improve junction properties of each contact as well as allow to high integration of these devices; and (3) low defects density and low residual impurities. Chemical vapor deposition (CVD), which was developed by a Japanese researcher group in 1981 [12], is a promising technique for diamond applicable to electronic devices. Attempts to obtain high quality homoepitaxial diamond films on (001) substrates have been made using both plasma enhanced CVD [13-20] and hot-filament assisted CVD [21-23].

Step-flow epitaxy observed in conventional semiconductors such as Si and GaAs is thought to be ideal. Recently, step-flow growth on vicinal (001) diamond substrates have been reported [14,16,18-20]; however, the presence of misoriented secondary crystals and the formation of hillocks have been generally observed for the low misorientation angle substrates [13,17,21-23] (see Fig. 4). Further, the main interest of these studies were focused on the morphology of the diamond films obtained and only little attention has been paid to other characteristics such as electrical properties and residual impurities.

In order to overcome this situation and to obtain high quality diamond films, two concepts were introduced. One is the "clean epitaxy", and the other is the "precise control of growth parameters", in particular, substrate temperature control independent from plasma power, which is difficult for conventional CVD reactors due to its configuration. This is because the surface migration of precursors and the nucleation are generally affected by impurity atoms on the surface as well as by substrate temperature. In the present study, we have tried the homoepitaxial growth of diamond films using an end-launch type microwave plasma CVD reactor [24] consisting of a 6 in. i.d. stainless-steel chamber [18-20]. Using this reactor, we can control the substrate temperature independently of the plasma power and keep plasma away from the chamber wall to reduce bombardment which results in unintentional doping of impurities. The procedure is also expected to realize low base pressure. Figure 5 shows a schematic diagram of the diamond deposition system. The microwave energy is launched into the growth chamber in a direction normal to the substrates surface through the quartz window. The substrate was set on a susceptor and heated inductively with a 60 kHz rf generator independently of plasma power. Diamond films were deposited using 0.5 % CH_4 diluted by H_2 gas and the substrate temperature was maintained at 800 °C as measured by a thermocouple attached to the backside of the susceptor. The gas pressure, the total gas flow rate, and the microwave power were 25 Torr, 400 sccm, and 750 W, respectively. The substrates used in this study were synthetic Ib diamond (001) substrates (4.0×4.0×0.3 mm^3) with misorientation angles of less than 3°. The deposition duration was typically 6 h and resulting film thickness was approximately 2.4 μm.

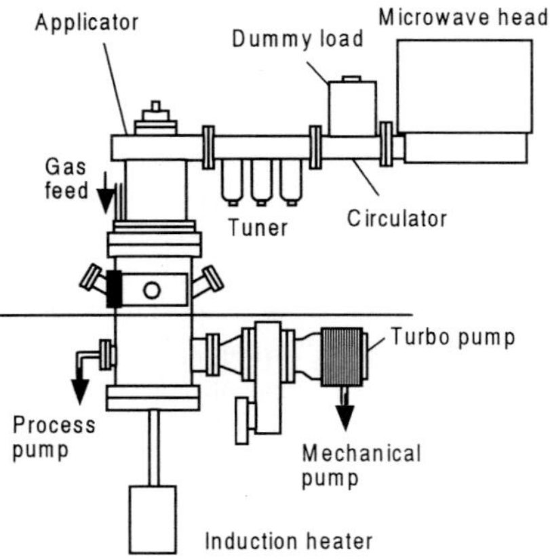

Figure 5. Schematic diagram of the diamond deposition system.

2.2 Step-flow growth and etching of diamond

In the case of epitaxial growth, the initial condition of the substrate surface including surface roughness and impurities on the surface, gives a great influence on the resulting surface. Because of the limitation of the polishing technique, however, commercially available single-crystal diamond substrates with reasonable cost possess polishing marks, which are visible in several analytical technique such as optical microscope, secondary electron microscopy, and AFM. In order to diminish substantially the surface roughness, we have tried the atomic scale control of diamond surfaces by hydrogen plasma treatment.

Figure 6 (a) shows a typical AFM image of an as-received diamond substrate. The corrugated structure with features about 5 nm high which was formed by the polishing process can be apparently observed. After 1 hour hydrogen plasma treatment, as shown in Fig. 6 (b), the corrugated structure observed before the treatment almost has disappeared and the surface of the film is found to be atomically flat within the vertical range of 1 nm. Further, the surface consists of successive terraces having almost the same width, and atomic steps of several atoms high running over the surface parallel to [110] direction were clearly observed. This indicates that the etching of the (001) diamond surface occurred dominantly by step-flow rather than by attacking the terrace.

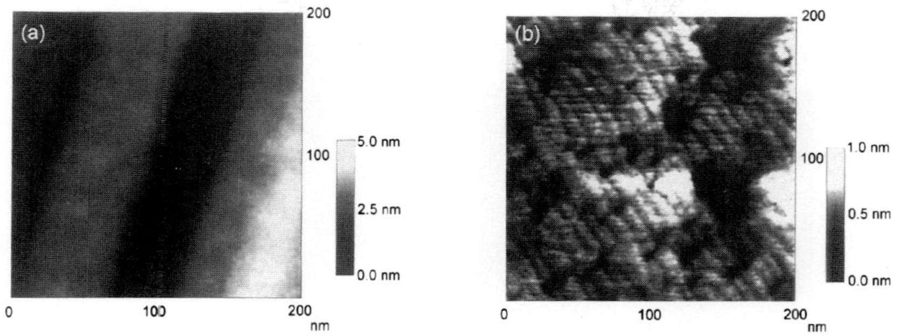

Figure 6. AFM images of (a) typical commercially-available diamond substrate (b) the diamond substrate after hydrogen plasma treatment on the area of 200 nm × 200 nm. Vertical ranges in the scan area for (a) and (b) are 5 nm and 1 nm, respectively.

Next, the epitaxial diamond films was grown on these plasma treated substrates. A typical AFM image of the 200 nm thick epitaxial films is shown in Fig. 7. The morphology corresponding to the multi atomic steps is again clearly observed. The average terrace width estimated from this image is approximately 8 nm, which corresponds to a misorientation angle of around 1°. This image indicates that the incorporation of precursors in atomic steps (step-flow growth) was dominant during the deposition on the vicinal substrates with a misorientation angle of around 1°, suggesting that the migration length of precursors under the present plasma growth conditions is longer than that previously reported.

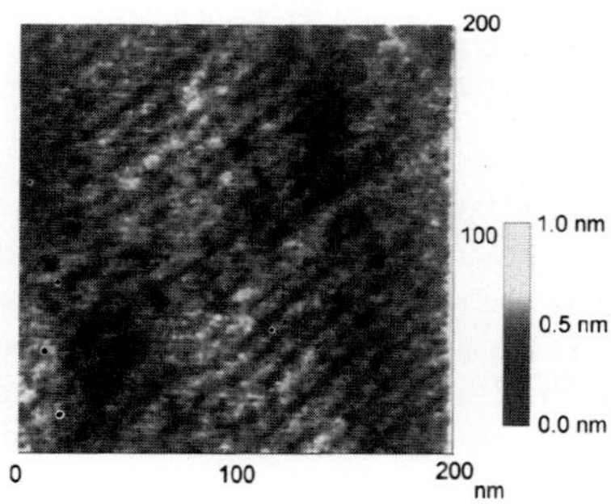

Figure 7. AFM image of the homoepitaxial films grown on hydrogen-plasma treated substrate.

2.3 Characterization of step-flow grown diamond films

As described in the pervious section, we have achieved the epitaxial growth of diamond films by step-flow mode, which is expected to produce higher quality films than previous cases. This is because, in this mode, the crystal growth takes place at the most stable point: atomic steps, rather than on the terraces. In this section, we show characterization of the step-flow grown diamond films in terms of the surface morphology and the film crystallinity. Secondary ion mass spectroscopy analysis was also performed to investigate the impurities in the films. Additionally, Al-Schottky barrier properties of these films were examined in order to evaluate the performances of the films in the functional aspect: the electronic properties of the diamond film.

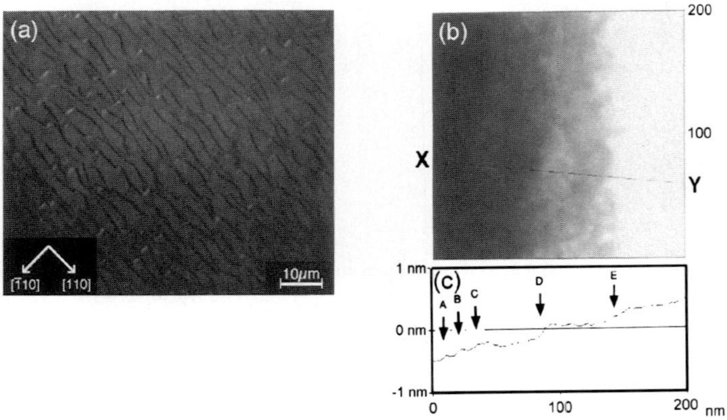

Figure 8. (a) Optical microscope image of surface morphology of typical 2.4 μm thick homoepitaxial diamond films. (b) AFM image of flat region of homoepitaxial diamond film on area of 200 nm × 200 nm. (c) Contour between X and Y in (b).

A typical optical microscope image of the 2.4 μm thick epitaxial films is shown in Fig. 8 (a). Macroscopic steps are clearly observed running parallel to [110] direction. It should be noted that the entire sample surface was covered with these macroscopic steps and no misoriented secondary crystals as shown in Fig. 4 were observed. The average width of the flat region was approximately 3 μm and the average height of the steps was approximately 0.05 μm. The misorientation angle of the substrate is estimated to be around 1° from this result, which is consistent with that obtained by X-ray analysis.

Figure 8(b) shows an AFM image of the flat region on an area of 200 nm × 200 nm. This image indicates that the diamond surface grown in the present reactor is atomically flat within the vertical range of 2 nm in the scan area. Contours between X and Y are shown in Fig. 8(c). The morphology which is thought to correspond to single atomic steps [for A – C as indicated by arrows in Fig. 8(c)] and double atomic steps (for D, E) can be observed. This result indicates that the obtained surfaces between macroscopic steps are extremely smooth. Thus, the morphologies observed in both optical microscope and AFM images suggest that the incorporation of precursors onto the atomic steps (*i.e.*, step-flow growth) rather than two-dimensional nucleation on the terrace was dominant during the 6 hour deposition.

The crystallinity of the deposited film was examined by Raman spectroscopy and RHEED. In the Raman spectra as shown in Fig.9, only 1332 cm^{-1} peak, corresponding to sp^3 bond, was observed and the full width at half-maximum of the film was as narrow as that of substrates used. RHEED patterns taken from [100] and [110] azimuths with an accelerating voltage of 40 kV shown in Fig.10 are narrow and streaky with strong Kikuchi bands. The half-order and first-order Laue rings are also observed. These patterns are consistent with the well known 2×1 and 1×2 double-domain structure which has been observed previously [25]. These narrow streaky patterns indicate that the surface of the epitaxial layer is smooth over a relatively wide area.

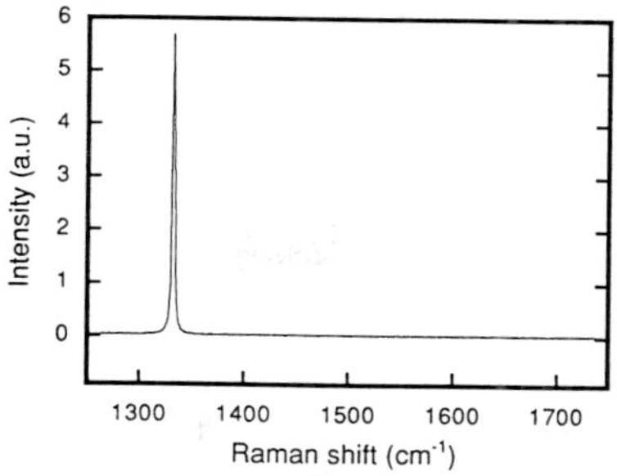

Figure 9. Raman spectrum of the homoepitaxial diamond film. The 1332 cm^{-1} peak corresponds to the sp^3 diamond bond.

Impurity concentrations of Fe, Si, and N, which might be incorporated in the films, were analyzed by SIMS using a primary beam of O_2^+ ions with acceleration energy of 15 kV. Positive secondary ions of masses 12 (C), 14 (N), 30 (Si), and 56 (Fe) were detected. As shown in Fig. 11, the concentrations of Fe and Si atoms in the film were found to be lower than the detection limit of the system. In the case of Si, the detection limit is determined by the secondary ions of NO, which have the same mass number of 30. This is because the plasma of about 2 in. diameter was confined away from the chamber wall having 6 in. inner diameter to prevent unintentional doping of impurities due to bombardment. This proves that this configuration has the advantage of reducing residual impurities.

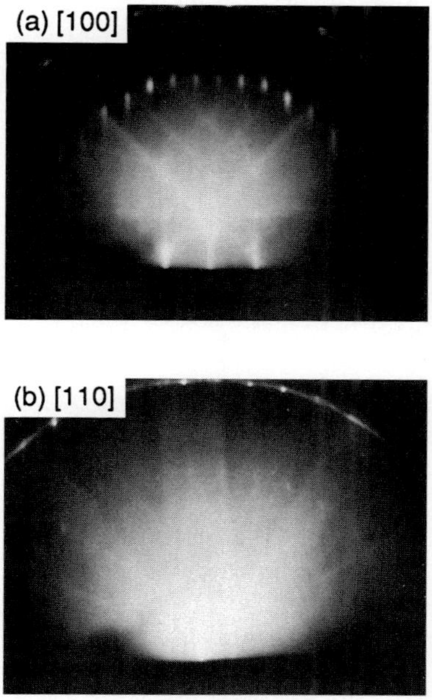

Figure 10. RHEED patterns of the film surface taken from (a) [100] azimuth and (b) [110] azimuth with accelerating voltage of 40 kV.

Figure 12 shows a typical forward *I-V* characteristic of Al-Schottky barrier of the as-deposited (hydrogenated) diamond surface of step-flow grown diamond films. The hydrogenated diamond films are known to have a high-conductivity semiconducting layer with p-type conduction [26-28]. A significant improvement is observed in both forward and reverse characteristics. The forward bias currents of the step-flow grown diamond films were found to show the conventional *I-V* characteristics of Schottky barrier with ideality factor [29] of 1.1 which is close to unity ever reported for hydrogenated films. It is also found that the reverse-bias leak currents of step-flow grown diamond films were found to be lower that the detection limit of the system ($\sim 10^{-13}$ A). These excellent rectification properties suggest the homoepitaxial diamond films obtained in the present study have a potential for electronic applications.

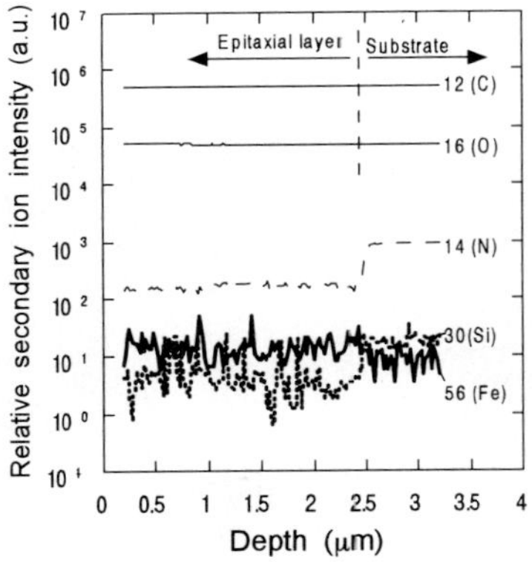

Figure 11. Depth profile of impurities in the film obtained by SIMS analysis.

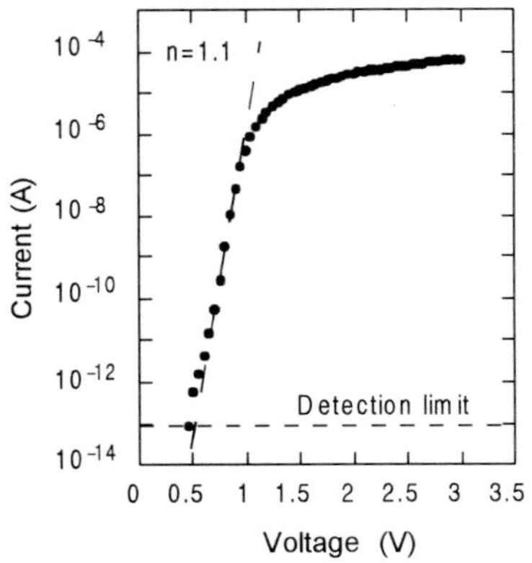

Figure 12. Forward *I-V* characteristics of Al-Schottky barrier to the step-flow grown diamond films.

3. Silicon Carbide

3.1 Historical background

The Schottky barrier heights of metal/semiconductor interfaces have historically been passive in terms of artificial controlling. This is mainly due to an interfacial Fermi level pinning which automatically determines the barrier height, resulting in the difficulty in controlling the barrier height. This low controllability has needed much efforts on Ohmic formation because the interfaces always have pinned Schottky barriers. The history of Ohmic formation was that of leaking currents through the barrier instead of lowering the barrier heights. A leaky Schottky barrier has complex mechanisms on current transports because of complex interface atomic and electronic structures caused by an interfacial reaction depending on an experimental condition. This difficulty in Ohmic formation will be basically solved if we can reduce the barrier height down to zero.

In our concept, Schottky and Ohmic formations are dealt with in the same category, where we control the density of interface electronic states to provide us controllable barrier heights as an active function. Also, these functional Schottky and Ohmic interfaces have a simple current transport mechanism of thermionic emission with a minor effect of tunneling current.

In this section, we propose a new concept to unify the Schottky and Ohmic formations as mentioned above and we demonstrate barrier height controls from a pinned Schottky barrier down to an Ohmic with a zero barrier height by changing the pinning rate.

3.2 Conventional Schottky and Ohmic formations

A Schottky barrier height ϕ_{bn} between a metal and an n-type semiconductor is classically expressed using Schottky-Mott rule,

$$\phi_{bn} = \phi_m - \chi_s, \tag{3}$$

where ϕ_{bn} is the Schottky barrier height for n-type semiconductor, ϕ_m is the metal work function, and χ_s is the electron affinity of the semiconductor. A practical system, however, has interface states which tend to pin the Fermi level within a band gap. When a pinning phenomenon is observed, ϕ_{bn} is expressed as,

$$\phi_{bn} = \frac{\partial \phi_{bn}}{\partial \phi_m}(\phi_m - \chi_s) + C, \tag{4}$$

where, $0 \leq \frac{\partial \phi_{bn}}{\partial \phi_m} \leq 1$. When we define the pinning rate p as $1 - \frac{\partial \phi_{bn}}{\partial \phi_m}$, the pinning rate becomes $0 \leq p \leq 1$. When p is 0, ϕ_{bn} satisfies eq.3 in principle because C is generally negligible. The situation with $p = 0$ is called the Schottky limit, where the Fermi level is free from pinning. When p is 1, ϕ_{bn} in eq.4 becomes constant, and is independent of the metal work function ϕ_m. This situation of the strongest pinning is the Bardeen limit. Since practical interfaces are present within the Schottky limit and Bardeen limit, we can control the barrier height to some extent by changing a metal. For instance, the maximum and minimum barrier heights for silicide/Si interfaces are around 0.5eV to 0.9eV and the width of controlling the barrier height is around 0.4eV[30], where no Ohmic interface is formed. So far, there has existed no technique to control the pinning rates; the Schottky barriers were out of control. As a result, the pinning rates have depended on their semiconductors.

Ohmic contacts had a more practical aspects. The most certain way to obtain the Ohmic property is a heavy impurity doping to reduce the Schottky barrier width so as to increase a tunneling current. This is inefficient in a new material because of the difficulty of high doping. Also, a minimum contact resistivity given by this way is limited by the doping concentration. The second way is to grow epitaxially another semiconductor with a narrower band gap which restricts the maximum barrier height within the band gap width. This is applicable to only systems where a heteroepixial technique is available. The most convenient way is to anneal the system. This, however, has ambiguous results in terms of

Schottky or Ohmic formation. Actually, annealing is a good way to form a stable Schottky barrier with a low leakage current in some cases, while annealing is the best way to form an Ohmic contact in other cases. The last one is to use a contact metal with a low work function for *n*-type semiconductor to reduce Schottky barrier height. For *p*-type, a high work function is needed. This way also often forms Schottky barriers with bad ideality factors because of no anneal resulting in contaminated interfaces which are formed by initial surfaces with contaminations and oxides.

As a general trend, a Schottky barrier has been formed for an ideal abrupt interface, whereas an Ohmic contact has been formed by losing the interface abruptness or by lowering the crystallinity of the semiconductor.

3.3 Unifying Schottky and Ohmic formations by controlling the Fermi level pinning

It is known that the pinning rate is dominated by the density of interface states D_{it} [31] as follows,

$$p = 1 - \frac{\varepsilon_i}{\varepsilon_i + q\delta D_{it}}, \tag{5}$$

where ε_i is the permitivity of the interfacial layer and δ its thickness[31]. Therefore, we can control the degree of the Fermi level pinning by changing D_{it}. Recent techniques using chemical solutions to terminate semiconductor surfaces progress to reduce the density of surface states and make their surfaces flat atomically. Such an ideal surface should result in forming an interface with a low D_{it}. We utilize two chemical solutions, *p*H-modified buffered HF or hot water to suppress D_{it}. *p*H-modified buffered HF has been found to terminate the Si(111) surface only by monohydrides[32]. Hot water also has the same effect[33]. In a semiconductor surface layer with a low crystallinity which is the origin of high D_{it}, removing the surface layer by an oxidation followed by a chemical etching is effective before the chemical termination.

Further, an actual barrier height is determined by the number of interface charges Q_{it} in D_{it}. Q_{it} is changeable if the metal work function ϕ_m is changed because the change in Fermi levels between the metal and the semiconductor yields a charge transfer. When $D_{it} \rightarrow 0$, the interface is in the Schottky limit. In the limit with zero D_{it}, the Ohmic property is obtained when ϕ_m is smaller than χ_s because C in eq.4 is generally small. Since this Ohmic interface has no Schottky barrier, the contact resistivity ρ_c is zero, independent of the donor concentration N_D. In contrast, in a common Ohmic contact, there exists a large D_{it} resulting in a Schottky barrier formation restricting the conductivity. In such a contact, the minimum value of ρ_c is limited by the Schottky barrier width determined by N_D because the current flows through the Schottky barrier by tunneling.

3.4 Demonstration of controlling barrier heights in metal/6H-SiC(001) interfaces

Unintentionally doped *n*-type 6H-SiC (0001) bulk substrates grown by modified Lely

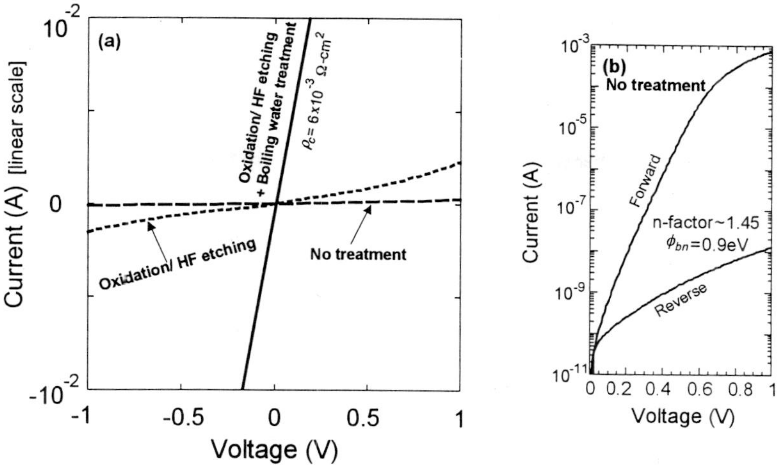

Figure 13. *I-V* characteristics between two Ti electrodes on 6H-SiC(0001) crystals (N_D ~2×10^{17}cm^{-3}). (a) linear *I-V* characteristics of no treatment, oxidation/HF etching treatment, and boiling water treatment. (b) a logarithmic *I-V* characteristic of no treatment.

method and nitrogen doped *n*-type 6H-SiC (0001) epitaxial layers were used. Si-faces of the SiC samples were used for all depositions and characterizations. The donor concentration N_D determined by capacitance vs. voltage (*C-V*) measurements was ~ 2×10^{17} cm^{-3} for the bulk substrates and ~ 5×10^{17} cm^{-3} for the epitaxial layers. Some of the samples were oxidized in a quartz tube furnace. The thickness of oxidized layers was about 30 Å. The oxidized layers were etched by dipping in 5 % HF solution. We refer to this oxidation and the successive HF etching as O/E treatment hereafter. Some of the samples with O/E treatment were further dipped in boiling water or in *p*H-modified buffered HF for 10 min. Finally, the samples were rinsed in deionized water. Ti and Ni were deposited with an electron beam evaporator to form contact electrodes. During the depositions, the sample temperatures were kept below 100 °C. No annealing was carried out for the Ti electrodes. Some of the Ni contacts were annealed at 1000 °C for 60 min in pure Ar to form Ohmic contacts.

Typical current vs. voltage (*I-V*) characteristics between two Ti contacts on the bulk substrates are shown in Fig.13(a). The contacts without O/E treatment show rectification properties as depicted by dashed line. The contacts are found to have Schottky properties with a bad ideality factor of around 1.45 as indicated in Fig.13(b). The barrier height is 0.9eV. After O/E treatment, the properties of Ti contacts on the bulk substrates change from rectification to Ohmic, showing good linearity (dotted line). However, ρ_C varies widely depending on electrodes. After dipping bulk substrates in boiling water, the average ρ_C of Ti contacts decreases to $(6\pm1)\times10^{-3}$ Ω-cm^2 (solid line) and shows good uniformity.

Figure14 shows the ϕ_{bn} as a function of ϕ_m in each treatment. Ni contacts in all treatments show Schottky properties and the heights are almost constant. Whereas in the Ti contacts ϕ_{bn} depends largely on the treatments. The pinning rate p obtained from the slopes of three straight lines connecting Ti and Ni in each treatment are 0.6 for no treatment, 0.3 for the O/E treatment and 0 for the boiling water treatment. It is found that the Fermi level is being released from pinning in order with proceeding the treatments. The cross point of the three lines in the figure, around 1.2eV of ϕ_{bn}, indicates the pinning position of the Fermi level because it is independent of the surface treatments. The corresponding ϕ_m in the pinning position is around 5.2eV in the figure. Since Ni has a work function of 5.35eV[34] which is almost the same as the pinning position, Ni shows the independency of the surface treatments. In the Ti contacts, they depend largely on the treatments because the work function of Ti is 4.33eV[34] which is much lower than the pinning position.

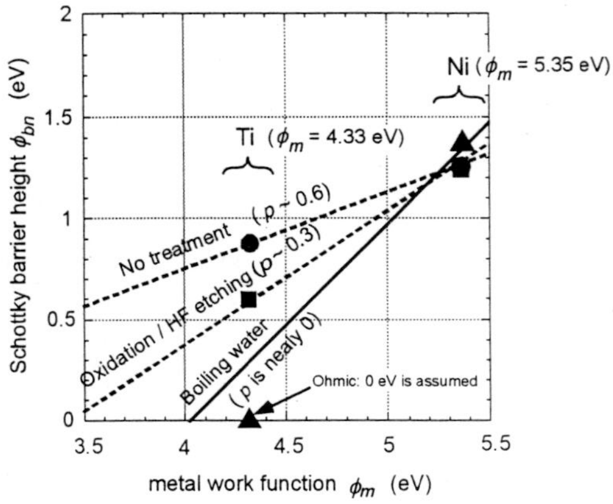

Figure 14. Schottky barrier heights as a function of metal work functions. No treatment (short dashed line), oxidation followed by HF etching treatment (long dashed line), and dipping treatment in boiling water (straight line) are indicated.

Ohmic contacts in the present experiments are formed on the lightly doped substrates (N_D~10^{17} cm^{-3}). In such an interface with a wide depletion layer, a tunneling current is negligible. Therefore, the Ohmic currents in the present experiments are generated by current flows over barriers with very low heights. Further, it should be noted that the Ohmic contacts are formed without post-annealing.

4. Conclusions

We have demonstrated the step-flow epitaxial growth of diamond films by introducing the two concepts of clean epitaxy and precise control of growth parameters. Al-Schottky barrier diodes using the step-flow grown diamond films show excellent rectification properties, indicating that the quality of these films is sufficient for electronic applications. Many technological issues, however, still remain. In particular, the basic research is needed on crystallinity, doping, energy states due to defects and impurities, and surface states. It is believed that this high-quality atomically flat diamond film contributes much not only to the realization of future electronic devices, but also to the scientific research of diamond. Also, we proposed a new method to control electric potential barriers by changing the Fermi level pinning and by changing the metal work function. This method unifies Schottky barriers and Ohmic conductivities in terms of their artificial control. We demonstrate this method on 6H-SiC substrates. The pinning control was achieved by dipping silicon carbide crystals into *p*H-modified buffered HF or boiling water before metallizations.

Acknowledgment

We thank T.Teraji and S.Yamanaka for their experimental cooperation. We also are grateful to Dr.N.Ohtani of Nippon Steel Co. for providing us 6H-SiC bulk wafers.

References

[1] E.O.Johnson, RCA Rev., 163 (1965).
[2] K.Shenai, R.S.Scott, and B.J.Baliga, IEEE Trans.Electron Devices, **36**, 1811 (1989).
[3] H.S.Kong, J.T.Glass, and R.F.Davis, J.Appl.Phys. **64**, 2672 (1988).
[4] T.Kimito, H.Nishino, W.S.Yoo, and H.Matsunami, J.Appl.Phys. **73**, 726 (1993).
[5] J.A.Powell, J.B.Petit, J.H.Edgar, I.G.Jenkins, L.G.Matus, J.W.Yang, P.Pirouz, W.J.Choyke, L.Clemen, and M.Yoganathan, Appl.Phys.Lett., **59**. 333 (1991).
[6] R.Kaplan, Surf.Sci. **215**, 111 (1989).
[7] M.A.Kulakov, G.Henn, and B.Bullemer, Surf.Sci., **346**, 49 (1996).
[8] F.Owman, Surf.Sci. **330**, L639 (1995).
[9] See, for example, S. Grot, in *Diamond: Electronic Properties and Applications*, edited by L. S. Pan and D. R. Kania (Kluwer Academic, Boston, 1995) chap. 9.
[10] J. T. Glass, D. L. Dreifus, R. E. Fauber, B. A. Fox, M. L. Hartsell, R. B. Henard, J. S. Holmes, D. Malta, L. S. Plano, A. J. Tessmer, G. J. Tessmer, and H. A. Wynand, *Proc. 4th Int. Conf. New Diamond Sci. Technol.*, edited by S. Sato, N. Fujimori, O. Fukunaga, M. Kamo, K. Kobashi, and M. Yoshikawa (MYU, Tokyo, 1994), p.355.
[11] Recent reports have described heteroepitaxial nucleation of diamond on SiC, Si, and Pt. See, for example, B. R. Stoner and J. T. Glass : Appl. Phys. Lett. **60**, 698 (1994). H. Kawarada, T. Suesada, and. H. Nagasawa : Appl. Phys. Lett. **66**, 583 (1995). X. Jiang, C.-P. Klages, R. Zachai, M. Hartweg, and H.-J. Russer : Appl. Phys. Lett. **62**, 3438 (1993). S. D. Wolter, B. R. Stoner, J. T. Glass, P. J. Ellis, D. S. Buhaenko, C. E. Jenkins, and P. Southworth : Appl. Phys. Lett. **62**, 1215 (1993). T. Tachibana, Y. Yokota, K. Nishimura, K. Miyata, K. Kobashi, Y. Shintani, Diam. Relat. Mater. (to be published).
[12] S. Matsumoto, Y. Sato, M. Kamo, and N. Setaka, Jpn. J. Appl. Phys. **21**, L183 (1982).
[13] M. Kamo, H. Yurimoto, and Y. Sato, Appl. Surf. Sci. **33**, 553 (1988).
[14] T. Tsuno, T. Tomikawa, S. Shikata, T. Imai, and N. Fujimori, Appl. Phys. Lett. **64**, 572 (1994).
[15] A. Badzian and T. Badzian, Diam. Relat. Mater. **2,** 147 (1993).
[16] N. Lee and A. Badzian, Appl. Phys. Lett. **66**, 2203 (1995).
[17] C. Wild, R. Kohl, N. Herres, W. Müler-Sebert, and P. Koidl, Diam. Relat. Mater. **3**, 373 (1994).
[18] K. Hayashi, S. Yamanaka, H. Okushi, and K. Kajimura, Appl. Phys. Lett. **68**, 1220 (1996).

[19] K. Hayashi, S. Yamanaka, H. Okushi, and K. Kajimura, Diam. Relat. Mater. (to be published).
[20] K. Hayashi, S. Yamanaka, H. Okushi, and K. Kajimura, Mater. Res. Soc. Symp. Proc. **416**, (to be published).
[21] W. J. P. van Enckevort, G. Janssen, W. Vollenberg, and L. J. Giling, J. Cryst. Growth **148**, 365 (1995).
[22] J.-P. Vitton, J.-J. Garenne, and S. Truchet, Diam. Relat. Mater. **2**, 713 (1993).
[23] W. J. P. van Enckevort, G. Janssen, W. Vollenberg, J. J. Schermer, and L. J. Giling, Diam. Relat. Mater. **2**, 997 (1993).
[24] L. S. Plano, in *Diamond: Electronic Properties and Applications*, edited by L. S. Pan and D. R. Kania (Kluwer Academic, Boston, 1995) p. 78.
[25] T. Tsuno, T. Imai, Y. Nishibayashi, K. Hamada, and N. Fujimori, Jpn. J. Appl. Phys. **30**, 1063 (1991).
[26] M. I. Landstrass and K. V. Ravi, Appl. Phys. Lett. **55**, 1391 (1989) .
[27] H. Kawarada, M. Aoki, H. Sasaki, and K. Tsugawa, Diam. Relat. Mater. **3**, 961 (1994).
[28] K. Hayashi, S. Yamanaka, H. Okushi, and K. Kajimura, Appl. Phys. Lett. **68**, 376 (1996).
[29] E. H. Rhoderick and R. H. Williams, *Metal-Semiconductor Contacts,* 2nd ed. (Clarendon, Oxford, 1988) chap. 3.
[30] S.M.Sze, *Physics of Semiconductor Devices, 2nd ed.* chap.5 (John Wiley & Sons, NY, 1981) 292.
[31] E.H.Rhoderick and R.H.Williams, *Metal-Semiconductor Contacts, 2nd ed.* (Oxford University Press, Oxford, 1988) 20.
[32] G.S.Higashi, Y.J.Chabal, G.W.Trucks, and K. Raghavachari, Appl.Phys.Lett. **56** (1990) 656.
[33] S.Watanabe, M.Shigeno, N.Nakayama, and T.Ito, Jpn.J.Appl.Phys. **12B** (1991) 3575.
[34] H.B.Michaelson, J. Appl. Phys. **48** (1977) 4729

Part II

Spectroscopy

Spectroscopy Methods for Low-Dimensional Systems

Minko Balkanski

Laboratoire de Physique des Solides, Université Pierre et Marie Curie, 4, Place Jussieu, Tour 13, 2ème étage, 75252 PARIS Cédex 05, FRANCE.

1. Introduction

Continuous progress in Solid State Physics and Materials Science has developed through a permanent interplay between theory and experiments in which spectroscopy is taking a major part. Absorption and Reflectivity spectroscopy is a key method for determining band gaps, necessary for establishing of electronic band structures and phonon dispersion relations which are the basic characteristics of materials.

The development of quantum theory of solids relied on a deep understanding of optical data. Unlike sharp atomic spectra, solid state spectra are broad. Energy band separations can not be easily extracted without a good theoretical calculation. Introduction of the Empirical Pseudopotential Method (EPM) allowed the interpretation of optical and photoemission spectra and provided accurate transition matrix elements. The actual behaviour of real solids at finite temperatures can be extracted from spectroscopic data, considering not only the frequency positions of the spectral lines but also their line shapes, by taking into account the many body aspects of solid state spectroscopy using Green's function techniques [1]. Light Scattering Spectroscopy and Resonant Light Scattering contain informations on the electronic structure, the collective excitation and the structure of materials.

Experimental feedback proved to be as important to establish *ab initio* approaches as improvements in formalism or in computing power.

In recent years, a completely new physics has emerged from the possibility to artificially create by MBE, or other analogous methods, structures of low dimensions. Limiting the dimensions in one of the three space directions confines the electronic states into a quantum box where the states of a continuous energy band becomes discrete energy levels. The fundamental theory, that of a particle in a one dimensional box has been known since the earliest days of quantum mechanics, but it tooks a new youth, once applied to the spectroscopy results of the new quantum structures of 2D, 1D and quantum dates. This field has now grown and matured to the point to become one of richest potentials and the basic promises for further development of the electronic and optoelectronic technologies.

More recently still new discoveries and further progress are appearing in low dimension structures for electronics : the C_{60}-related balls and fibers [2].

Parallely the EPM and *ab initio* approaches evolved together toward the establishment of *ab initio* pseudopotentials. The use of *ab initio* pseudopotential and density functional theory to calculate electronic energy structures, phonon spectra, mechanical properties, superconductivity, and a host of other measured properties of the ground and excited states of a wide variety of solids was recently discussed by Marvin L.Cohen [3].

A general approach sometimes referred to as « standard model of solids » is now making its way. This approach has been used to calculate band structure of fullerenes, the C_{60} molecule. This theory offers the possibility to explain electronic and optical properties, of this new types of molecules for electronics.

The discovery of carbon nanotubes (bucky tubes) by S. Iijima [4] has triggered a very broad interest. The electronic properties of these tubes change with the chirality introduced when the sheet is rolled into a tube. Semiconductor and semimetal systems can be obtained by purely geometric changes in the tubes. It is interesting to focus on the unusual possibility of varying electronic properties of tubules by altering their structure.

Introducing other than 6-fold rings in C tubules creating 5-fold and 7-fold « defect », not only changes the geometric structure by creating positives and negatives tubule curvatures, but also alters the electronic properties of the tube. This defects allows the joining of a chiral tube and non-chiral tube. The result is a nano-heterojunction with two semiconductors of different band gaps joined in a small « interface » created by « 5-7 » defect. This shows that nanodevices on a scale, just about as small as one can picture when using groups of atoms are possible to make.

Of particular interest would be the study of the nature and electronic properties of tubes when they are filled with atoms such as K, inducing charge transfer from K atoms to the inter tube walls. Charge transfer from Li atoms inserted in the van der Waals gap of layered compounds has been successfully investigated by infrared reflectivity spectroscopy [5] following the evolution of the plasmon mode frequency which measures the amount of charges transferred into the layers. This method can be used to study the charge transfer resulting from the insertion of alkali atoms in the tubes grown with open ends.

Another interesting area of investigation that of nanotubes involving systems based on $B_xC_yN_z$. The BN tubes are expected to be semiconducting with fairly large band gaps because of the ionic nature of this system. In some of these materials it has been shown that the anisotropic conductivity leads to chiral currents when the tube is formed. This raises the possibility of creating nanocoils based on BC_2N [6].

2. Spectroscopy of Low Dimensional Systems

2.1. Bulk and 2D Structures

In an ideal crystal the discrete electronic levels of the infinite number of atoms constituting the crystal are degenerated and form broad energy bands. The optical spectra represent absorption edges and broad emission bands. Isolated atoms introduced as impurities in semiconductor for example form hydrogen like energy spectra resulting in sharp absorption or emission lines characteristic of the nature of impurity. In this case the electron still feels the nuclear charge but the Coulomb interaction is screened by the dielectric constant of the medium and the resulting spectrum corresponds to very small binding energies, of the order of tens of mV.

Hydrogen like spectra result also from the Coulomb interaction between electrons and holes seen each other through the dielectric susceptibility of the crystal and hence having very small binding energies. Under optical excitation when the resulting carriers remain strongly coupled to the radiation field, the state of the system is just an oscillation between exciton and photon state [7] and remains such until an external perturbation : field fluctuation or phonon, contributes the value of binding energy and forces the dissociation of the excitonic state into separate free electrons and holes. The exciton is the prime and essential effect of the interaction of radiation with matter and it controls all optoelectronic phenomena and applications.

The developments of modern technology calls for higher and higher density of informations and consequently for smaller and smaller systems with high density of states, which in term of spectroscopy means sharp lines. A new era in solid state physics was opened some twenty years ago with the advent of technologies which allowed the reduction of the dimensions of solids down to atomic scale. The resulting quantum structures have thicknesses of few atomic layers along the growth direction.

The simplest quantum structure consists of a single layer of material B between sheets of material A. The material A has a band gap larger than B and the band discontinuities are such that both types of carriers are confined in the B material. In an approximation of infinitely deep well the electron wave function is completely confined in the well and the energy spectrum is a set of discrete levels. In the final-well case the wave function spread out over the boundary and extend into material A. A periodic array of quantum wells forms a superlattice where the discret energy levels of the individual quantum wells form narrow energy bands.

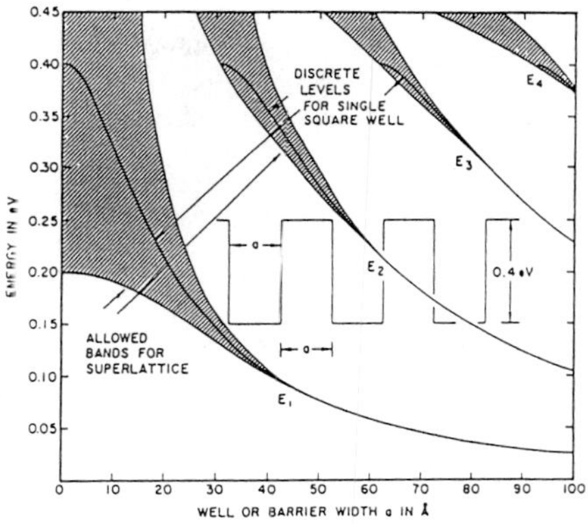

Figure 1 : Allowed energy bands E_1, E_2, E_3, and E_4 (hatched) calculated as a function of well or barrier width ($L_z = L_B = a$) in a superlattice with a barrier potential $V = 0.4$. Note the existence of forbidden gaps even above the barrier potential (Reprinted with permission from World Scientific Pub. Co., L. Esaki, « Recent Topics in Semiconductor Physics » (H. Kamimura and Y. Toyozawa, eds., 1983).

The 2D quantification has a drastic effect on the excitonic spectrum. Comparison of the absorption coefficients due to 3D and 2D (Figure 2) shows that the characteristic binding energy is 4 times larger for 2D excitons. Excitons are much more stable in 2D quantum structure which is of great importance in lowering the threshold energy of semiconductor lasers.

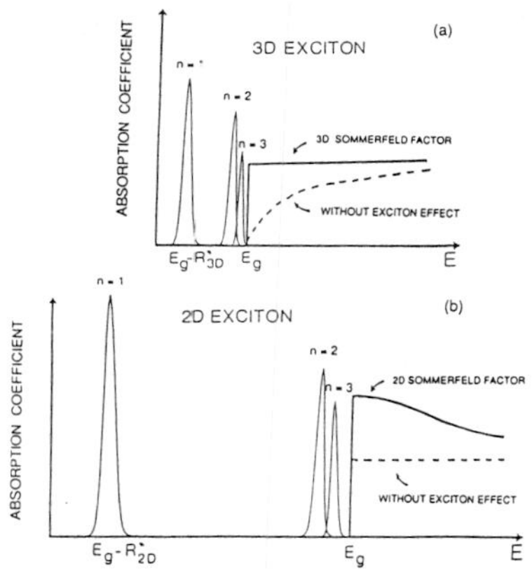

Figure 2 : Comparison of the absorption coefficients due to (a) 3D or (b) 2D excitons. The characteristic energy is ~ 4 times larger for 2D excitons. Oscillator strengths are increased (~ a_B^{-3} in 3D, ~ a_B^{-2} in 2D). For continuum states, the absorption coefficient is increased over the excitonless value (---) by the Sommerfield factor, determined by the continuum wave functions of the hydrogen atom, which represents the effect of electron-hole correlation in unbound states.

2.2. Confined Electrons and Photons in 1D and 0D [8]

In the search for sharp atom-like transitions in solids today's efforts are in two directions :

i) Confining electrons in smaller and smaller quantum structures resulting into quantizing electron motion. Different solutions are examined : quantum dots, II-VI particles dispersed in a dielectric medium, porous Si, Fullerenes and related balls, tubes and fibers.

ii) Quantizing photons. Here also diverse solutions are envisaged : microcavities, photonic bandgap materials, coherent coupling, controlled spontaneous emission.

The concept of *photon mode control* applied in semiconductor optics allows one to significantly change the parameters of the photon-matter interaction. Emission lines in solids are usually broad due on the one hand to the simultaneous occurrence of a continuum of electronic excited states populated by thermal excitation and on the other to a continuum of energy and momentum-matched photon states. By singling-out specific photon modes, microcavities lead to sharp emission lines. In addition, the emission is directional. This could eventually lead to a thresholdless laser if all the spontaneous emission took place in a single photon mode. The reader can refer to Yokoyamma for an introduction to optical microcavities[9].

Another way to control spontaneous emission has recently developed, thanks to the concept of photonic bandgap (PBG) materials [10] by structuring the dielectric functions of a material in 3-D, it has been shown possible to prevent the propagation of light in all directions due to coherent Bragg reflection, in a manner similar to the creation of electron energy bandgaps in solids due to the periodic atomic potential. Photon propagation is forbidden within that phonon energy gap and therefore any active « matter » with emission energy within the optical bandgap will have its spontaneous emission strongly suppressed. By modifying locally the dielectric constant, one is able to create a phonon localized state (similar to the localized electronic states of chemical impurities or defects in solid state physics). This localized photon state situated in a photonic bandgap can lead to sharp optical features, when that state is resonant with some active matter within the PBG material.

2.3. Light Scattering - Many Body Aspects

One of the most powerful spectroscopy methods informing on different kinds of elementary excitations is light scattering. Light scattering informs not only on the structure of materials but also on its electronic and vibrational processes. The successive steps in a light scattering processes are first absorption of a photon creating an electronic excited state which after scattering on a collective excitation of the material such as phonon, magnon, or plasmon, returns to the ground state by emitting a photon. The difference between the absorbed and emitted photons measures the energy of collective mode. The intermediate electronic excited state can be put in resonance with the incident beams which produces the resonant light scattering with characteristics enhancements on the scattering intensity. The line shape of the Raman band is rich of informations of statistical nature. Taking into consideration the many body aspects one can gain detailed informations on the actual behaviour of real solids at finite temperatures. One can extract information on the actual population density from spectroscopic data taking into consideration not only the frequency position of the spectral lines but also their lineshape. In the case of coincidence in the spectral range of a single mode excitation, say a phonon, with a broad continuum of electronic states, one observes [11] spectacular interference distortions in terms of antiresonance and line shape asymmetry. An example of interference distortion of Raman scattering line shape is given in Figure 3.

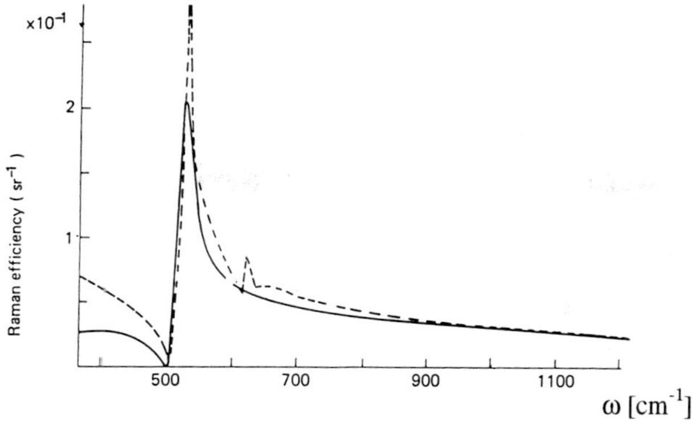

Figure 3 : Theoretical (solid curve) and experimental (dashed curve) results for Raman spectra in doped silicon, 1.1×10^{20} boron atoms per cm^3, at 2.4 K for incident laser wavelength 647.1 nm. The small peaks around 620 and 649 cm^{-1} in the experimental curve are due to boron impurity modes [ref. 1].

Raman intensity profiles are also reach of information on the crystal structures. The Raman intensity analysis of crystal structures is analogous to X-ray diffraction analysis in which the structures are determined from fitting the theoretical intensity profiles based on model structures to the experimental ones. Recently S. Nakashima [12] has developed a careful investigation on Raman intensity of folded modes for the identification of the stacking structures of SiC and CdI_2 polytypes. It is found that for the FTO and FTA modes in SiC and CdI_2 the relative Raman intensity does not depend on the value of band polarizability parameters but depends on the atomic displacements alone. This enables to confirm models of lattice dynamics from the Raman intensity analysis. The force constants and the Raman polarizabilities can be determined from the fitting of the theoretical intensity profiles and frequencies. From this procedure the identification of crystal structures can be done. Investigations on Raman intensity profiles have been also fruitful in the studies of superlattices.

2.4. Electron Transfer from Inserted Atoms

An other basic phenomenon of the low dimensional structures is the intercalation process. For layered compounds foreign atoms can be inserted in the interlayer space and produce significant changes in the electronic and vibrational structure of the host. The most striking effect is the fact that alkali metals induce an electron transfer from the inserted atom to the host to the point of transforming an isolator host into a metal.

Not always transferred electrons result into free charge carriers in the host lattice. It is important to exactly evaluate the charge transfer and this is best done by infrared reflectivity and magnetoreflectivity spectroscopy.

At sufficiently high carrier concentration N, the dielectric constant of the material becomes strongly affected by the free carrier susceptibility. In the infrared region where $\omega\tau >> 1$ the reflectivity is

$$R = \frac{(n-1)^2}{(n+1)^2} \quad \text{and} \quad n^2 = \varepsilon_\infty - \frac{4\pi N e^2}{m^* \omega^2} = \varepsilon_\infty \left(1 - \frac{\omega_p^2}{\omega^2}\right)$$

where ω_p is the plasma frequency determined by the equation

$$\omega_p^2 = \frac{4\pi N e^2}{m^* \varepsilon_\infty}$$

Clearly, when ω becomes equal to ω_p, $n = 0$ and $R = 1$, giving total plasma reflection. One then observes a reflectivity minimum at slightly higher frequencies [1] where $R \sim 0$ and $n \sim 1$:

$$\omega_{min} = \omega_p \left(\frac{\varepsilon_\infty}{\varepsilon_{\infty - 1}} \right)^2$$

An example of the measure of electron transfer during intercalation [13] is shown in Figure 4 in which are compared the infrared reflectivity spectra of pure TiS_2 and Li-intercalated TiS_2. The reflectivity minimum gives directly the amount of charges transferred. In this case one can assume that all charge carriers are transferred to empty band supposing a rigid band model with no major perturbations on the structure of the host.

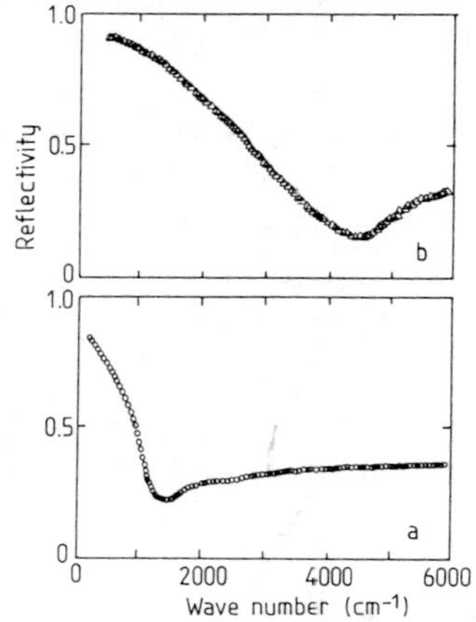

Figure 4 : Room temperature infrared reflectivity of pure TiS_2 (curve a) and electrochemically lithium-intercalated $Li_{1.0}TiS_2$ (curve b) single crystals [ref. 13].

Completely different is the case of Li intercalation in InSe. Experimental and theoretical investigations [14] indicate that the electronic band structure and lattice dynamics are specifically modified by the intercalation process.

The excitonic transitions persist after Li insertion, which suggests that all of the Li-2s electrons do not transfer to the conduction band and thus do not transform semiconducting InSe into a metal. If we had a metallic transition, the Coulomb interaction between the electron and hole of the exciton would be screened and the excitonic state would be washed out. The persistence of the excitonic transitions in highly intercalated InSe suggests that the Li-2s electrons form a low mobility impurity band or are efficiently trapped into localised states.

When the Li concentration in InSe samples is increased we observe that the direct absorption gap at 1.3 eV increases with Li content whereas the second absorption gap decreases with increasing Li concentration. These results are shown in Figure 5.

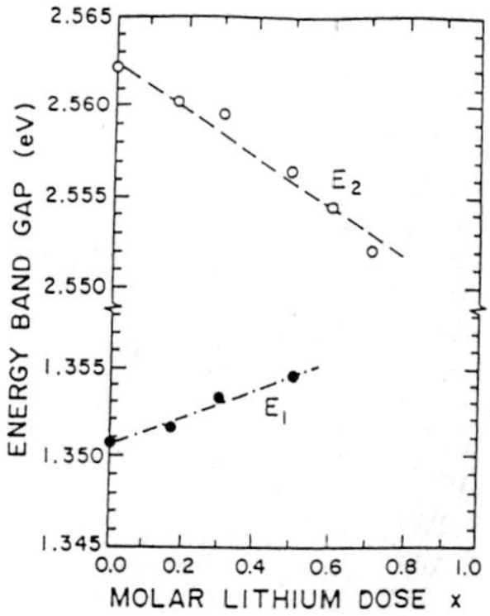

Figure 5 : Energy gap versus lithium content in γ-InSe for the smallest and next-to-smallest energy gaps E_1 and E_2 [ref. 14].

Figure 6 demonstrates that the photoluminescence spectrum of InSe is significantly modified by the intercalation of Li. A new photoluminescence peak appears at a photon energy somewhat less than that of the fundamental exciton peak of pure InSe. This new peak can be due to recombination of an electron at the lower edge of the Li-2s band with a hole at the valence band edge. The insertion also induces significant modifications in the phonon spectrum and the plasmon phonon coupling.

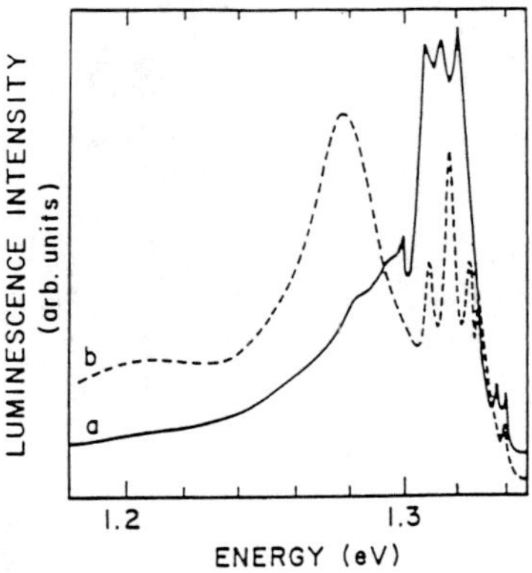

Figure 6 : Photoluminescence spectra for γ-InSe before (a) and after (b) lithium intercalation at 5 K under 1.916 eV excitation by a Kr^+ laser [ref. 14].

Figure 7 shows the experimental Raman spectra for pure γ-InSe and γ-InSe intercalated with lithium when excited with laser light having a wavelength of 514 nm. The arrows specify the extra modes due to the presence of Li, as indicated by their absence in the case of pure InSe. L^- and L^+ indicate LO-plasmon coupled modes.

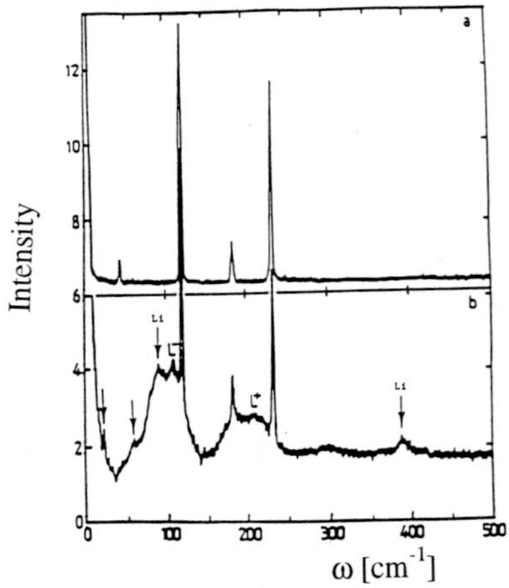

Figure 7 : Raman spectra of pure InSe (a) and Li_xInSe (b) at 5 K excited with 514 nm wavelength light. The modes labelled L^+ and L^- are thought to be coupled LO-phonon-plasmon modes [ref. 14].

3. Effects of Insertion in C_{60}-Related Balls and Fibers [2]

Intercalation of alkali metals into C_{60} to a stoichiometry M_zC_{60} (where M = K, Rb, Cs) has yielded relatively high T_c superconductors, and this has spanned a great deal of interest in C_{60}-related materials. In the C_{60} structure each carbon atom has its valence requirements fully satisfied, hence this structure is expected to be an insulator or a semiconductor. The semiconductor behaviour is stabilized by the distortion in the band lengths, yielding a band gap of 1.8 eV. In the solid phase C_{60} is a molecular solid with narrow electronic bands (~0.6 eV band) width for the conduction bands.

There are different ways to dop the C_{60} molecules and to convert the material into a conductor. One method is the addition of rare earth transition metal into the interior of the C_{60} ball. The second is the substitution of a carbon atom on the C_{60} ball by an impurity atom with a different valence state. In fact, the only smaller atom than C which can be substituted for a carbon atom on the bucky ball surface is boron which makes the charged ball p-types.

The third method of doping a bucky ball is to place the dopant (e.g. an alkali metal) between two adjacent balls, so that charge transfer takes place between the guest species and the balls. This method of doping closely parallels the charge transfer that occurs when an alkali metal is intercalated into a layered material. Many similarities of behaviour justify the comparison of doped C_{60} with layered and particularly graphite intercalation compounds.

Just as for graphite and other layered compounds Raman spectroscopy provides valuable informations about solid C_{60} and C_{60}-related compounds. There are only 10 Raman-active mode frequencies for an isolated C_{60} molecule despite of the 180-6 = 174 degree of freedom of the C_{60} molecule and the 46 distinct phonon modes.

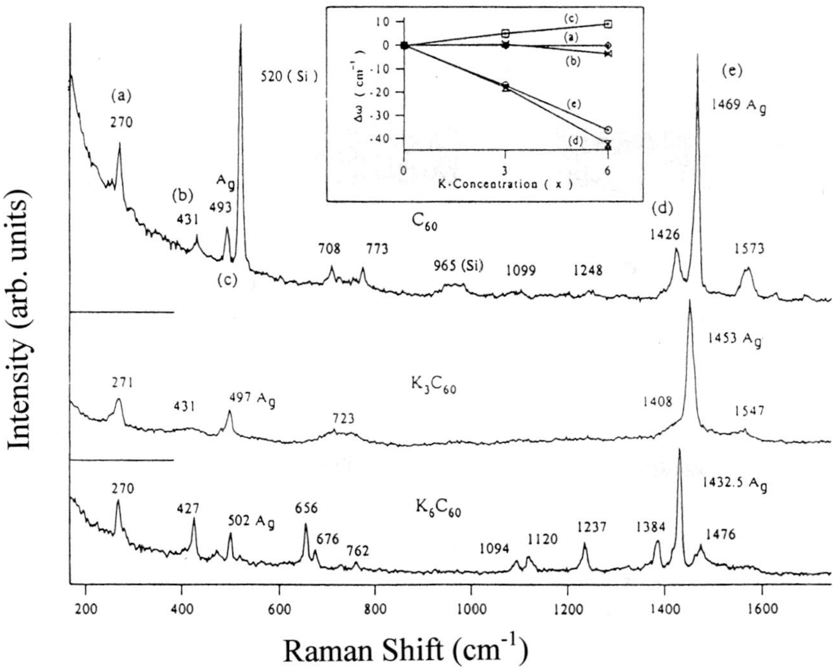

Figure 8 : Room temperature Raman spectra from films of (a) undoped C_{60}, (b) K_3C_{60}, and (c) K_6C_{60}. Si denotes the silicon substrate. The inset shows the potassium concentration x dependence of five representative Raman modes in K_xC_{60} $(0 \leq x \leq 6)$ [ref. 2].

In the Raman spectrum shown in Figure 8 one observes ten strong lines exactly as predicted for an isolated ball. The addition of alkali metal dopants to saturation M_6C_{60} (for M = K, Rb, Cs) perturbs the Raman spectra only slightly. A remarkable feature is that the lowest, frequency mode suffers an interference distortion of the Raman lineshape in the case of K_3C_{60} due to a coupling of the phonon with a continuum associated either with electronic state or with multiphonon processes related to the alkali metal ions [11].

Because of the weak interaction of the balls with each other and with the alkali metal dopants, solid M_xC_{60} can be viewed as a molecular solid having energy levels with little dispersion, giving rise to a very high density of states near the Fermi level.

Doping with alkali metal increases the electrical conductivity. The stoichiometry K_3C_{60} half fills the lowest unoccupied molecular orbitals and gives the highest conductivity - up to nearly metallic conduction.

4. C_{60}-Related Fibers - Bucky Fiber

Carbon fibers are industrially important for their extraordinarily high modules and strength. When activated carbon fiber can become highly porous. Activated carbon fibers have been used to absorb various gazes and as a double layer capacitor for electrical applications. Having made a new form of carbon fiber, a « bucky fiber » based on the C_{60} molecule.

By adding a ring of five armchair hexagous to C_{60} along the equator normal to a fivefold axis one gets a rugby-ball shopped C_{70} molecule. This suggests that one might add j rows of such armchair hexagous to C_{60} and obtain a $C_{60+10\ j}$ molecule which would be in the form of a fiber with a cylindrical sheath of carbon atoms one layer thick terminated on either end by a C_{30} cap formed by bisecting the C_{60} molecule normal to its five fold axis along armchair hexagons, thereby fitting these caps perfectly to the cylindrical sheath.

The nucleation of a bucky fibre instead of a bucky ball would be stabilized by a defect in the cap region during the early formation stage of the cluster. In general these defective caps will introduce some chirality which is propagated in the cylindrical tubule nucleated by the cap. Iijima [4] reports that the majority of the carbon tubules that he observed had screw axes. The chirality and the fiber diameter of any graphene tubule can be specified by the vector **AA'** or $C_h = n\hat{a}_1 + m\hat{a}_2$ which connects two crystallographically equivalent sites on a 2D graphen sheet [11] (see Figure 9).

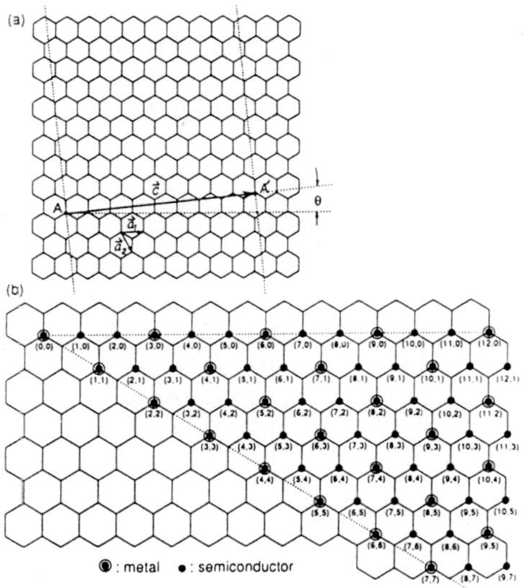

Figure 9 : (a) The vector AA' or $C_h = n\hat{a}_1 + m\hat{a}_1$ on the honeycomb defined by unit vectors $\hat{a}_1$ and $\hat{a}_2$ and the chiral angle θ with respect to the zigzag axis. (b) Possible vectors for general fibers, including zigzag, armchair and chiral fibers. The circled dots and bullets, respectively, denote metallic and semiconducing behavior for each of the possible fibers [ref. 2].

The cylinder is formed by superimposing the two ends of the vector AA', and the joint is made along the two lightly dotted lines which are perpendicular to AA' at points A and A' in Figure 9.

The chiral fiber thus generated has no distortion of band angles other than distortions caused by the cylindrical curvature of the fiber. Differences in chiral angle θ provide a basis for differences in the properties of the various types of fibers of very small diameter. The two limiting cases are defined by the vectors AA' extending from the zigzag fiber, $\theta = 0°$, to the armchair fiber $\theta = 30°$. All other vectors (n, m) correspond to chiral fibers. Both right- and left-banded chirality is possible, therefore it is expected that chiral fibers are optically active to either right or left circularity polarized light, depending on their chirality. This property could lead to practical applications for chiral fibers.

4.1. Electronic Structure

One direct method for obtaining the electronic dispersion relations for a very small diameter fiber is based on folding the electronic dispersion relations of a 2D graphite sheet [17]. A set of 1D energy dispersion relations is obtained by slicing up the 2D energy bands structure of graphite in the circonferencial direction. For both armchair and zigzag type fibers one has two 1D energy bands which cross at the Fermi energy giving rise to a metallic conduction.

For the case of chiral fiber one can have metallic and semiconducting fibers.

5. Theory Predictions and Prospectives

The wealth of experimental results and knowledge on materials accumulated in recent years has helped our microscopic view of solids to mature up to the point to construct for a large class of materials an useful first-principles or « standard model » sufficiently powerful to explain and predict many physical properties.

The fundamental concepts for density functional theory an pseudopotentials established in the 1930's by Dirac [18] and Fermi [19] have gained creative and useful refinements up to the capacity to provide explanations and predictions owing to close collaborations between theory and experiments.

In solid state physics as in atomic physics the great advances of those who developed quantum theory relied on a deep understanding of data, in principle optical data. However unlike sharp atomic spectra, solid state spectra are broad. It is difficult to extract energy band separations without a good theoretical calculation. Empirical pseudopotentials were used to decipher solid state spectra especially for semiconductor [20]. This empirical pseudopentential method not only allowed an interpretation of an optical or photoemission spectrum in terms of energy level separations, it also provides accurate dipole transition matrix elements.

With accurate EPM bands, density of states, electron densities and response spectra, tests could be done on *ab initio* theories to determine their worth. This feedback proved to be as important as improvements in formalism or in computing power. An excellent example of this view is the development of Angular Resolved Photoemission Spectroscopy (ARPES) which gives the band structure $E_n(\mathbf{k})$ directly. The first convincing results were on 2D materials, layered semiconductors.

The next step was the use of *ab initio* pseudopotentials and density functional theory. The *ab initio* pseudopotentials were obtained using approaches requiring only the imput of the atomic number. By combining the pseudopotential and a local density approximation (LDA) for the density functional theory (DFT), it becomes possible to do precise calculations for ground state properties from *ab initio* theory.

The theory is now on the stage where it is possible to calculate to a good approximation ground state properties such as electronic energy structure, phonon spectra, mechanical parameters, superconductivity, and a host of other measurable properties of the ground and excited states of a wide variety of solids.
This general approach is referred to as a standard model of solids, it does apply for a variety of metals, semiconductors, semimetals, and insulators when the electrons are not too localized.

Recently Marvin L. Cohen [3] has presented a remarkable overview illustrating the power of the standard model and giving the prospectives of the possible uses of this approach. We shall discuss here only a few examples concerning the applications to nanotubes.

6. Prospective Applications of Nanotubes

The electronic properties of carbon nanotubes (bucky tubes) are predicted [21] to change with the chirality introduced when the sheet is rolled into a tube. Semiconductor and semimetal systems can be obtained with purely geometric changes in the tubes - without the necessity of doping.

A particularly interesting question is the study of the nature of the electronic properties of the tubes when they are filled with atoms such as K.

Calculations indicate that there should be significant charge transfer from the K atoms to the inter tube walls. This situation reminds of electron transfer from Li intercalated in InSe and could be studied by the same spectroscopy method as those described in section 2.4.

The theoretical approaches are also parallel. Using density functional theory within the local density approximation one has determined the sites of lowest potential energy occupied by Li in the van der Waals gap de InSe. Those sites are determined by comparing the total energy of different trial configurations [22]. Total energy calculations comparing the energy of an empty tube and a tube filled with a linear arrangements of K atoms suggests a lower energy for the latter [23].

To insert experimentally K in the tubes it will be necessary to grow tubes with open ends. Recent experiments indicate that this might be possible. If we achieve K insertion in carbon tube the question would be on the possibility to achieve metal conductivity as one would expect in an intercalation process. One might even expect superconductivity. This opens very exciting experimental challenges.

K insertion in carbon tubules is schematically represented in Figure 10.

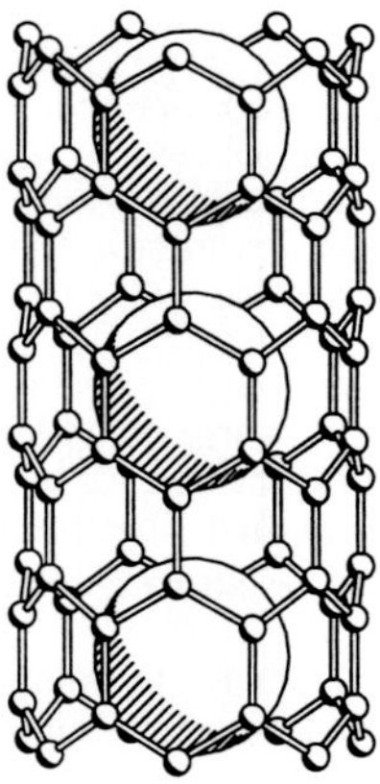

Figure 10 : Schematic picture of a (7.0) carbon tubule having K atoms inside. Large and small circles indicate K and C atoms, respectively [ref. 23].

Very recent results on a multiwalled carbon nanotube reminds again of the analogy of electron transfer in Li intercalated InSe. The conductance of an individual carbon nanotube exhibits a lnT dependence [24] just as in Li-intercalated InSe [25]. In the two cases the data can be interpreted in terms of two-dimensional weak localization.

The doping of nanotubes adds significantly to the range of properties which can be explored ; however it is interesting to focus on the unusual possibility of varying the electronic properties of tubules by altering their structure. The caps on the ends of tubes and the shape of the narrowing down of tubes near the ends requires different rings of atoms other than 6. Using C as prototype tubule, positive curvature can be obtained, as in the C_{60} molecule, by introducing 5-fold rings. Negative curvature is achieved with 7-fold rings and 4-fold and 8-fold rings can cause other changes in the tubule curvatures.

Introducing rings other than 6-fold rings not only changes the geometric structure and structural properties, it also alters the electronic properties of a tube. An interesting situation arises when the two 6-fold rings are replaced by 5-fold - 7-fold defect. This defects allows the joining of the chiral tube and non-chiral tube [26]. The result is a nano-heterojunction with two semiconductors of different band gaps joined in a small « interface » created by the « 5-7 » defect .

A schematic representation of the structure of a (8.0) tube joined to a (7.1) tube forming a semiconductor-metal junction is given in Figure 11.

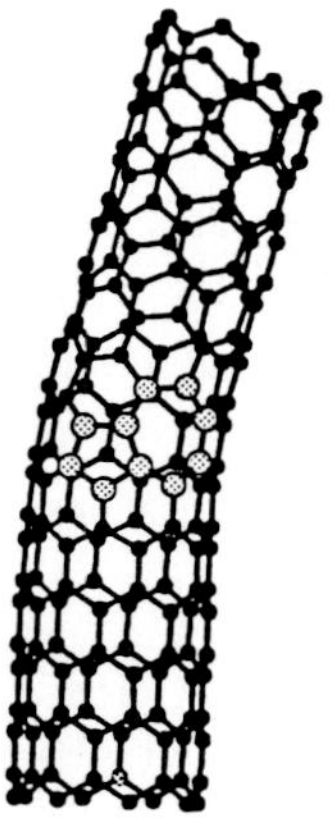

Figure 11 : Atomic structure of an (8.0/(7.1) tube, large light-grey balls denote the atoms forming the septagonal pentagon pair (ref. 26].

In a similar manner Schottky barriers can be formed. These studies may make nanodevices possible where the scale is just about as small as one can picture when using groups of atoms. These predictions, open a wide, completely new field of research and applications in molecular electronics where spectroscopy and the methods described in the previous sections will have a major role to play. Here again the association of theory and experiments is bound to open an extremely exciting, and full with application, new area of physics.

References

[1] R.F. Wallis and M. Balkanski, *Many Body Aspects of Solid State Spectroscopy,* North-Holland (Amsterdam), 1986.

[2] M. Dresselhaus, G. Dresselhaus, R. Saito and P.C. Ekland, C_{60}-Related Balls and Fibers, in *Elementary Excitations in Solids* - A Special Volume in Honor of Professor Minko Balkanski, edited by J.L. Birman, C. Sébenne and R.F. Wallis, North-Holland (Amsterdam), 1992.

[3] Marvin L. Cohen, Density Functional Theory and Pseudopotentials : a Panacea for Calculating Properties of Materials, *Int. J. of Quantum Chem.* (to be published).

[4] S. Iijima, *Nature* **354**, 1991, pp. 56.

[5] C. Julien and M. Balkanski, Is the Rigid Band Model Applicable in Lithium Intercalation Compounds, *Solid State Ionics III, Materials Research Society Symposium Proceedings,* Vol. **293**, 1992, pp. 27-37.

[6] Y. Miyamoto, A. Rubio, M.L. Cohen and S.G. Lonie, Chiral Tubules of Hexagonal BC_2N, *Phys. Rev.* **B 50**, 1949, pp. 4976.

[7] M. Balkanski, Energy Transport in Semiconductors, *J. Phys. Chem. Solids,* Vol. **8**, 1959, pp. 179-181.

J.J. Hopfield, Aspects of Polaritons, Proc. Int. Conf. Phys. of Semicond., Kyoto, 1966 : *J. Phys. Soc. Japan* **21**, 1966, pp. 77-88.

[8] E. Burstein and C. Weisbuch, Confined Electrons and Photons, Plenum Press, New-York, 1995.

[9] H. Yohoyama, Physics and Device Applications of Optical Microcavities, *Source* **256**, 1992, pp. 66.

[10] E. Yablonovitch, Photonic Bandgap Structures, *J. Opt. Soc. Am.* **B 10**, 1993, pp. 283-297.

[11] M. Balkanski, K.P. Jain, R. Beserman and M. Jouanne, Theory of Interference Distortion of Raman Scattering Line Shapes in Semiconductors, *Phys. Rev.*, Vol. **B 12** (1975), pp. 4328-4337.

[12] S. Nakashima, Raman Intensity Profiles and Crystal Structures, in *Elementary Excitations in Solids,* Special Volume in Honor Professor Minko Balkanski, edited by J.L. Birman, C. Sébenne and R.F. Wallis, North-Holland (Amsterdam), 1992, pp. 167-195.

[13] C. Julien, I. Samaras, O. Gorochov and M. Ghorayeb, Optical and Electrical-Transport Studies on Lithium-Intercalated TiS_2, *Phys. Rev.* Vol. **45** (1992), pp. 13390-13395.

[14] M. Balkanski, P. Gomes da Costa and R.F. Wallis, Electron Energy Bands and Lattice Dynamics of Pure and Lithium Intercalated InSe, *Phys. Status Solidi* (to be published).

[15] Ping Zhou, K.A. Wang, A.M. Rao, P.C. Ekland, G. Dresselhaus and M.S. Dresselhaus, *Phys. Rev.* Vol. **B 45**, 1992, pp. 10838-10840.

[16] M.S. Dresselhaus, G. Dresselhaus and R. Saito, *Phys. Rev.*, Vol. **B 45**, 1992, pp. 6234-6242.

[17] R. Saito, M. Fujita, G. Dresselhaus and M.S. Dresselhaus (unpublished).

[18] P.A.M. Dirac, *Proc. Cambridge Philos. Soc.,* Vol. **26**, 1930, pp. 376.

[19] E. Fermi, *Nuovo Cimento*, Vol. **11**, 1934, pp. 157.

[20] M.L. Cohen and J.R. Chelikowsky, *Electronic Structure and Optical Properties of Semiconductors,* Springer-Verlag, Berlin, 1988.

[21] N. Hamada, S. Sawada and A. Oshiyama, *Phys. Rev. Lett.,* Vol. **68**, 1992, pp. 1579.

[22] K. Kunc and R. Zeyher, *Europhys. Lett.* **7**, 1988, pp. 611.

[23] Y. Miyamoto, A. Rubio, X. Blase, M.L. Cohen and S.G. Lonie, Ionic Cohesion and Electronic Doping of thin Carbon Tubules with Alkali Atoms, *Phys. Rev. Lett.,* Vol. **74**, 1995, pp. 2993.

[24] L. Langer, V. Bayot, E. Grivei, J.P. Issi, J.P. Heremans, C.H. Ock, L. Stockman, C. Van Hacsendonck and Y. Bruynsaraede, Quantum Transport in Multiwalled Carbon Nanotube, *Phys. Rev. Lett.* **76**, 1996, pp. 479-482.

[25] D. El-Khatouri, A. Khater, M. Balkanski, C. Julien and J.P. Guesdon, Two-Dimensional Conductivity in the Layered Semiconductor InSe at Low Temperatures Owin to Weak Localisation, *J. Appl. Phys.* **66**, 1989, pp. 2049-2051.

D. El-Khatouri, A. Khater, M. Balkanski and J. Tuchendler, Two-Dimensional Quantum Corrections to the Magnetoconductance of InSe at Low Temperatures Owing to Weak Localization, *J. Appl. Phys.* **6**, 1989, pp. 5409-5411

[26] L. Chico, V.H. Grespi, L.X. Benedict, S.G. Lonie and M.L. Cohen, Pure carbon nanoscale devices : nanotube heterojunctions (to be published).

Time-resolved Spectroscopic Studies of Photoexcited Retinal Isomers in Solution and the Mechanism of Photoisomerization

Hiro-o Hamaguchi

Department of Basic Science, Graduate School of Multidisciplinary Sciences, The University of Tokyo, 3-8-1 Komaba, Tokyo 153, Japan
and
Kanagawa Academy of Science and Technology (KAST), KSP East 301, 3-2-1 Sakato, Kawasaki 213, Japan

Abstract

The photophysiscs and photochemistry fololwing the photoexcitation of retinal isomers in hydrocarbon solutions have been studied by nanosecond time-resolved spontaneous Raman, nanosecond time-resolved infrared absorption, picosecond time-frequency two-dimensional CARS, picosecond time-resolved fluorescence, and femtosecond visible absorption spectroscopies. The mechanism of the cis-trans photoisomerization of retinal is discussed on the basis of the accumulated time-resolved spectroscopic data.

1. Introduction

Photoisomerization of retinoid plays crucial roles in photobiology [1]. The elementary process of vision stars with the 11-cis to the all-trans photoisomerization of the retinyl chromophore (1) in rhodopsin. The proton-pumping photochemical cycle in bacteriorhodopsin is triggered by the all-trans to the 13-cis photoisomerization of the same chromophore. Photoisomerization in either case transforms a photon signal into spatial information that is to be received by the protein. Such transformation is carried out in a unique predeterminate way with a high quantum efficiency. On the contrary, photoisomerization of retinal (2) in solution is much less selective.

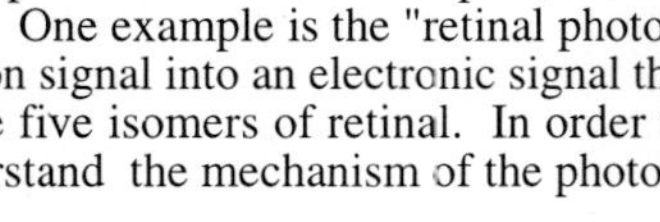

(1)

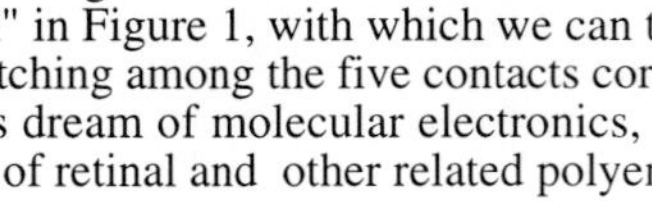

(2)

If we are able to manipulate the photoisomerization pathway of retinal as nature does it so well in rhodopsin and bacteriorhodopsin, we may be able to design a molecular electronics device based on it. One example is the "retinal photo-rotary switch" in Figure 1, with which we can transform a photon signal into an electronic signal through the switching among the five contacts corresponding to the five isomers of retinal. In order to achieve this dream of molecular electronics, we need to understand the mechanism of the photoisomerization of retinal and other related polyenes.

In the following, I review our efforts in the past thirteen years for elucidating the mechanism of the photoisomerization of retinal in solution. The method we used includes nanosecond time-resolved spontaneous Raman spectroscopy, nanosecond time-resolved infrared absorption spectroscopy, picosecond time-frequency two-dimensional CARS (Coherent anti-Stokes Raman Scattering) spectroscopy, picosecond time-resolved fluorescence spectroscopy, and femtosecond time-resolved visible absorption spectroscopy. All of the apparatus used in these studies were developed in our laboratory.

Springer Proceedings in Physics, Vol. 81
Materials and Measurements in Molecular Electronics
Editors: K. Kajimura · S. Kuroda

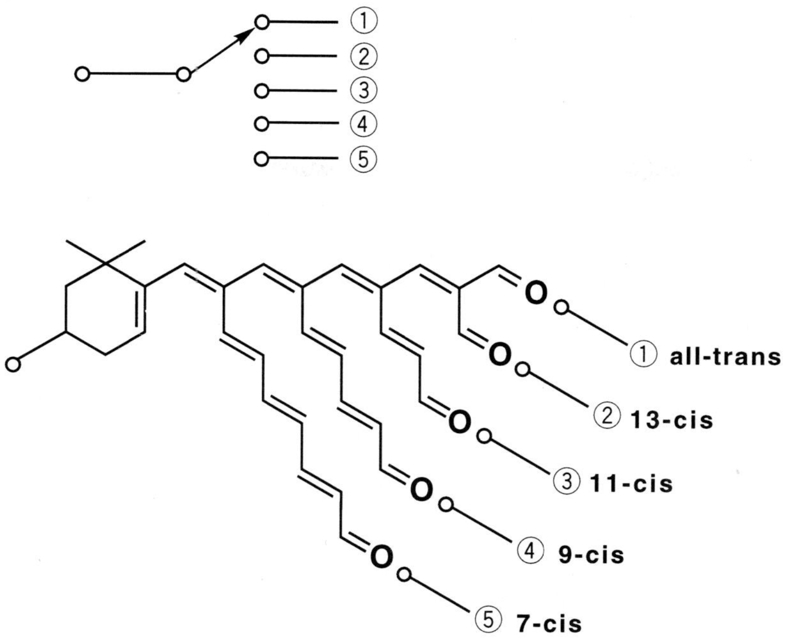

Figure 1: Retinal photo-rotary switch: a dream of molecular electronics using photoisomerization.

2. The Photoisomerization of Retinal

Retinal has four mono-cis isomers, the 7-, 9-, 11-, and 13-cis isomers, in addition to the all-trans. The scheme of photoisomerization among these five isomers is shown in Figure 2, where the length of the arrow shows a rough measure of the photoisomerization quantum yield in hydrocarbon solutions. For all the four mono-cis isomers, the cis to trans photoisomerization is very efficient. On the other hand, the photoisomerization from the all-trans is much less efficient. The photoexcitation of the all-trans isomer results in the formation of small amounts of the 9-cis and the 13-cis isomers. The 7-cis and the 11-cis isomers are not formed. The all-trans/7-cis and the all-trans/11-cis photoisomerization are therefore categorized to the so called "one-way" photoisomerization [2]. For the all-trans/9-cis and the all-trans/13-cis pairs, the quantum efficiency of photoisomerization is much higher for the cis to trans than for the trans to cis. This tendency of the retinal photocismerization contrasts sharply with the ordinary photoisomerization of olefins in which the photoexcitation of either of the trans and cis isomers results in a one to one mixture of the two isomers. In polar solvents, the trans to cis photoisomerization quantum yield is twice or three times larger than that in hydrocarbon solutions. Furthermore, both the 11-cis and 7-cis isomers are formed by the photoexcitation of the all-trans. Such a marked solvent dependence is another key aspect of the retinal photoisomerization.

In this review, we focus on the photoexcited retinal isomers in hydrocarbon solutions for which more spectroscopic data have been accumulated. We note that we recently gave a qualitative account for the solvent dependence of the retinal photoisomerization using the exchange polarization model of photoisomerization[3].

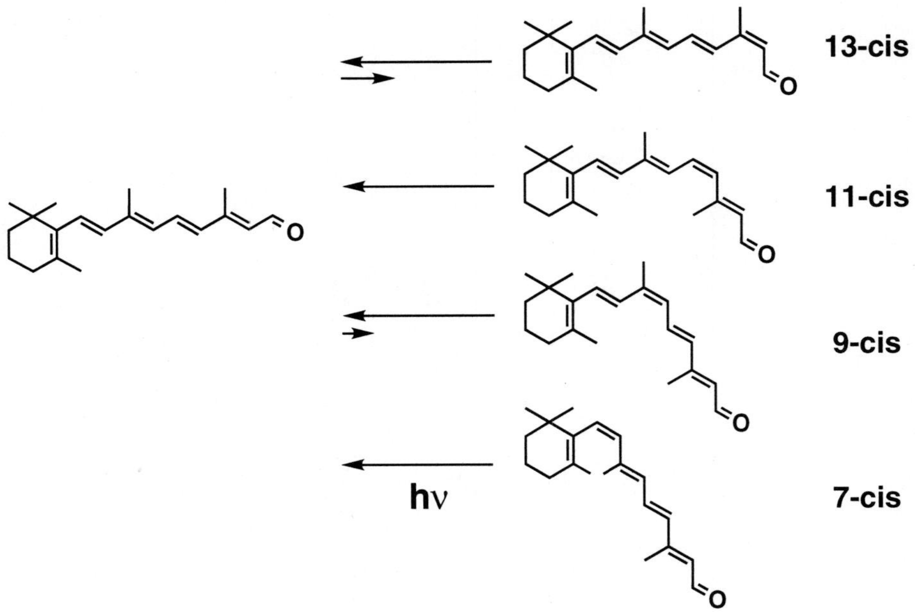

Figure 2: Photoisomerization among retinal isomers.

3. Nanosecond Time-resolved Raman Spectroscopy: Photoexcitation of the All-trans and the Four Mono-cis Isomers [4,5]

Time-resolved Raman spectra of the excited triplet states generated from the all-trans, 7-cis, 9-cis, 11-cis, 13-cis isomers of retinal are shown in Figure 3, together with the corresponding ground state Raman spectra. All the spectra were measured 20 ns after the phtoexcitation. Though the Raman spectra in the ground state differ from one another reflecting the structural difference in the cis-trans configuration, the excited triplet Raman spectra are identical except for the one generated from the 13-cis. This common triplet Raman spectrum is assigned to the all-trans triplet state for a number of reasons [5]. The Raman spectra in Figure 3 then means that the photoexcitation of the 7-, 9-, 11-cis isomers results in the formation of the all-trans triplet states, after 20 ns from the excitation, and that the photoisomerization from the three mono-cis isomers to the all-trans is complete in the excited triplet manifold. It is therefore concluded that the excited triplet potential surface of retinal has a deep minimum at the all-trans configuration and shallow or virtually no minima at the 7-, 9-, and 11-cis configurations. This shape of the excited triplet potential surface explains well the "one way" characteristic of the retinal photoisomerization.

The photoexcitation of 13-cis retinal need a further consideration. Detailed analysis of the Raman spectra indicates that two distinct triplet species, one with the all-trans configuration and the other most probably with the 13-cis configuration, are generated by the photoexcitation of the 13-cis isomer in the ground state. It is presumed that the 13-cis configuration has a potential minimum that is slightly deeper than those of the other three mono-cis configurations.

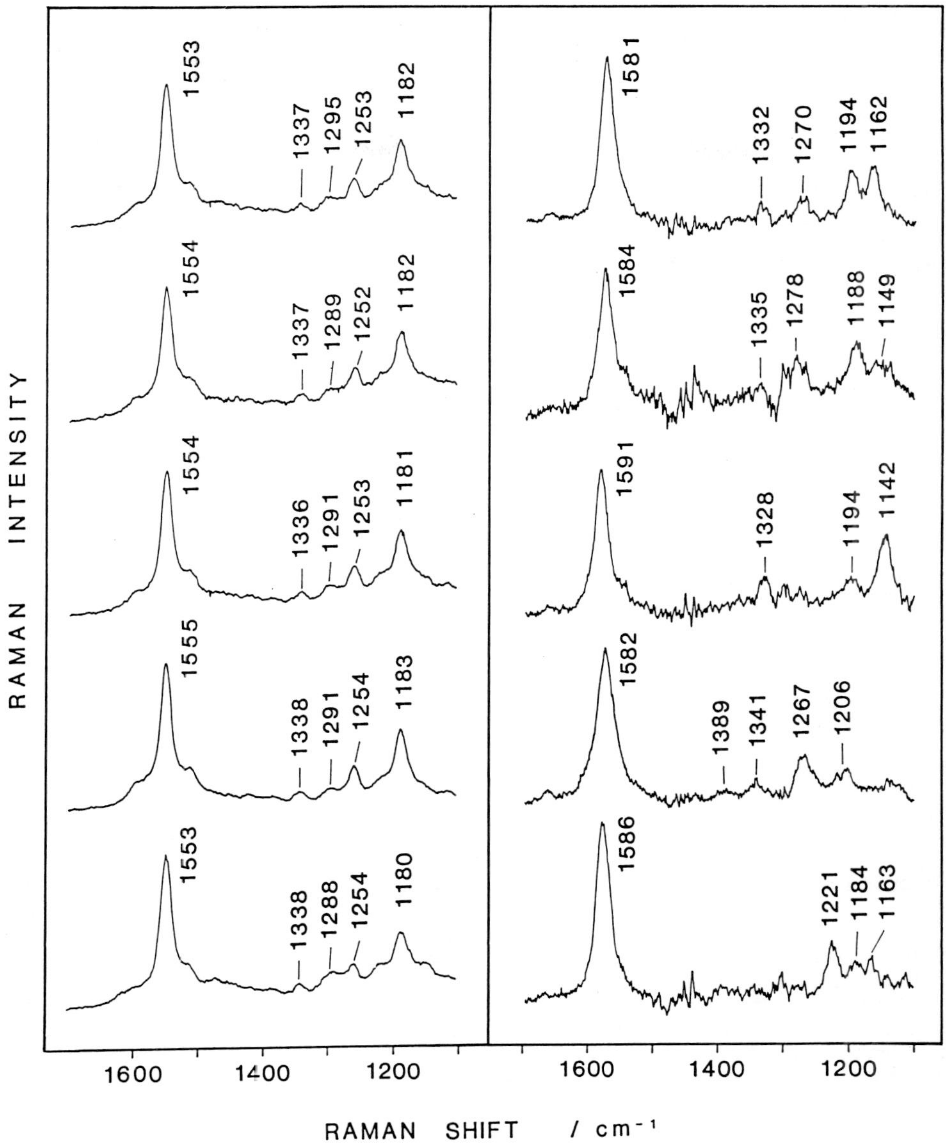

Figure 3: Time-resolved triplet Raman spectra (left hand) and the corresponding ground-state Raman spectra (right hand) of retinal isomers in hexane. From the top to the bottom, all-trans, 7-cis, 9-cis, 11-cis and 13-cis retinal.

4. Nanosecond Time-resolved Infrared Absorption Spectroscopy: Photoexcitation of the All-trans Isomer [6]

Nanosecond time-resolved infrared system used in the present study [7] is based on a dispersive infrared spectrometer equipped with a fast photovoltaic MCT (Mercury Cadmium Tellurium) detector and an AC-coupled low-noise wide-band preamplifier (Figure 4). It detects the change (only the AC component) of the transmitted infrared light intensity which is induced by the photoexcitation. This AC-coupling scheme enables us to achieve a high sensitivity which detects a signal of one part in million in the induced intensity change. This sensitivity is more than two orders of magnitude higher than ordinary FT-IR spectrometers. The time response of the spectrometer is about 50 ns. A full infrared wavenumber range from 4000 cm^{-1} to 700 cm^{-1} is covered. All-trans retinal in cyclohexane is photoexcited by the third harmonic (349 nm) of a diode-pumped Q-switch Nd:YLF laser at a repetition rate of 200 Hz. The sample solution is circulated through a flow cell with Ar bubbling in the reservoir.

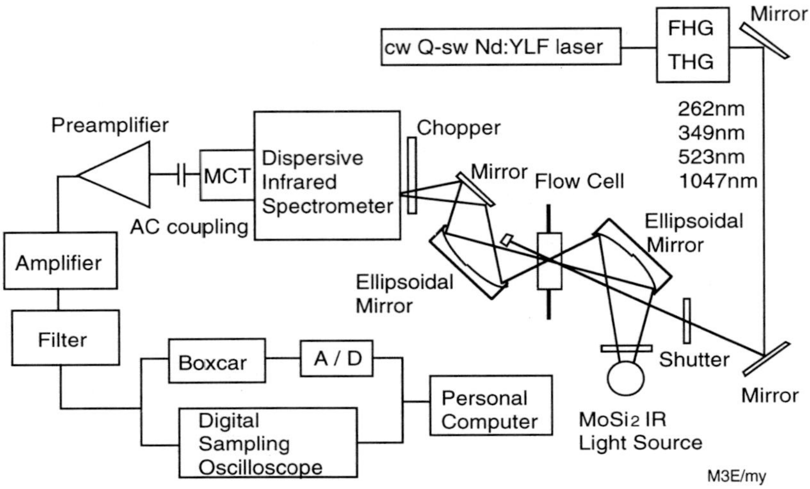

Figure 4: Nanosecond time-resolved dispersive infrared spectrometer.

Figure 5 shows the time-resolved infrared spectra after 0 μs to 45 μs of the photoexcitation of all-trans retinal in cyclohexane [6]. There, negative peaks represent the depletion of the all-trans isomer in the ground state and the positive peaks are due to the photogenerated species. The spectra consists of more than one components with different time behavior. In order to sort out these spectral components, we use the SVD (Singular Value Decomposition) analysis. With a few assumption of the decay dynamics of the transient species, we obtain two SVD component (a) and (b) as shown in Figure 6. The component (a) has the spectrum consisting of the depletion of the ground state all-trans band and the generation of the excited triplet all-trans band. It rises within the limit of the spectrometer response and decays with a time constant of a few microsecond. This indicates that the all-trans triplet state is generated from the all-trans ground state by the photoexcitation and that it relaxes back to the ground state with the same configuration (all-trans). No isomerization is involved in the component (a). The component (b) has a more complicated spectrum which can be well reproduced by the depletion of the all-trans ground state and the generation of the 9- and 13-cis ground states with a ratio of 1:3. The temporal behavior of the component (b) is a step function which rises within the time response of the spectrometer and stays constant afterwards. This fast rise suggests that the 9- and 13-cis isomers are formed via the singlet excited manifold and not via the triplet. Thus, nanosecond time-resolved infrared study has proved that the photophysics and photochemistry of all-trans retinal in cyclohexane consists of three different pathways; 1) formation of the all-trans triplet state and its relaxation back to the all-trans ground state, 2) isomerization to the 9-cis isomers via the singlet manifold, 3) isomerization to the 13-cis isomer via the singlet manifold.

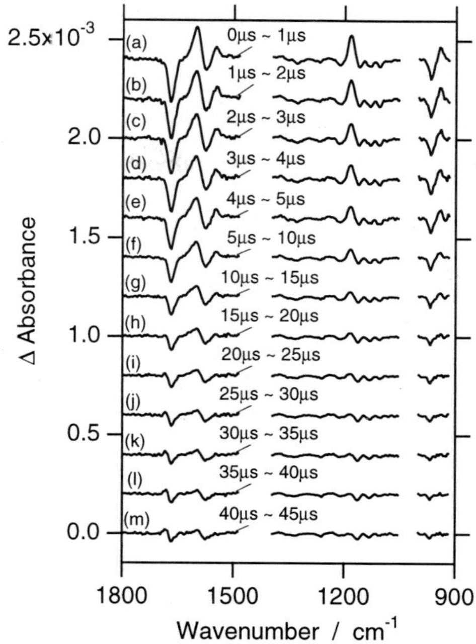

M3E/my

Figure 5: Time-resolved infrared absorption spectra of photoexcited all-trans retinal in cyclohexane.

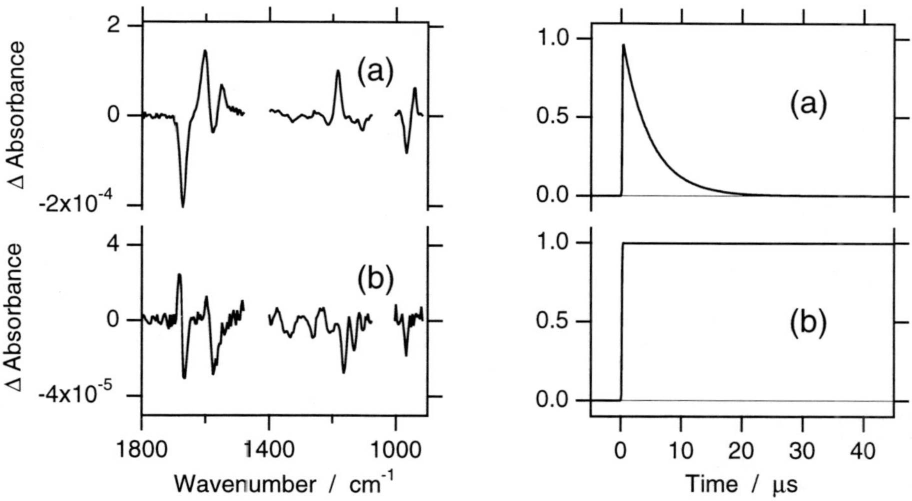

M3E/my

Figure 6: The spectral (left hand) and temporal (right hand) components obtained by the SVD analysis of the time-resolved infrared absorption spectra in Fig. 5.

5. Picosecond Time-frequency Two-dimensional Multiplex CARS Spectroscopy; Photoexcitation of the All-trans and 9-cis Isomers [8]

Picosecond two-dimensional (2d) multiplex CARS (Coherent Anti-Stokes Raman Scattering) spectroscopy [9] uses three laser pulses; one pico/femtosecond pulse for the photoexcitation of the molecule and two nanosecond pulses with angular frequency ω_1 and ω_2 for CARS probing of the photoexcited species (Figure7). The photoexcitation pulse is obtained from the third harmonic of a regeneratively amplified mode-locked Nd:YAG/Ti:Sapphire laser. The ω_1 pulse for the CARS probing is obtained from a diode-pumped Q-switch Nd:YLF laser and the ω_2 pulse from a broad-band dye laser pumped by the Nd:YLF laser. The spectrum of the broad-band ω_2 pulse covers from 900 cm^{-1} to 1600 cm^{-1} of the anti-Stokes side of the ω_1 pulse. The generated CARS signal is first spectrally resolved by a polychromator and then temporally resolved by a streak camera. The resultant time-frequency two dimensional CARS image is detected by a CCD detector. The time resolution of the system is determined primarily by the triggering jitter of the streak camera. In the case of femtosecond photoexcitation with a Ti:Sapphire laser, time-resolution as high as 15 ps is achieved. The efficiency of this 2-D CARS system is much higher than the conventional scanning CARS and this has made us possible to obtain high-quality CARS spectra of fluorescent electronically excited species in solution. Note that retinal, which is often referred to as "non-fluorescent", is strongly fluorescent in the scale of very low light level detected in Raman spectroscopy. Spontaneous time-resolved Raman spectroscopy of retinal just after the photoexcitation is prohibitingly difficult due to the coexisting fluorescence.

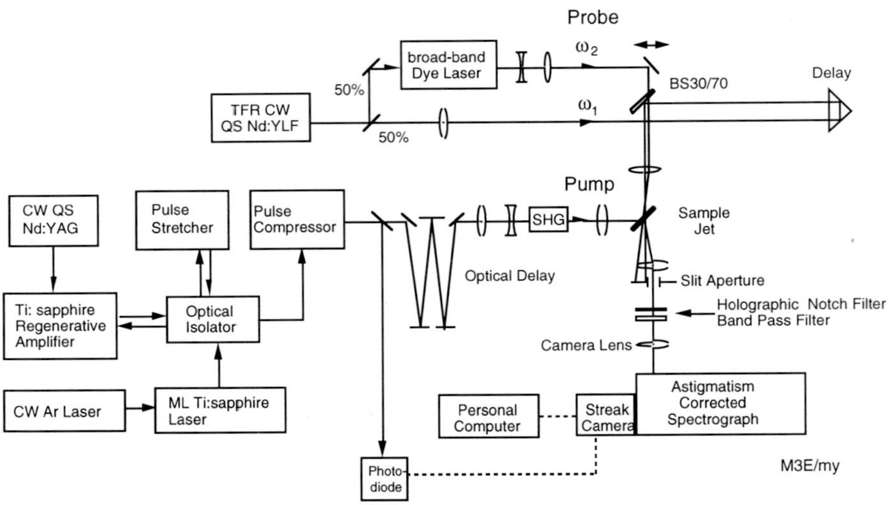

Figure 7: Picosecond time-frequency two-dimensional multiplex CARS system.

Picosecond 2-D CARS spectroscopy of either of the all-trans and the 9-cis photoexcitation detects the formation of the all-trans triplet state which has exactly the same Raman frequencies as those observed in spontaneous Raman spectroscopy [8]. The difference found between the two isomers is the rise time of the all-trans triplet population (Figure 8). For the all-trans photoexcitation (Figure8-A), the population of the all-trans triplet state rises within the response of the spectrometer which is about 90 ps in this experiment. On the other hand, for the photoexcitation of the 9-cis isomer (Figure 8-B), the all-trans triplet rises with a time constant of about 900 ps. This delay in the rise means that there exists a precursor with 900 ps lifetime for the all-trans triplet formation. Since the lifetime of 900 ps is too long for a singlet state, we conclude that this precursor is the 9-cis isomer in the triplet state. In other words, the isomerization from the 9-cis to all-trans configuration occurs in the triplet manifold with a time constant of 900 ps. The 9-cis to all-trans photoisomerization pathway is thus elucidated by picosecond 2-D CARS spectroscopy.

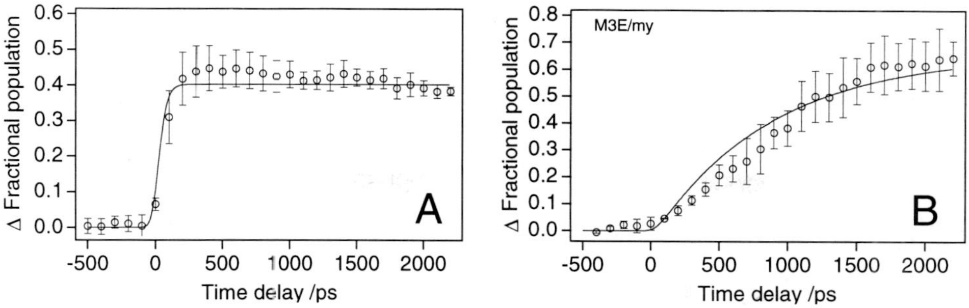

Figure 8: The rise of the population of the all-trans triplet state following the photoexcitation of all-trans (A) and 9-cis retinal (B).

6. Picosecond Time-resolved Fluorescence Spectroscopy; Photoexcitation of the All-trans Isomer [10]

Combining a steak camera with a stable femtosecond Ti:Sapphire laser system, we are able to perform high efficiency time-resolved fluorescence spectroscopy. Advantage of this method lies in the fact that not only the temporal information (rise and decay curves) but also the spectral information is obtained with a single measurement. Note that other competent methods, the time-correlated single photon counting and the frequency up-conversion, can give accurate temporal profile of the fluorescence intensity. However, they need many separate measurements with changing the laser wavelength before the time-resolved fluorescence spectra are constructed.

Figure 9 shows the time-resolved fluorescence spectra of all-trans retinal in hexane. Just after the photoexcitation, the fluorescence maximum is located around 460 nm and, after 15 ps, it shift to a longer wavelength. The analysis of the decay curves at four different wavelengths indicates that two fluorescent excited singlet states are detected. One has a very short lifetime which decays within the time response (10 ps) of the spectrometer and the other has a lifetime of 33 ps. The species with 33 ps lifetime is assigned to the S_1 state because the lifetime agrees with reported rise time of the all-trans triplet state. Picosecond fluorescence spectroscopy has revealed that there is an additional singlet excited state that may participate in the process of photoisomerization.

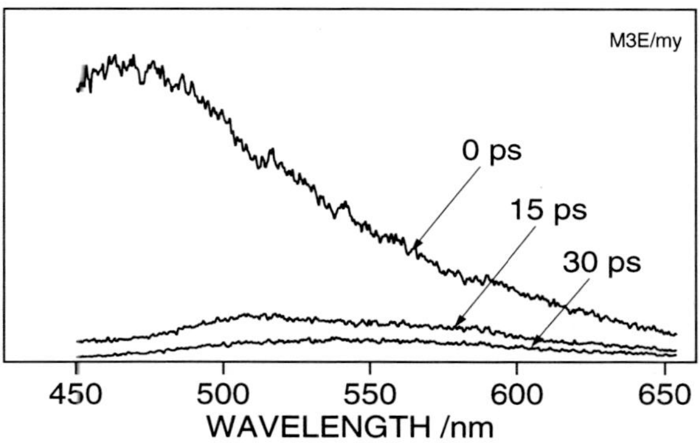

Figure 9: Picosecond time-resolved fluorescence spectra of all-trans retinal in hexane.

7. Femtosecond Time-resolved Visible Absorption Spectroscopy; Photoexcitation of the All-trans Isomer [11]

Our most recent work on the retinal photoisomerization uses a femtosecond time-resolved absorption spectrometer which is based on a femtosecond Ti:Sapphire oscillator/regenerative amplifier system. Photoexcitation is made with the second harmonic of the regeneratively amplified femtosecond pulse (400 nm) at a repetition rate of 1 kHz. Femtosecond white light for probing the transient absorption is generated using self phase modulation by focusing the fundamental of the amplified pulse into water. The effect of the chirp of the probe white light is corrected with a new method recently developed by us [12]. The measured cross correlation time is a few hundred femtosecond.

Time-resolved visible absorption spectra of all-trans retinal in cyclohexane in the time range between 0 ps to 80 ps are shown in Figure 10. The SVD analysis indicates that, in addition to the triplet state for which the transient absorption is well established, there are at least three singlet excited state that contribute to the time-resolved absorption spectra in the femto/picosecond regime [11]. These excited singlet state, tentatively called S_3, S_2, and S_1 states, have the lifetimes of 100 fs, 1.6 ps, and 26 ps, respectively. All these singlet states can contribute to the photoisomerization.

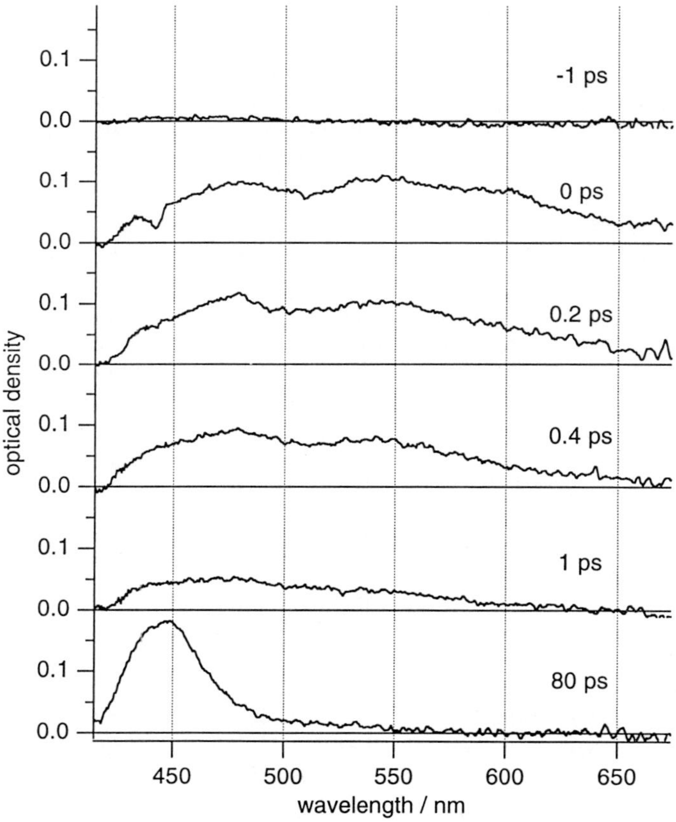

Figure 10: Femtosecond time-resolved visible absorption spectra of photoexcited all-trans retinal in hexane.

8. The Mechanism of the Retinal Photoisomerization: What Has Been Elucidated up to the Present Time

Cmbining all the time-resolved spectroscopic data presented in Sections 3-7, we are able to construct a model that explains some of the characteristics of the retinal photoisomerization. The model is schematically depicted in Figure 11 for the all-trans/9-cis photoisomerization.

The photoexcitation of the all-trans isomer first generates the all-trans S_3/S_2 state(s) which populate the the all-trans S_1 state within a few picosecond. The all-trans S_1 state is then converted with a time constant of 26 ps to the all-trans T_1 state by intersystem crossing. The all-trans T_1 state relaxes back to the all-trans ground state with a time constant of a few microsecond. No isomerization takes place in this pathway. The isomerization does take place from the singlet excited states. Most probably, it proceeds via the perpendicular state which, after a diabatic conversion to the ground state potential, generates eqaul amounts of the all-trans and the 11-cis isomers. This isomerization pathway from the singlet state(s) seems to universal for polyenes.

The photoexcitation of the 9-cis isomer also generates the 9-cis S_3/S_2 state(s). The 9-cis S_1 state is populated within 1 ps of the photoexcitation, which transformes into the 9-cis T_1 state with a time constant of 26 ps [13]. The 9-cis T_1 state is then converted to the all-trans T_1 state with a 900 ps lifetime. The photoisomerization takes place on the excited triplet potential surface. The 9-cis to all-trans isomerization may also take place, though in a much smaller quantum efficiency, through the singlet pathway.

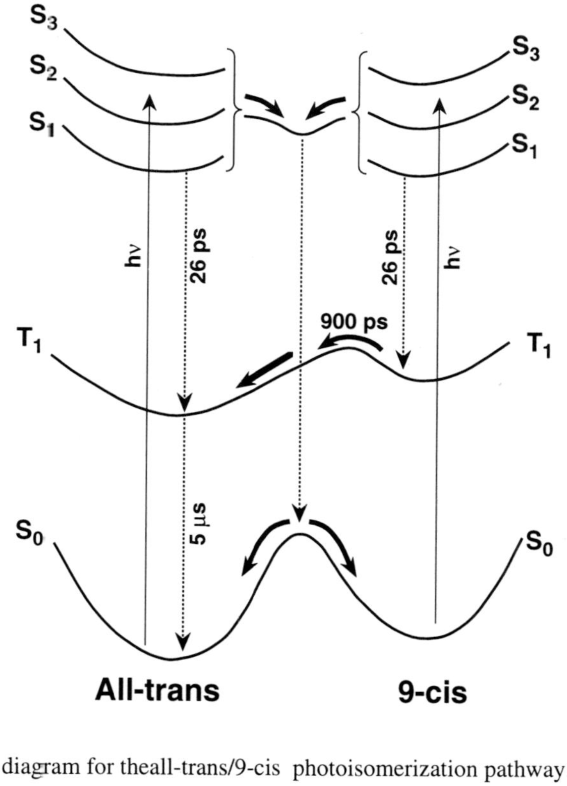

Figure 11: Schematic diagram for theall-trans/9-cis photoisomerization pathway.

Acknowledgement

I would like to thank all the collaborators of the work included in this review.

References

[1]. M. Ottolenghi, *Adv. Photochem.,* vol. 12, 1970, p. 249 .
[2]. T. Arai and K. Tokumaru, *Chem. Rev.,* vol. 93, 1993, pp. 23-39.
[3]. V. K. Deckert, K. Iwata, and H. Hamaguchi, *J. Photochem.,* in press.
[4]. H. Hamaguchi, in *Vibrational Spectra and Structure*, J. R. Durig ed., Elsevier, Amsterdam, 1987, pp. 227-309.
[5]. H. Hamaguchi, H. Okamoto, M. Tasumi, Y. Mukai, and Y. Koyama, *Chem. Phys. Lett.,* vol. 107, 1984, pp. 355-359.
[6]. T. Yuzawa and H. Hamaguchi, *J. Mol. Structure*, vol. 352/353, 1995, pp.489-495.
[7]. T. Yuzawa, C. Kato, M. W. George, and H. Hamaguchi, *Appl. Spectrosc.,* vol. 48, 1994, pp. 684-690.
[8]. T. Tahara, B. N. Toleutaev, and H. Hamaguchi, *J. Chem. Phys.,* vol. 100, 1994, pp. 786-796.
[9]. T. Tahara and H. Hamaguchi, *Rev. Sci. Instruments,* vol. 65, 1994, pp. 3332-3338.
[10]. T. Tahara and H. Hamaguchi, *Chem. Phys. Lett.,* vol. 234, 1995, pp.275-280.
[11]. S. Yamaguchi and H. Hamaguchi, *J. Mol. Structure,* in press.
[12]. S. Yamaguchi and H. Hamaguchi, *Appl. Spectrosc.,* vol. 49, 1995, pp. 1513-1515.
[13]. S. Yamaguchi and H. Hamaguchi, paper in preparation..

Structure Determination by Using Pulsed ESR

Junichi Isoya[1,2] and Satoshi Yamasaki[2,*]

[1]University of Library and Information Science, 1-2 Kasuga, Tsukuba-city, Ibaraki 305 Japan
[2]Joint Research Center for Atom Technology, National Institute for Advanced Interdisciplinary Research, 1-1-4 Higashi, Tsukuba-city, Ibaraki 305 Japan

Abstract

Electron spin echo envelope modulation (ESEEM) method of pulsed electron spin resonance (ESR) technique is able to resolve weak hyperfine interaction which is hidden underneath the linebroadening in conventional ESR of continuous-wave (cw) mode. Several examples in which the ESEEM method is applied to elucidate the local structure of impurities and point-defects which exist in low concentrations are demonstrated.

1 Introduction

Impurities and point-defects, even in low concentrations, might strongly affect the electronic and optical properties of device materials. In the case of the solid-state lasers and the doping of semiconductors, the useful properties in practical applications are created by impurities. On the other hand, physical properties of device materials are, quite often, severely spoiled by impurities and point-defects. For example, the applicability of hydrogenated amorphous silicon (a-Si:H) based solar cells is, at present, severely limited by the presence of the light-induced degradation which is attributed to an increase in the concentration of the dangling bond defects. Characterization of impurities and point defects is important in improving device properties from material side.

Even in the case that an impurity atom is replacing a single atom of crystal lattice, both the wavefunction of the impurity state and the structural relaxation are expected to be extending to a considerable spatial range. To understand the behaviors of impurities and the mechanism of formation of point-defects, local structure determination which contains much more detailed information than simple identification as "chemical species" is required.
ESR, although its applicability is restricted to a species with unpaired electron(s), is a powerful tool for identifying impurities and point-defects which exist in low concentrations by supplying detailed microscopic information of structure and electronic wavefunction. Definite identification of point-defects in semiconductor crystals such as impurities, vacancies, interstitials and anti-site-defects have been given by using ESR.

The ESR spectrum is described by using the spin Hamiltonian. For S=1/2, the spin Hamiltonian is:

$$\mathcal{H} = \beta_e \mathbf{S} \cdot \mathbf{g} \cdot \mathbf{B} + \Sigma_i \left[\mathbf{S} \cdot \mathbf{A}_i \cdot \mathbf{I}_i - (g_n)_i \beta_n \mathbf{I}_i \cdot \mathbf{B} + \mathbf{I}_i \cdot \mathbf{Q}_i \cdot \mathbf{I}_i \right].$$

Springer Proceedings in Physics, Vol. 81
Materials and Measurements in Molecular Electronics
Editors: K. Kajimura · S. Kuroda © Springer-Verlag Tokyo 1996

where the successive terms denote the electron Zeeman interaction, hyperfine interaction, nuclear Zeeman interaction, and nuclear quadrupole interaction which is non-zero only for $I \geq 1$. The g-tensor is useful in identifying the symmetry of paramagnetic species and the electronic configuration, and in distinguishing different paramagnetic species. In microscopic structural identification, definite information is often obtained from the observation of the hyperfine interactions. From the hyperfine interaction, we identify the kind of the impurity atom and determine the extent of the wave function of the unpaired electron, the *sp* hybrid ratio and the geometrical structure. Although we observe the microwave-induced flipping of electron spin in ESR experiments, the most reliable structural information is obtained by picking up the information of nuclei coupled to the electron spin by hyperfine interaction.

Synthetic diamond crystals grown by the high pressure, temperature-gradient method, usually, exhibit bright yellow color which is caused by isolated nitrogen (20~100 ppm) substituting for carbon. Why the nitrogen, which has an extra electron in the lattice of diamond, does not give n-type semiconducti iiamond ? ------ we get answer from its ESR spectrum [1,2]. The spin densities and the *sp* hybrid ratio which were determined from the ^{14}N and ^{13}C hyperfine interactions observed serve as information of the structure relaxation. The unpaired electron enters into anti-bonding orbital of one of four N-C bonds. The anti-bonding character is significantly decreased by a distortion involving elongation of the N-C bond which lowers the symmetry from T_d to C_{3v}. The spin density is larger on the carbon atom than on the nitrogen atom, with the displacement from the normal lattice position larger for the carbon atom than for the nitrogen atom. The localized character of the unpaired electron is confirmed by the observation of ^{13}C hyperfine interaction of 9 carbon atoms surrounding the N-C bond.
In a conventional ESR of cw-mode, with the lack of sufficient resolution, a wealth of information is often hidden underneath the line broadening. The ESR signals in solid samples are usually inhomogeneously broadened, consisting of many overlapping spin packets with different resonant frequencies. Small hyperfine splitting is easily obscured by the line broadening caused by other sources. When many nuclei are coupled to the same electron spin, hyperfine interaction itself is often the major source of the inhomogeneous broadening by splitting the signals into too many lines to be resolved.

To extract hyperfine interactions and nuclear quadrupole interaction which are hidden underneath the inhomogeneous broadening, such methods as conventional ENDOR (electron nuclear double resonance), pulsed ENDOR of pulsed ESR technique, and ESEEM (electron spin echo envelope modulation) have been successfully applied. In these methods, we observe the hyperfine interaction from the side of nuclei. When an electron spin (S =1/2) is coupled to N non-equivalent nuclei with nuclear spin I, the ESR spectrum is splits into $(2I+1)^N$ lines of ($\Delta M_S=\pm1$, $\Delta m_I=0$) transitions. In the same system, if we observe nuclear transitions ($\Delta M_S=0$, $\Delta m_I=\pm1$), the spectrum (ENDOR spectrum) consists of $4 \times I \times N$ lines. When electron spin and nuclear spin are coupled by hyperfine interaction, the shift of the nuclear frequencies (ENDOR frequencies) from its free precession frequency $\nu_n = g_n \beta_n B$ is determined by hyperfine interaction and nuclear quadrupole interaction.The first-order ENDOR frequencies, in the case of weak hyperfine interaction, for I=1/2 and for I=1 are

$$\nu_{\mathrm{ENDOR}} = \nu_n + M_S A_{\mathrm{eff}} / h,$$

$$\nu_{\mathrm{ENDOR}} = \nu_n + M_S A_{\mathrm{eff}} / h \pm \nu_Q,$$

respectively, where A_{eff} / h is the hyperfine splitting and $2\nu_Q$ is the quadrupole splitting. The first-order quadrupolar correction for I=1, with asymmetric parameter $\eta = 0$ is

$$\nu_Q = (3/8)\, e^2 qQ\, (3\cos^2\theta_Q - 1)/h .$$

In cw-ENDOR and pulsed ENDOR experiments, the nuclear transitions are observed by monitoring the change in the ESR signals as a function of the radiofrequency. The ESEEM method of pulsed ESR technique, which is an alternative to ENDOR, is particularly fitted to extract weak hyperfine interactions and nuclear quadrupole interactions such as one arising from magnetic nuclei of surrounding atoms nearby paramagnetic species [3-5].

In pulsed ESR, transient signals produced by coherent excitation using strong and short microwave pulses (typically, with the power of 100W and the duration of 20 ns) are observed in time domain. The signal is measured, normally with the magnetic field fixed, as a function of time after the pulse or as a function of the interpulse delay. The microwave pulse, which is characterized by the turning angle θ_p through which the microwave field B_1 turns the electron spins, affects the spins within $\sim B_1$ of resonance simultaneously. The free-induction-decay (FID), which is the decay of the transverse magnetization following a single pulse (normally with θ_p=90°) and which corresponds to the Fourier transform (FT) of the continuous-wave ESR spectrum, decays rapidly since the time constant of the decay is determined by the total linewidth of the inhomogeneous broadening (the lifetime of the FID is determined by B_1, if the linewidth exceeds B_1). Thus, we do not gain the resolution from FT-ESR which utilizes the free-induction decay. Moreover, in most samples of solid state, the most part of the FID disappears within the deadtime of the spectrometer. In two-pulse Hahn echo sequence (90°- τ - 180° - τ - echo) , we use the echo signal which appears after two pulses. If we follow the echo intensity as a function of the interpulse delay τ, the signal is observed in a time range much longer than the free induction decay. In the dangling bond defects (ΔB_{pp}=0.7 mT at 9.5 GHz) of deuterated amorphous silicon (a-Si:D), the two-pulse echo is observable for τ well beyond 100 μs, while the FID decays with the lifetime of ~20 ns. Since the echo decay in the time domain (the time constant of the decay is called phase memory time T_M) corresponds to the lineshape of the spin packet in the frequency domain, the two-pulse method supplies a resolution of the linewidth of spin packet by overcoming inhomogeneous line broadening.

When nuclear spin is located in such a position that the dipolar hyperfine interaction is comparable in magnitude to the nuclear Zeeman interaction, nuclear spin is influenced by two magnetic field from two different directions, the dipolar hyperfine field from electron spin and the external magnetic field. In this case, since m_I is not a good quantum number, in addition to allowed transitions (ΔM_S=±1, Δm_I=0), formally forbidden transitions (ΔM_S=±1, Δm_I=±1) are partially allowed. In pulsed ESR, strong microwave pulse(s) excites both allowed and semi-forbidden transitions simultaneously. The ESEEM which arises from an interference effect between allowed and semi-forbidden transitions appears as a periodic variation (modulation) of the echo intensity superimposed on the slow echo decay in time domain.The modulation frequencies consist of ENDOR frequencies and sum and difference of ENDOR frequencies. If the echo decay is sufficiently slow, the modulation can be observed for many cycles. With the two-pulse ESEEM, ENDOR frequencies can be measured with an accuracy basically determined by the homogeneous linewidth of the spin packets. A three-pulse stimulated echo sequence (90°- τ - 90° - T - 90°- echo, τ fixed, T scanned) is also frequently employed in the ESEEM measurements. The three-pulse echo decay, which is mainly caused by the spin-lattice relaxation, persists over a time scale much longer than T_M. The modulation frequencies of the three-pulse ESEEM consists of the ENDOR frequencies without involving their sum(s) and difference(s).

In both two-pulse and three-pulse echo decays, the ESEEM spectrum observed is a product of the slow echo decay due to spin relaxation and the normalized modulation. Both the modulation frequencies and the modulation amplitudes can be calculated in a straightforward manner from the spin-Hamiltonian parameters. In single crystal, the ENDOR frequencies obtained as the modulation frequencies are used to determine the A-tensor and the **Q**-tensor. For a randomly-oriented system, usually, simulation of the normalized modulation is carried out. When the anisotropic parts of the hyperfine interactions are describable in terms of electron-nuclear dipolar interactions, the positions of magnetic nuclei nearby electron spin are determined with high accuracy. Since the origin of ESEEM is the state-mixing caused by weak anisotropic hyperfine interaction, the method is suited to locate magnetic nuclei sited at a relatively remote distance (2~ 6 Å).

2 Site determination of Ni impurity in synthetic diamond crystals

Synthetic diamond crystals are grown by using metal solvent such as iron, cobalt, nickel and their alloys. The assignment of the ESR signal with an isotropic g-value of 2.0319 to be arising from Ni, which is inferred from the fact that the signal is absent for crystals grown from a metal solvent without Ni, is confirmed by the observation of the ^{61}Ni hyperfine structure (I=3/2, natural abundance 1.2%). The tetrahedral symmetry is confirmed by the observation that the ^{13}C hyperfine tensor of the nearest-neighbor carbons has trigonal symmetry (the 111 class) [6]. There are two tetrahedral sites for impurities, the substitutional site and the tetrahedral interstitial site (Fig. 1). Both sites are tetrahedrally surrounded by four carbon atoms 1.54Å distant. Thus, the determination of the arrangement of the nearest-neighbors is not sufficient for assigning the site. The substitutional site has twelve next-nearest-neighbors 2.52Å apart belonging to monoclinic-I symmetry (the 110 class), while the interstitial site has six next-nearest-neighbors 1.78 Å apart belonging to tetragonal symmetry (the 100 class). The assignment of the site requires to determine the arrangement of atoms up to the second-nearest-neighbors. The linewidth of cw-ESR of the substitutional Ni^- center strongly depends on both the concentration of the center and that of isolated nitrogen. In the cw-ESR measurements, the angular dependence of the ^{13}C hyperfine splitting of the nearest-neighbor carbons and the observation of the ^{61}Ni hyperfine structure were achieved by obtaining a crystal which gave the linewidth of 0.025 mT (700 kHz) at 77 K by varying the composition of the metal solvent [6]. Even with the narrow linewidth, the ^{13}C hyperfine structure of the next-nearest carbons was only partially resolved.

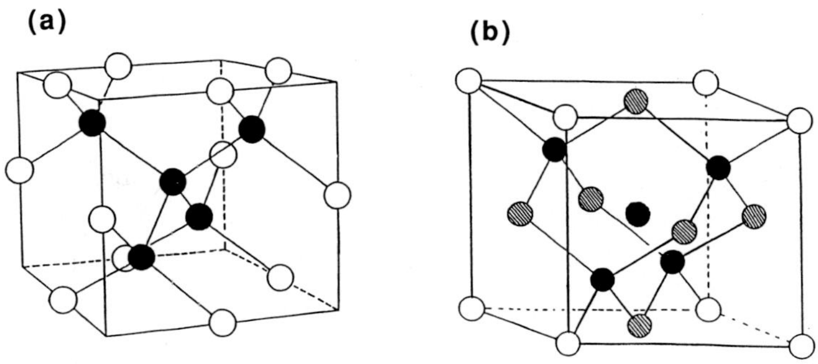

Fig.1 (a) Substitutional site. (b) Tetrahedral interstitial site of diamond.

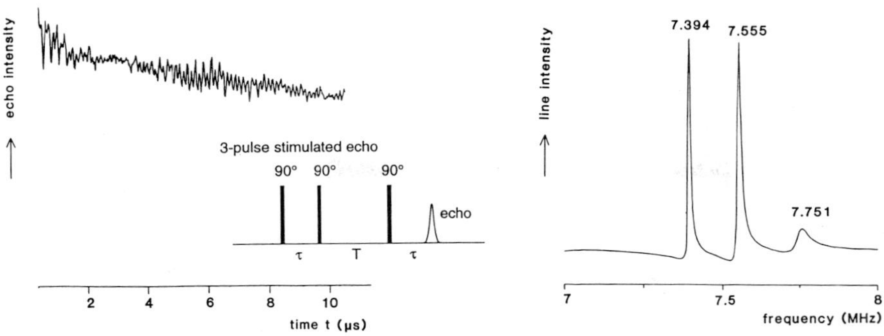

Fig. 2 Three-pulse ESEEM spectrum of the substitutional Ni^- center in synthetic diamond crystal (10% ^{13}C) at 45 K. The time-domain spectrum and the portion of the frequency-domain spectrum obtained by LP-ZOOM method are shown.

By applying three-pulse ESEEM, the angular dependence of the ^{13}C ENDOR frequencies was measured with a rotation of the crystal around the [011] axis, which could resolve all seven sets as expected from the symmetry of the next-nearest-neighbors of the substitutional site [6]. The linewidth of the frequency domain spectra which was obtained by LP-ZOOM method was 20 kHz (Fig.2). The ESEEM spectra were taken by using a crystal 10% enriched in ^{13}C (I=1/2, natural abundance 1.1%). The charge state -1 was determined from the effective spin S=3/2 which was measured by transient nutation method of pulsed-ESR.

Perhaps, it is somehow surprising that Ni could be incorporated into the lattice of diamond (C-C bond length : 1.54 Å) as dispersed impurities. So far incorporation of dispersed impurities into the lattice of diamond crystals at the growth stage are limited to only a few elements (N, B, Si, Ni). While p-type semiconducting diamond crystals are easily obtainable by incorporating boron, the growth of n-type crystals has not been established yet. The fact that nickel really occupies the substitutional site should be encouraging to try to incorporate new impurities, especially those which are promising candidates for n-type dopants. The ESR signals assigned to interstitial Ni^+ in diamond has been also reported [7].

3 Dangling-bond defects in *a*-Si:H

Unlike unhydrogenated amorphous silicon (*a*-Si), hydrogenated amorphous silicon (*a*-Si:H) exhibits photoconductivity and doping and has become a technologically important material for applications such as solar cells and thin-film transistors. The concentration of dangling-bond defects, which amounts to ~10^{19} cm^{-3} in *a*-Si, is reduced to ~10^{15} cm^{-3} in *a*-Si:H of device quality. Typically 10% H is contained in *a*-Si:H (5×10^{22} Si cm^{-3}) films of device quality. The covalently bonded hydrogen plays a key role in improving the electronic properties by passivating the dangling bond defects . However, prolonged light illumination increases the concentration of the dangling-bond defects by more than one order of magnitude and causes a decrease in the photoconductivity and dark conductivity. The dangling bond defects, which give an ESR signal of g=2.0055 (S=1/2) , are the dominant recombination centers for excess electrons and holes. Thermal annealing, typically 150°C for 1 h, restores the original low defect concentration. It has not been established whether the modification of the amorphous network by the incorporation of hydrogen causes the metastability or not. To clarify the microscopic creation mechanism of the photocreated metastable dangling bond defects, the experimental determination of the spatial relationship between the dangling bond and the hydrogen is crucial.

In a conventional ESR of cw-mode, weak hyperfine interaction, which gives information of the distance between the hydrogen and dangling bond is completely hidden underneath the inhomogeneous broadening (ΔB_{pp} ~ 0.7 mT at 9.5 GHz) which is mainly caused by the distribution of the g-value, arising from both the random orientation and site-to-site variation in structure. Since there is essentially no difference in linewidth between *a*-Si:H and *a*-Si:D, hydrogen atoms are located away ($r > 3$Å) from the unpaired electron, at least for the majority of the dangling-bond defects.

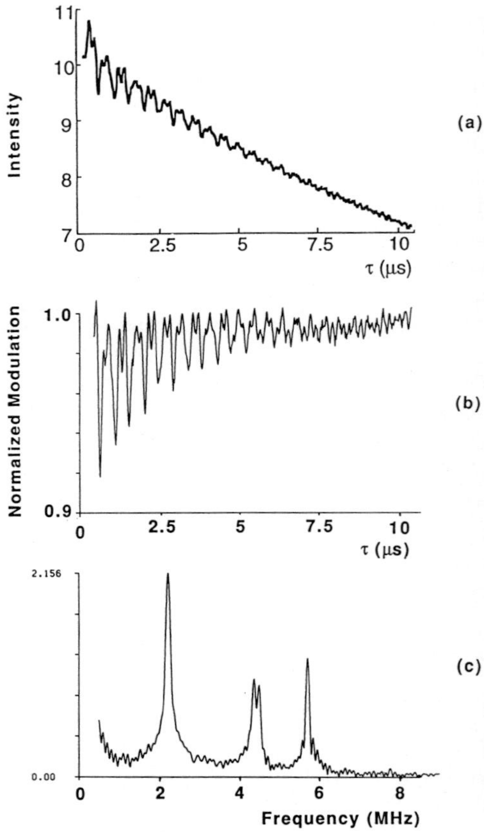

Fig.3 Two-pulse ESEEM spectrum of photocreated dangling bond of *a*-Si:D (9.54 GHz, 56 K).
(a) Two-pulse echo decay.
(b) Normalized modulation.
(c) FT-ESEEM spectrum.

We have applied the ESEEM method to determine both the bonding nature and the spatial distribution of deuterium nearby dangling bonds in deuterated amorphous silicon (*a*-Si:D) for both the native dangling bonds which remain after annealing and the metastable dangling bonds which are created by prolonged light illumination [8]. We have used *a*-Si:D instead of *a*-Si:H, since D-ESEEM requires less excitation bandwidth of microwave pulses, since the modulation depth which is proportional to $I(I+1)$ for small nuclear quadrupole interaction is larger in D-ESEEM than in H-ESEEM, since the modulation in randomly-oriented system persists for a longer time scale for D-ESEEM than for H-ESEEM, and since the two-pulse echo decay is slower for a-Si:D than for a-Si:H. In Fig.3, two-pulse ESEEM spectrum is shown. In the frequency domain ESEEM spectrum which was obtained by Fourier transform of the normalized modulation, the nuclear quadrupole splitting is noticed in the sum frequency ($2\nu_n$) peak of deuterium. The sum frequency peak is free from hyperfine broadening in the first-order. The quadrupole interaction of deuterium located in the vicinity of the electron spin is similar to that of the majority of deuterium in the a-Si:D film measured by NMR. We note that information obtained by ESEEM corresponds to particular deuterium atoms which are located in the vicinity of the unpaired electron. It was determined that the bonding nature of the deuterium atoms sited near by the unpaired electron is similar to that of Si-D bonding in bulk, for both native and metastable dangling bonds.

The spatial distribution of deuterium nearby dangling bonds was determined by simulation of the normalized modulation of the three-pulse ESEEM. In the simulation, the normalized modulation for each orientation (θ, θ_Q, ϕ_Q) was calculated from the eigen values and eigenvectors obtained by diagonalization of two 3×3 submatrices of the spin-Hamiltonian (S =1/2, I =1) matrices. The normalized modulation from a single D nucleus at a distance r is calculated as an spatial averaging of all possible orientations (θ, θ_Q, ϕ_Q), where the angle θ is the angle between **B** and the vector (**r**) connecting the electron spin and the D nucleus and (θ_Q, ϕ_Q) is the direction of the nuclear quadrupole axis. The overall normalized modulation for the case that several D are coupled to the same electron spin is obtained as the product of the modulation functions of the individual D nucleus. It was determined that the distance to the closest deuterium atom is 4.2Å for both native and metastable dangling bonds [8].

4 Trapping site of hydrogen atom in zeolite

Zeolites, which consists of a regular three-dimensional framework of corner-sharing SiO_4 and AlO_4 tetrahedrons, provides regular arrays of cavities of different shape and angstrom-to-nanometer extensions. By using these cavities as microscopic reactor vessels, , novel species such as small clusters of II-VI semiconductors, one-dimensional chain of Se, clusters of alkali atoms are synthesized. To understand the unique properties of these novel species that are confined in the cavities of zeolites, determination of both the location within the cages and the interaction with the cage walls is important.

As a model case of detailed structure determination by applying the ESEEM method, we have determined the arrangement of water molecules around hydrogen atoms trapped in the cavity of zeolite [9]. Single crystal of natural zeolite (scolecite, $Ca_8(Al_{16}Si_{24}O_{80})\cdot 24H_2O$, F1d1) was irradiated by γ-rays (1×10^7 R) from ^{60}Co source at 77 K. When the crystal was transferred to the cryostat without warm-up after γ-irradiation , the ESR spectrum exhibited a doublet (~50.5 mT) arising from atomic hydrogen. Althogh the shoulders arising from nuclear-spin flip transitions appeared in the ESR signals, hyperfine interaction(s) due to surrounding magnetic nuclei were not resolved in a conventional ESR of cw-mode.

An example of three-pulse ESEEM spectrum recorded at 20 K by using a three-pulse stimulated echo sequence is shown in Fig.4. The frequency domain spectrum was obtained by Fourier transform of the time domain spectrum after the slow echo decay was subtracted. The modulation contains the ENDOR frequencies of ^{27}Al nuclei forming the cage of the trapping site of atomic hydrogen and those of ^{1}H nuclei of H_2O molecules nearby atomic hydrogen.
The local arrangement of ^{1}H nuclei of H_2O molecules surrounding atomic hydrogen is determined by analyzing the angular dependence of the ^{1}H-ENDOR frequencies. When the anisotropic parts of the hyperfine interactions can be described by point-dipole-point-dipole approxi-

mation, the effective hyperfine splitting ($\nu_{ENDOR}(+1/2) - \nu_{ENDOR}(-1/2)$) is

$$A_{eff} = (\mu_0/4\pi)(g\beta_e g_n \beta_n / r^3)[3\cos^2\alpha - 1] + a_{iso},$$

where α is the angle between **B** and the vector (**r**) connecting the electron spin (hydrogen atom) and the ^1H nucleus of H_2O. The resolution of the frequency-domain spectrum in the present case is 0.04 MHz since the modulation is measured over the time range of 25 μs. This resolution corresponds to an accuracy of 0.01Å in the determination of distance r for hydrogen nuclei with r ~ 3 Å.

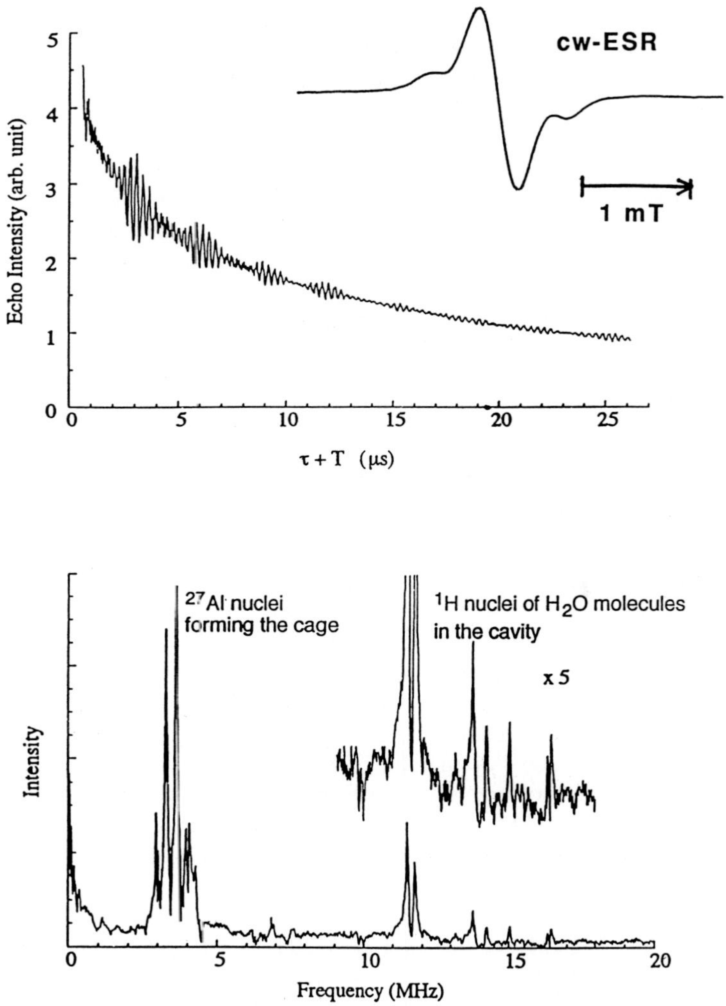

Fig.4 Three-pulse ESEEM spectrum of atomic hydrogen trapped in the cavity of scolecite (τ=400 ns, 9.36 GHz, 20 K). Time-domain spectrum and frequency-domain spectrum are shown, The low field component of the cw-ESR spectrum is also shown.

The ESEEM spectra were measured as a function of angle between the magnetic field and the crystal axis. The angular dependence was measured by rotating the crystal with a step of 2.5° with the c-axis perpendicular to the external magnetic field (**B**). Since some of the modulation frequencies ($\nu\tau \sim n$, n = 1, 2, 3, ·) are suppressed in the three-pulse ESEEM, the ESEEM spectra were recorded at two different τ values at each orientation. The effective hyperfine splitting in the crystal coordinate system (x//a, y//b, z//c) is

$$A_{\text{eff}} = (\mu_0/4\pi)\, g\beta_e g\beta_n/r^3\, [\, 3\sin^2\theta \cos^2(\varphi - \omega) - 1\,] + a_{\text{iso}}\ .$$

where (θ, φ) is the direction of **r** and ω is the angle between **B** and a-axis. The principal values and the principal directions of ^{1}H hyperfine tensors were determined by the least-squares fitting to the hyperfine splittings observed (Fig.5). In Table 1, the local arrangement of hydrogen nuclei of H_2O molecules surrounding atomic hydrogen in scolecite is given. The positions of ^{27}Al nuclei in the cage walls, which give the position of the atomic hydrogen in the cavity were also determined [10].

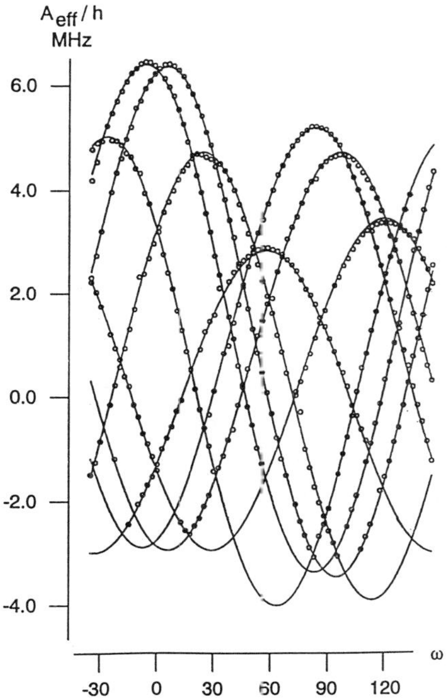

Fig. 5 Angular dependence of the ^{1}H hyperfine splitting of the surrounding H_2O molecules. Observed and calculated splittings are plotted as a function of the rotation angle.

Table 1 Local arrangement of ^{1}H nuclei around atomic hydrogen trapped in scolecite

A_{dip}/h (MHz)	r (Å)	θ	ϕ
4.00	2.70	60.2	-27.1
3.91	2.72	59.3	23.4
3.46	2.83	77.0	4.3
3.37	2.86	80.1	-7.4
3.00	2.98	53.8	56.9
2.95	2.99	58.5	119.1
2.94	3.00	69.1	95.7
2.89	3.01	75.4	82.4

5 Size of micro-pores in the random-network of *a*-SiO_2

Amorphous silicon dioxide (*a*-SiO_2), which is called in such terms as silica-glass, vitreous-silica, and fused quartz, is widely used as the low-loss optical fibers and the optical materials, especially in the uv region. In the most significant role of *a*-SiO_2 in silicon LSI technology is to serve as the gate-oxide layer in MOS devices and as the masking layer for doping. The structure of α-quartz, which is the most common phase of crystalline SiO_2, contains large open channels (the small O-O diameter of 3.55 Å) along the c-axis. Since the density (2.20 g/cm^3) of silica-glass is smaller than that (2.65 g/cm^3) of α-quartz, it is expected that the structure of silica glass should contain micro-pores (small cavities) in the random network. However, it has been difficult to characterize the medium-range structural order of the amorphous network such as the size of silica rings, the silica chain structure, and the size of micro-pores.

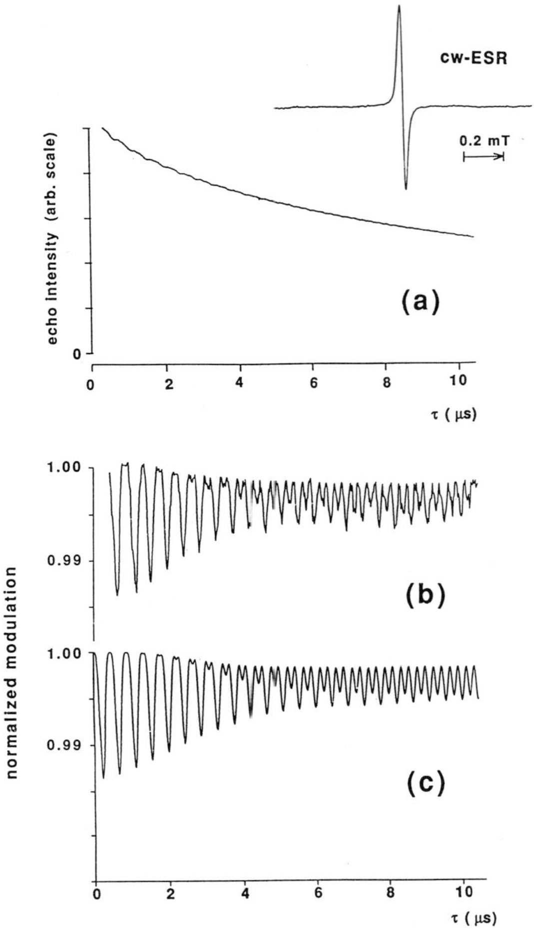

Fig.6 Two-pulse ESEEM spectrum of atomic hydrogen trapped in micro-pores of random-network of a-SiO_2 (the low field component of the hyper-fine structure, 8.21 GHz, 50 K)
(a) Two-pulse echo decay. The high field component of the cw-ESR spectrum is also shown.
(b) Observed normalized modulation.
(c) Calculated normalized modulation with 0.41 ^{29}Si nuclei at 4.5Å in addition to 1.0 ^{29}Si nuclei at 7.0Å.

Hydrogen is among the most common impurities in silica glass. The ESR signals of atomic hydrogen in silica-glass are observed after γ-irradiation at 77 K. Since the hydrogen atoms in silica glass decay above ~100 K, the samples irradiated by γ-rays at 77 K needed to be transferred to a cryostat for ESR measurements without warm-up. Hydrogen atoms, which are produced by γ-ray irradiation at 77K from precursors such as Si-OH and Si-H, diffuse in the lattice, and are trapped in the micro-pores in the network.. The ESR signals of the hydrogen atoms serve to estimate the concentration of the precursors. We could easily detect the ESR signals of atomic hydrogen even for the silica glass samples used for optical communications in which the hydrogen content should be decreased to a minimum level achievable by the present process technology.

We use hydrogen atoms produced by γ-irradiation at 77 K as probe to determine the size of micro-pores in the random network of a-SiO_2 [11]. The ESEEM method has been used to extract weak dipolar hyperfine interaction between the electron spin of the atomic hydrogen trapped in a micro-pore and the ^{29}Si (I = 1/2, natural abundance 4.7 %) nuclei in the cage wall of the micro-pore. In the cw-ESR spectrum of silica-glass irradiated by γ-rays at 77K, a doublet with the splitting of ~50.5 mT (the hyperfine interaction with ^1H with I = 1/2 and natural abundance 99.985 %) characteristic of atomic hydrogen was observed. In cw-ESR, even with the line-width of 0.034 mT (Fig.6), the ^{29}Si hyperfine interaction could not be resolved.

An example of the two-pulse echo decay of hydrogen atoms in silica glass is shown in Fig.6. Superimposed on the slow echo decay, presence of a weak modulation is noticed. When the modulation is normalized to the echo amplitude, the modulation consisting of the free precession frequency ν_n of ^{29}Si and $2\nu_n$ is clearly seen (Fig.6). The modulation arises from dipole-dipole interaction between the electron spin of the hydrogen atom and the nuclear spin of ^{29}Si which is determined by both the distance r between the spins and the angle α between the external magnetic field **B** and the vector **r** connecting the electron spin and the nuclear spin. In a randomly-oriented system, the ESEEM spectrum is a superposition of spectra at many different orientations. In simulation of the ESEEM spectra, the normalized modulation from a single ^{29}Si nucleus at a distance r is calculated as an spatial averaging of all possible orientations (α). The overall normalized modulation for the case that several ^{29}Si are coupled to the same electron spin is obtained as the product of the modulation functions of the individual ^{29}Si nucleus.

If we assume two sets of silicon nuclei, closer ones and distant ones, a good fitting is achieved with 0.41 ^{29}Si nuclei at 4.5 Å in addition to distant silicon nuclei as shown in Fig.3. The small number (0.41) of ^{29}Si nuclei, which comes from the occupation of silicon sites by ^{29}Si with the natural abundance 4.7 %, corresponds to 8 to 9 as the number of the nearest-neighbor silicon atoms. With a simple model of spherical shape for the micro-pores, the number and the distance of ^{29}Si nuclei in the cage wall of the trapping site of hydrogen atom was determined.

6 Summary

The ESEEM method, which has proven to be useful in determining the distance and the number of magnetic nuclei nearby the paramagnetic species, even for a disordered system, has been developed through applications mainly in chemistry and in biology. Here, we have demonstrated several examples in which the ESEEM method was successfully applied for material science.

ESEEM has a limited applicability since there is no modulation effect for a strong hyperfine interaction system in which no significant forbidden nuclear spin-flip transitions occur. The ESEEM method is powerful in determining the local geometry surrounding a paramagnetic probe, rather than in determining the arrangement of atoms on which the unpaired electron is mainly localized. The ESEEM method has a high sensitivity in locating the magnetic nuclei of those atoms which are sited in a specific range of distance from electron spin. In the cases of dangling bonds in a-Si:D and the hydrogen atoms trapped in a-SiO_2, the number of nuclei (D in a-Si:D and ^{29}Si in a-SiO_2) should increase with the distance r, since a-Si:D film contains 10 at.% D and since silicon sites are occupied by ^{29}Si with the probability of the natural abundance (4.7 %). However, the contribution to the modulation from distant nuclei diminishes rapidly with the increase of r, since the modulation intensity is sensitive to the distance, being proportional to r^{-6} for $r > 3.6$ Å. The ESEEM method is suited for identifying and locating impurity atom(s) and charge compensating ion(s) associated with the paramagnetic species. In the dangling-bond defects in a-Si:D, the distance between the dangling bond and the closest deuterium atom was determined to be 4.2Å by using the point-dipole-point-dipole approximation. Among the surrounding D nuclei, the closest one contributes dominantly to the modulation intensity. The distance of 4.2Å seems to be too large for the defect-creation model in which the dangling bond should accompany a deuterium atom at a specific close distance.

When the ESEEM method is applied to single crystal samples, local structure of the paramagnetic species of very low concentration might be determined with an accuracy (0.01Å) compatible to one obtainable for the host lattice by X-ray diffraction. Weak hyperfine interactions due to ^1H nuclei of H_2O molecules within a sphere of diameter of ~10 Å of atomic hydrogen which was trapped in the cavity of the zeolite-network were measured. The accuracy of the local structure determination obtainable depends on the resolution of the modulation frequency which depends on the time span of the time-domain spectrum. Although both the modulation frequency and the modulation intensity which are determined by the spin-Hamiltonian [12] is independent of the spin relaxation rate, the accuracy obtainable depends on the spin relaxation. When the concentration of magnetic nuclei is high, both in single crystals and in disordered systems, the two-pulse echo decay might be fast since those nuclei which are sited at too remote distance to contribute to the modulation do contribute to the phase relaxation by dipolar field fluctuation arising from nuclear spin flip-flops [13]. In these samples, as in our case of hydrogen atoms in zeolite, three-pulse ESEEM which supplies slow echo decay is preferable in gaining resolution.

In single crystal, the modulation amplitude is strongly angular dependent, since the intensity of the forbidden transitions depend on the angle α. One of disadvantages of the ESEEM method is that the modulation is extremely weak for the principal axis directions (α=0°, 90°). In ESEEM measurements of a single crystal sample, the rotation of the sample with precise alignment and with a small step of rotation angle might be required.

In a randomly oriented system, the ESEEM spectrum is a superposition of spectra at many orientations with different modulation frequencies (ENDOR frequencies) and with different modulation amplitudes. The frequency domain spectrum usually shows featureless peaks centered at the free precession frequency ν_n and at $2\nu_n$ in two-pulse ESEEM, and at ν_n in three-pulse ESEEM. To determine the distance and the number of magnetic nuclei in a randomly oriented system, since the lineshape of the frequency-domain spectrum is not like one from which the principal values of the hyperfine tensor is easily assigned, simulation of the damping behaviour of the modulation in the time domain or simulation of the lineshape of the frequency domain spectrum is required. On the other hand, the fact that the modulation frequency agrees with the free precession frequency ν_n in the randomly-oriented sample unless the nuclear quadrupole interaction is large is convenient in assigning the kind of nucleus.

The ESEEM method is one of examples that specific information can be selectively extracted by choosing a pulse sequence. Now, pulsed-ESR has emerged into a new stage in which variety of pulse sequences are selectable, even with a commercial instrument [14]. We believe that importance of ESR spectroscopy in local structure determination should be enhanced by use of various methods of pulsed techniques.

Acknowledgements

This project is partly sponsored by Special Coordination Funds of the Science and Technology Agency of the Japanese Government. The part of this work was performed under the management of Joint Research Center for Atom Technology (JRCAT) partly supported by New Energy and Industrial Technology Development Organization (NEDO). The construction of our pulsed ESR spectrometer benefited greatly from what J.I. learned from collaboration with Prof. J. R. Norris and Dr. M. K. Bowman.

* Permanent address : *Electrotechnical Laboratory, Tsukuba, Ibaraki 305 Japan*

References

[1] W.V. Smith, P. P. Sorokin, I. L. Gelles, and G. J. Lasher, Electron-spin resonance of nitrogen donors in diamond, Phys. Rev. 115, 1959, pp.1546-1552
[2] R. C. Baeklie and J. Guven, ^{13}C hyperfine structure and relaxation times of the P1 centre in diamond, J. Phys. C: Solid State Phys. 14, 1981, pp.3621-3631
[3] W. B. Mims, Electron spin echoes, in *Electron Paramagnetic Resonance*, edited by S. Geschwind, Plenum, New York, 1972, pp.263-351
[4] L. Kevan, Modulation of electron spin-echo decay in solids, in *Time Domain Electron Spin Resonance*, edited by L. Kevan and R. N. Schwartz, Wiley, 1979, pp.279-341
[5] W. B. Mims and J. Peisach, ESEEM and LEFE of metalloproteins and model compounds, in *Advanced EPR, Applications in Biology and Biochemistry*, edited by A. J. Hoff , Elsevier, 1989, pp.1-57
[6] J. Isoya, H. Kanda, J. R. Norris, J. Tang, and M. K. Bowman, Fourier-transform and continuous-wave EPR studies of nickel in synthetic diamond: Site and spin multiplicity, Phys. Rev. B 41, 1990, pp.3905-3913
[7] J. Isoya, H. Kanda and Y. Uchida, EPR studies of interstitial Ni centers in synthetic diamond crystals, Phys. Rev. B 42, 1990, pp.9843-9852
[8] J. Isoya, S. Yamasaki, H. Okushi, A. Matsuda, and K. Tanaka, Electron-spin-echo envelope modulation study of the distance between dangling bonds and hydrogen atoms in hydrogenated amorphous silicon, Phys. Rev. B 47, 1993, pp.7013-7024
[9] J. Isoya, S. Yamasaki, J. K. Lee, T. Umeda, and K. Tanaka, Determination of atom positions of nano-scale local structure by pulsed EPR, in *Extend Abstracts of '95 JRCAT International Symposium on Nanoscale Self- Organization*, 1995, pp.225-228
[10] J. Isoya, S. Yamasaki, J. K. Lee, T. Umeda, K. Tanaka, and Y. Morita, to be published
[11] J. Isoya, S. Yamasaki, K. Tanaka, H. Kanda, Y. Morita, and T. Ohshima, Network structure of a-SiO_2 probed by hydrogen atom, in *Extend Abstracts of '96 JRCAT International Symposium on Atom Technology*, 1996, to be published
[12] M. K. Bowman and R. J. Massoth, Nuclear spin eigenvalues and eigenvectors in electron spin echo modulation, in *Electron Magnetic Resonance of the Solid State*, edited by J. A. Weil, M. K. Bowman, J. R. Morton, and K. F. Preston , The Canadian Society for Chemistry, Ottawa, 1987, pp.99-110
[13] J. Isoya, S. Yamasaki, A. Matsuda, and K. Tanaka, Time-domain measurements of spin relaxation processes of dangling-bond defects in hydrogenated amorphous silicon, Phil. Mag. B 69, 1994, pp.263-275
[14]A. Schweiger, Pulsed electron spin resonance spectroscopy: basic principles, techniques, and examples of applications, Angew. Chem. Int. Ed. Engl. 30, 1991, pp.265-292

ELECTRONIC AND CHEMICAL STRUCTURE OF CONJUGATED POLYMER SURFACES AND INTERFACES: A REVIEW AND SOME NEW RESULTS

K. Xing, M. Fahlman and W. R. Salaneck

Department of Physics, IFM, Linköping University, S-581 83 Linköping, Sweden

Abstract

Conjugated polymers have emerged as viable electronic materials for numerous applications. In the context of polymer electronic devices, it is of critical importance to understand the nature of the electronic structure of the polymer surface and the interface with metals. It has been shown that, especially for conjugated polymers, photoelectron spectroscopy provides a maximum amount of both chemical and electronic structural information in one (type of) measurement. An overview of some details of the early stages of interface formation with metals on the surfaces of conjugated polymers and model molecular solids, especially in connection with polymer-based LED devices, is presented. Materials involved include poly(*p*-phenylenevinylene), or PPV, as well as a series of substituted PPV's, and a diphenylpolyene molecule, namely α, ω-diphenyl-tetradecaheptaene. Some general trends in the behaviour of light-metal atoms on the clean surfaces of conjugated polymers will be pointed out. Finally, a series of new results will be discussed, one in limited detail, the others more briefly, in order to indicate some recent new developments in this area.

1 Introduction

In the field of π-conjugated polymers [1-4], one of the recent breakthroughs is the discovery by the Cambridge group [5,6] that poly(*p*-phenylenevinylene), or PPV, and related polymers can be used as the active component in polymer-based light-emitting diodes [7-16], or polymer-LEDs. The PPV films provide high quantum yield for electro- or photoluminescence in the yellow/green portion of the visible spectrum [8,9,13]. In addition, other π-conjugated polymers may be used to provide light from somewhat different portions of the visible spectrum: poly(*p*-phenylene), or PPP [11], which emits in the blue, and a derivative of PPV, poly(2-methoxy-5-(2'-ethyl-hexoxy)-1,4-phenylenevinylene), or MEH-PPV [7,12,16], and a cyano-substituted PPV, or CN-PPV [17], both with emission more-or-less in the red/orange part of the spectrum. A major factor determining the quantum yield for luminescence is the competition between radiative and nonradiative decay of the electron-hole pairs created within the polymer layer. These pairs can migrate along the chains and are therefore susceptible to trapping at quenching sites where nonradiative (*e.g.*, multiphonon) processes can occur [6].

Another factor in determining the over-all quantum efficiency of polymer-LEDs is the injection of electrons and holes at the respective metal-polymer interfaces [16,18]. Typically, but not exclusively, indium-tin-oxide-coated glass (ITO) is used as the hole injecting contact; one which also allows visible light to escape from the active light-generation medium [16]. The counter electrode is usually a metal with low work function, enabling electron injection into the conduction band of the polymer medium. In this context, the injection efficiency of the minority carrier may be one major limiting factor in over-all device efficiency.

Springer Proceedings in Physics, Vol. 81
Materials and Measurements in Molecular Electronics
Editors: K. Kajimura · S. Kuroda

2 Review

Over the past four years, the combined experiment-theory approach to the study of polymer surfaces and interfaces has been applied to a wide variety of π-conjugated polymers and model molecules, as well as the early stages of metal-on-polymer (and model molecule) interface formation [18,19]. During the course of these studies, certain general trends have emerged.

Broadly, the work has focused upon poly(*p*-phenylenevinylene) [20], poly(2,5-diheptyl-1,4-phenylenevinylene) [21,22], poly(2,5,2',5'-tetrahexyloxy-8,7'-dicyano-diphenylenevinylene) [23], which is a CN-substituted PPV, and several alkyl-substituted polythiophenes [24-28]. In addition, certain molecules were studied in the condensed molecular state as model systems for the conjugated polymers: a diphenylpolyene model molecule for PPV, namely α, ω-diphenyltetradecaheptaene [29-31], abbreviated as "DP7", and a model molecule for the polythiophenes, α-sexithiophene, or α6T . The metal atoms on surfaces of ultra-thin polymer films, or of the condensed molecular solids, include aluminium, sodium, calcium, potassium, and rubidium. In all cases, experiments were carried out using ultra violet and X-ray photoelectron spectroscopy (UPS and XPS) in ultra-high vacuum ($p \leq 10^{-10}$ Torr) [18,19,32]. Quantum chemical modelling used in the interpretation of spectra [33,34] was done using the valence effective Hamiltonian (VEH) calculations; the VEH method has an excellent record of providing reliable estimates of ionisation potentials, bandwidths, and bandgaps for a wide variety of conjugated polymers. The geometries used for the VEH calculations [35] are obtained from full molecular-geometry optimisations with the Modified Neglect of Differential Overlap (MNDO) or Austin Model 1 (AM1) Hartree-Fock semiempirical techniques. In addition, when appropriate for treating metal atoms or ions, the local spin density (LSD) model was employed [36].

2.1 Aluminium

Starting with aluminium [37], in general, at the initial stages of interface formation, atoms deposited upon essentially oxygen-free surfaces lead to the formation of covalent bonds: at the carbon atoms of the α-linkages in polythiophenes [37,38] and α6T [38]; with any of the carbon atoms along the polyene portion of the DP7 molecule; and at the vinylene-carbon atoms of PPV (and alkyl-substituted PPV), although slightly more elaborate configurations are possible in MEH-PPV, DMeO-PPV [39], and CN-PPV [23].

A careful study of the surface localisation of the aluminium atoms indicated that, although clustering occurs on the surface just as in the case of aluminium atoms on the surfaces on inorganic semiconductors, diffusion takes place into the near surface region. Although a diffusion depth distribution profile could not be obtained from the measurements, an estimate of the scale of the diffusion was obtained. The aluminium atoms forming covalent bonds with the molecular or polymer systems were localised to the near surface region within a characteristic length scale on the order of an electron tunnelling distance, *i.e.*, on the order of 20 to 30 Å. Note that these are not exact numbers; they are, however, indicative of the *scale* of the interfacial region.

2.2 Sodium

In all cases studied by the present authors and co-workers, sodium diffuses uniformly throughout the polymer or condensed molecular solid film, and donates electrons to the π-

system, forming Na^+-ions; this leads to the generation of bipolarons and the appearance of bipolaron states in the original energy gap. For the special case of the DP7 molecule, two solitons are generated, which are confined to the (finite sized) polyene portion of the molecule by the phenyl rings which cap the molecule (for environmental stability); these two confined solitons appear formally equivalent to a single bipolaron, with the exception that if the polyene portion of the molecule was infinitely long, the two observed states in the energy gap would coalesce into one degenerate state [31,39]. In all of the other cases, the two bipolaron states within the energy gap are a consequence of the non-degenerate ground state of the π-system in all works referred to in here [40].

2.3 Rubidium and potassium

In these cases, essentially the same behaviour as that of sodium was observed, with the exception that, following vapour deposition, longer times were required to obtain equilibrium, leading there by to somewhat greater detail in the observed spectra than in the case of sodium. For potassium, although again very subtle differences occur, essentially the same behaviour as for sodium or rubidium was obtained [40,41]. Thus for Na, Rb and K, bulk diffusion and n-type doping of the entire polymer or molecular film are observed.

2.4 Calcium

The case of calcium vapour-deposited upon *clean surfaces* in UHV is of particular interest. Observed clearly first for DP7, and subsequently for DHPPV [42], calcium diffuses into the near surface region, donates electrons to the π-system, and forms Ca^{++}-ions. The interfacial region between the Ca-metal contact and the polymer has an approximate scale in the range of 20 to 30 Å (similar to the case of Al atoms). In contrast, cases where there are large numbers of *oxygen-containing species at the surface* of, e.g., PPV [43] or substituted PPV's [21], an interfacial layer of an oxide of calcium is formed initially upon the deposition of calcium atoms in UHV, followed by the deposition of calcium metal after the oxygen-containing species have been consumed by the initial calcium atoms. The scale of this interfacial oxide (insulating) layer also is on the order of 20 to 30 Å, depending upon the details of the surface contamination, chemical impurity of the polymer, and/or the vapour-deposition environment.

3 New developments

3.1 "Dirty calcium" electrodes

Using spectroscopically clean (essentially oxygen-free) surfaces of PPV and CN-PPV, single-layer LED's, consisting of calcium electrode on a single polymer film on ITO glass, have been constructed. The vapor deposition of the calcium was done in the presence of a varying background of partial pressures of oxygen ranging from less then 10^{-12} mbar to 10^{-4} mbar in (otherwise) UHV. Although it might be considered difficult to define an unambiguous figure-of-merit, considering factors such as device yield, performance life-time, and initial luminescence intensity, an optimum occurs for about 10^{-6} mbar of O_2 background pressure. This environment results in a metallic oxide contact which provides "better" devices [44]. In addition, even if an oxidized layer of calcium is deposited first, and then covered with pure calcium, the devices develop shorts relatively soon after initial operation. Apparently calcium atoms diffuse into clean PPV's, doping the conjugated polymer to a state of relatively high electrical conductivity.

The possibility that certain oxides of calcium are electrically conductive, there by providing proper electrical contact and charge injection in the devices, has a basis in recent work on the oxides of calcium, magnesium and other metals [45,46]. Results of published studies of the electronic structure of the oxides of calcium, aluminium and strontium have shown that neither an ionic $Ca^{2+}O^{2-}$ model nor a homopolar $Ca^{+}O^{-}$ model are satisfactory. Rather, the potential energy barrier inside the oxygen sphere for the ground state is such that only one of the electrons of the O^{2-} is localised within the oxygen sphere. The second electron is delocalised in the interstitial space and at neighbouring cation sites. The model which applies is more like $Ca^{2+}O^{-}e^{-}$, where the e^{-} is a delocalised electron. This model is proposed to hold both for CaO and MgO. Although the precise chemical composition of the "dirty calcium" electrodes has not been studied, it appears that oxidised calcium is not a detriment to the injection of electrons.

The over-all implications of the results of the "dirty calcium" work are important. First, it seems clear that the use of UHV in the fabrication of polymer-LED's is not necessary. On the contrary, the use of UHV when clean surfaces of PPV are involved is detrimental to device performance.

3.2 "Dirty magnesium" electrodes

Studies similar to those mentioned for "dirty calcium" above, are in the process of being carried out using magnesium instead. Although these studies are not finished, initial indications are that magnesium does not behave as calcium. The best devices seem to be prepared in UHV on clean surfaces of PPV. Initially higher current densities, at a given voltage, are obtained. The light output is proportional to the current. Thus higher turn-on voltages occur, if "dirty magnesium" electrodes are used [47].

3.3 A blue LED

Using a regular alternating copolymer of di-heptyl substituted benzene and thiophene, namely poly(2,5-diheptyl-1,4-phenylene-alt-2,5-thienylene) [48], or PDHPT, single-polymer-layer LED's have been prepared [49]. The PDHPT exhibits a rather large band gap, because of steric effects which cause the torsion angles between the phenyl rings and the thiophene units to be about 50°, leading to an optical absorption edge near 3 eV and a photoluminescence which peaks near 2.5 eV. Polymer LED's were fabricated using "dirty calcium" electrodes (deposited in the presence of 10^{-6} Torr of O_2) on PDHPT spin-coated on ITO glass. Since the electro-luminescence peaks near 2.4 eV, the colour of the emitted light is blue-green. The turn-on electric filed is about 1.3×10^{8} V/m, and the device appears very bright in room light at 1.7×10^{8} V/m. This translates to bright light at a turn-on voltage of 5 V for a thickness of 300 Å, a film thickness that would be practical in commercial devices. The polymer used in these devices is inherently stable in the presence of the light.

3.4 Water in PPV

The performance of polymer LED's based upon the PPV's is well known to depend upon the presence of water vapor. A study has been undertaken to examine the effect of water on the electronic structure of PPV [50]. Although many different types of water influence may be studied, a short description of part of this study is contained here.

Studies were carried out using both UPS and XPS, taking the appearance of a well-defined π-band edge in the UPS spectrum as the best indication of a clean PPV surface [19]. Initially, the PPV used exhibited the presence of a small amount of residual oxygen-containing species, which

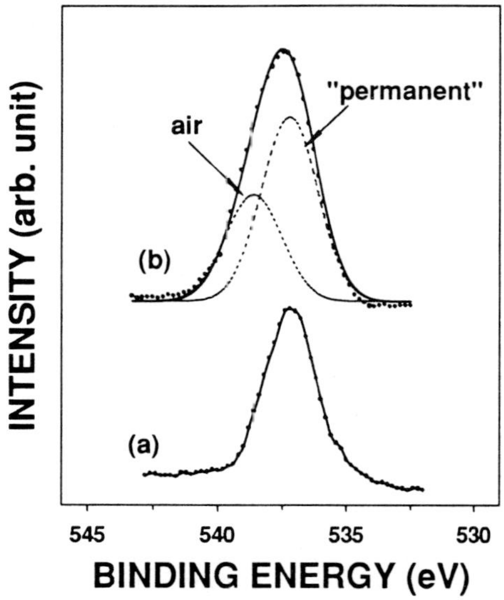

Fig. 1. O1s peak of (a) clean PPV and (b) PPV exposed to air for 13 hours.

are impossible to remove using conventional *in situ* heating processes. About 3 atomic percent (or a little less) of oxygen is seen in the O(1s) XPS spectra, as shown in Fig. 1. Upon exposure to air (a typical practical situation), an additional peak appears in the O(1s) spectra (Fig. 1). This new peak, clearly corresponding to a different oxygen-containing species, may be removed by heating in UHV to about 200 C for a number of hours. The effect of the exposure of PPV to air may be seen in the UPS spectra of Fig. 2. The well-defined π-band edge, the signature of a clean PPV surface, disappears upon air exposure, but returns upon heating in UHV. When exposure experiments were repeated using pure oxygen instead of air, no effects were observed in either the UPS or the XPS spectra, indicating that the oxygen-containing species which affects the PPV upon air exposure is indeed water vapor. Studies using optical absorption spectroscopy and wave-length-dependent ellipsometry also have been employed. The results of the optical measurements are consistent with the effects seen at the π-band edge in the UPS spectra.

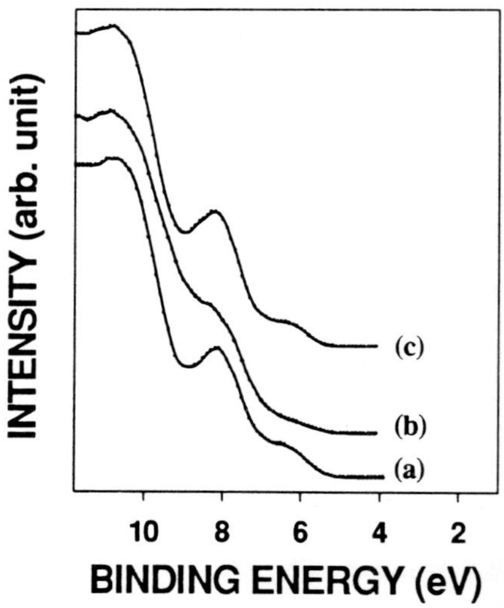

Fig. 2. HeI spectra of : (a) clean PPV, (b) PPV after 13 hours of exposure to air, and (c) PPV after heating in vacuum at 200C for 2 hours.

The affect of water in PPV is likely due to combined effects of swelling and hydrogen-bonding. Using simulations, carried out with the Austin Model One (AM1) model on PPV oligomers, it appears that water molecules have a tendency to be weakly hydrogen-bonded to the oligomers as indicated in Fig. 3 [51].

A main point of the above studies is that water does affect the electronic structure of PPV, right at the π-band edge which is important in luminescence. Fortunately, however, the water may be removed by proper heating in UHV. The details of these studies will be reported elsewhere [50].

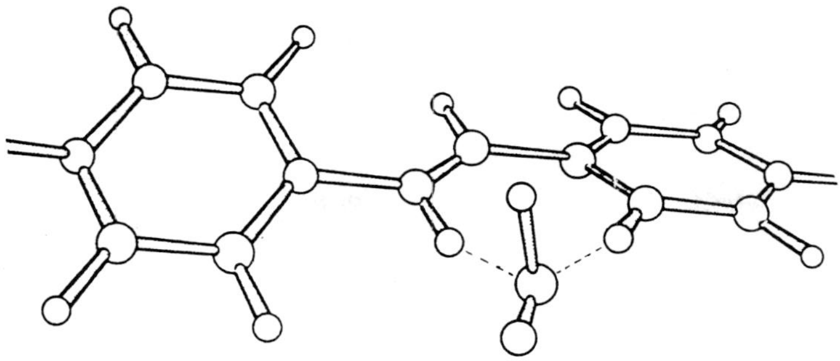

Fig. 3. A water molecule interacting with a segment of a PPV oligomer by forming weak hydrogen bonds with a vinylene and a phenylene hydrogen.

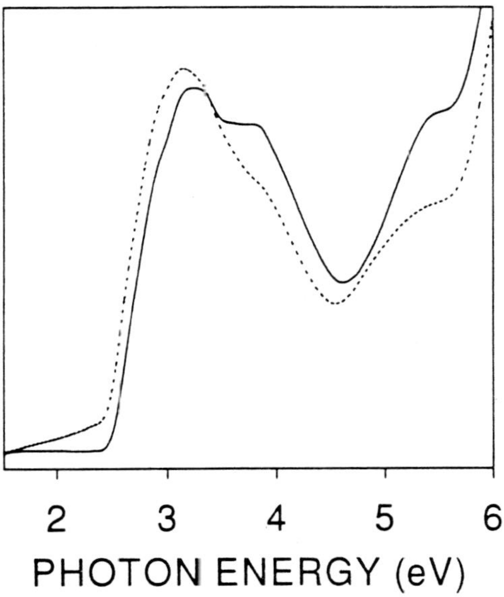

Fig. 4: The optical absorption spectra of PPV are shown, containing absorbed water (dashed line) and after heat reatment to remove water (solid line).

The UPS and the modelling discussed above imply that there should be observable changes in the optical absorption spectra in connection with the presence of water in PPV [50]. In Fig. 4 is shown an optical absorption spectrum of PPV, in UHV, but immediately following a long exposure to water, and after heat treatment in UHV to remove the water (to the extent possible, as discussed above). It is clear that there is an effect, increased absorption, in the low photon energy region below that of the band gap absorption edge in clean PPV. Although these effects are not yet understood in terms of the molecular and electronic structural modelling, the point is made; the presence of water has a pronounced effect on the electronic properties of PPV.

4 Summary

Some results, previously published as well as in progress, have been presented to illustrate the importance of the metal-polymer interface in determining device performance in polymer LED's. In every instance studied, there is *always* some chemistry that occurs at the interface. There is no such thing as an ideal metal-polymer contact without interfacial chemistry of one sort or another. In this context, it is very clear that ultimately control of the interface will be one of the major determining steps in the commercialization of polymer-based electronic devices.

5 Acknowledgements

Research leading to the results reported was supported by the Commission of the European Union, within the ESPRIT program (project number 8013 LEDFOS), as well as Philips Research, NL and Hoechst AB, FRG (in connection with the Brite/EuRam project PolyLED, 0592), and the Human Capital and Mobility program (project 0669 PANET). Research on conjugated polymers in Linköping is supported in general by grants from the Swedish Natural Sciences Research Council (NFR), the Swedish Research Council for Engineering Sciences (TFR), the Swedish National Board for Industrial and Technical Development (NUTEK), and the ESPRIT network of excellence NEOME.

6 References

[1] J. L. Brédas and R. Silbey, "*Conjugated Polymers*," (Kluwer Academic, Dordrecht, 1991).

[2] S. Stafström, W. R. Salaneck, O. Inganäs, and T. Hjertberg, *Proc. 1992 Intern. Conf. Synth. Met., Synth. Met.* **55 - 57** (1993).

[3] W. R. Salaneck, I. Lundström, and B. Rånby, "*Nobel Symposium in Chemistry: Conjugated Polymers and Related Materials; The Interconnection of Chemical and Electronic Structure*," (Oxford University Press, Oxford, 1993).

[4] W. R. Salaneck and J. L. Brédas, "Conjugated polymers," *Solid State Communications, Special Issue on "Highlights in Condensed Matter Physics and Materials Science"* **92**, 31 (1994).

[5] J. H. Burroughes, D. D. C. Bradley, A. R. Brown, R. N. Marks, K. Mackay, R. H. Friend, P. L. Burn, and A. B. Holmes, "Light emitting diodes based upon conjugated polymers," *Nature* **347**, 539-541 (1990).

[6] R. H. Friend, in *Nobel Symposium in Chemistry: Conjugated Polymers and Related Materials; The Interconnection of Chemical and Electronic Structure*, edited by W. R. Salaneck, I. Lundström, and B. Rånby (Oxford Sci., Oxford, 1993), pp. 285.

[7] D. Braun and A. J. Heeger, *Appl. Phys. Lett.* **58**, 1982 (1991).

[8] P. L. Burn, A. B. Holmes, A. Kraft, D. D. C. Bradley, A. R. Brown, and R. H. Friend, *J. Chem. Soc., Chem. Commun.* **32** (1992).

[9] P. L. Burn, A. Kraft, D. D. C. Bradley, A. R. Brown, R. H. Friend, R. W. Gymer, R. H. Friend, and R. W. Jackson, *J. Am. Chem. Soc.* **115**, 10117 (1993).

[10] R. E. Gill, G. G. Malliaras, J. Wildeman, and G. Hadziioannou, *Adv. Mat.* **6**, 132 (1994).

[11] G. Grem, G. Leditzky, B. Ultrich, and G. Leising, *Adv. Mat.* **4**, 36 (1992).

[12] G. Gustavsson, Y. Cao, G. M. Treacy, F. Klavetter, N. Colaneri, and A. J. Heeger, *Nature* **357**, 477 (1992).

[13] D. A. Halliday, P. L. Burn, D. D. C. Bradley, R. H. Friend, O. M. Gelsen, A. B. Holmes, A. Kraft, J. H. F. Martens, and K. Pichler, *Adv. Mat.* **5**, 40 (1993).

[14] Y. Ohmori, M. Uchida, K. Muro, and K. Yoshino, *Jap. J. Appl. Phys.* **30**, L1938 (1991).

[15] Y. Ohmori, M. Uchida, K. Muro, and K. Yoshino, *Solid State Commun.* **80**, 605 (1991).

[16] I. D. Parker, *J. Appl. Phys.* **75**, 1656 (1994).

[17] N. C. Greenham, S. C. Moratti, D. D. C. Bradely, R. H. Friend, and A. B. Holmes, "Efficient Light Emitting Diodes Based on Polymers with High Electron Affinities," *Nature* **365**, 628-630 (1993).

[18] W. R. Salaneck, S. Stafström, and J. L. Brédas, *Conjugated Polymer Surfaces and Interfaces* (Cambridge University Press, Cambridge, 1996).

[19] M. Lögdlund, P. Dannetun, and W. R. Salaneck, in *Handbook of Conducting Polymers*, edited by T. Skotheim, J. Reynolds, and R. Elsenbaumer (Marcel Dekker, New York, 1996).

[20] M. Fahlman, D. Beljonne, M. Lögdlund, P. L. Burn, A. B. Holmes, R. H. Friend, J. L. Brédas, and W. R. Salaneck, "Experimental and theoretical studies of the electronic structure of Na-doped poly(*p*-phenylenevinylene)," *Chem. Phys. Lett.* **214**, 327 (1993).

[21] P. Dannetun, *Private communication.*

[22] M. Fahlman, J. Rasmusson, K. Kaeriyama, D. T. Clark, G. Beamson, and W. R. Salaneck, "Epitaxy of poly(2,5-diheptyl-p-phenylene) on ordered polytetrafluoroethylene," *Synt. Met.* **66**, 123 (1994).

[23] M. Fahlman, W. R. Salaneck, and J. L. Brédas, "Experimental and theoretical studies of the interaction between aluminum atoms and CN-substituted poly(*p*-phenylenevinylene)," *submitted* (1995).

[24] W. R. Salaneck, O. Inganäs, B. Thémans, J. O. Nilsson, B. Sjögren, J.-E. Österholm, J.-L. Brédas, and S. Svensson, "Thermochromism in Poly(3-Hexylthiophene) in the Solid State: A spectroscopic Study of Temperature-dependent Conformational Defects," *J. Chem. Phys.* **89**, 4613 (1988).

[25] M. Lögdlund, R. Lazzaroni, S. Stafström, W. R. Salaneck, and J.-L. Brédas, "Direct Observation of Charge-Induced π-Electronic Structural Changes in a Conjugated Polymer," *Phys. Rev. Lett.* **63**, 1841 (1989).

[26] R. Lazzaroni, M. Lögdlund, S. Stafström, W. R. Salaneck, and J. L. Brédas, "The poly-3-hexylthiophene/NOPF6 system: A photoelectron spectroscopy of electronic structural changes induced by the charge transfer in the solid state," *J. Chem. Phys.* **93**, 6 (1990).

[27] R. Lazzaroni, M. Lögdlund, S. Stafström, W. R. Salaneck, D. D. C. Bradley, R. H. Friend, N. Sato, E. Orti, and J. L. Brédas, in *Conjugated Polymeric Materials: Opportunities in Electronics, Optoelectronics, and Molecular Electronics, NATO ASI Series E, Vol. 182*, edited by J.-L. Brédas and R. Chance (Kluwer Academic, Dordrecht, 1990), pp. 149.

[28] R. Lazzaroni, M. Lögdlund, A. Calderone, J. L. Brédas, P. Dannetun, C. Fauquet, C. Fredriksson, S. Stafström, and W. R. Salaneck, "Chemical and Electronic Aspects of Metal/Conjugated Polymer Interfaces. Implications for Electronic Devices," *Synth. Met.* **71**, 2159 (1995).

[29] C. W. Spangler, E. G. Nickel, and T. J. Hall, *Am. Chem. Soc., Div. Polym. Chem.* **28**, 219 (1987).

[30] M. Lögdlund, P. Dannetun, B. Sjögren, M. Boman, C. Fredriksson, S. Stafström, and W. R. Salaneck, "The electronic structure of a,ω-diphenyltetradecaheptaene, a model molecule for polyacetylene, as studied by photoelectron spectroscopy," *Synth. Met.* **51**, 187 (1992).

[31] M. Lögdlund, P. Dannetun, S. Stafström, W. R. Salaneck, M. G. Ramsey, C. W. Spangler, C. Fredriksson, and J. L. Brédas, "Soliton pair charge storage in doped polyene molecules: Evidence from photoelectron spectroscopy studies," *Phys. Rev. Lett.* **70**, 970 (1993).

[32] W. R. Salaneck and J. L. Brédas, *Adv. Mat.* (in press) (1996).

[33] J. M. André, J. Dekhalle, and J. L. Brédas, *Quantum Chemistry Aided Design of Organic Polymers* (World Scientific, Singapore, 1991).

[34] J. L. Brédas, R. R. Chance, and R. Silbey, *Phys. Rev. B* **26**, 5843 (1982).

[35] J. L. Brédas and W. R. Salaneck, in *Organic Electroluminescence*, edited by D. D. C. Bradley and T. Tsutsui (Cambridge University Press, Cambridge, 1996).

[36] C. Fredriksson and J. L. Brédas, *J. Chem. Phys.* **98**, 4253 (1993).

[37] R. Lazzaroni, J. L. Brédas, P. Dannetun, C. Fredriksson, S. Stafström, and W. R. Salaneck, "The chemical and electronic structure of the interface between aluminum conjugated polymers," *Electrochim Acta* **39**, 235 (1994).

[38] P. Dannetun, M. Boman, S. Stafström, W. R. Salaneck, R. Lazzaroni, C. Fredriksson, J. L. Brédas, R. Zamboni, and C. Taliani, "The chemical and electronic structure of the interface between aluminum and polythiophene semiconductors," *J. Chem. Phys.* **99**, 664 (1993).

[39] M. Lögdlund and J. L. Brédas, "Theoretical Studies of the Interaction Between Aluminum and Poly(*p*-phenylenevinylene) and Derivatives," *J. Chem. Phys.* **101**, 4357-4364 (1994).

[40] J. L. Brédas and G. B. Street, *Adv. Chem. Res.* **18**, 319 (1985).

[41] G. Iucci, X. K., M. Lögdlund, M. Fahlman, and W. R. Salaneck, *Submitted.* (1995).

[42] P. Dannetun, M. Fahlman, C. Fauquet, K. Kaerijama, Y. Sonoda, R. Lazzaroni, J. L. Brédas, and W. R. Salaneck, in *Organic Materials for Electronics: Conjugated Polymer Interfaces with Metals and Semiconductors*, edited by J. L. Brédas, W. R. Salaneck, and G. Wegner (North Holland, Amsterdam, 1994), pp. 113.

[43] Y. Gao, K. T. Park, and B. R. Hsieh, "X-ray photoemission investigations of the interface formation of Ca and poly(*p*-phenylenevinylene)," *J. Chem. Phys.* **97**, 6991 (1992).

[44] P. Bröms, J. Birgersson, N. Johnsson, M. Lögdlund, and W. R. Salaneck, *Synth. Met., in press* (1995).

[45] A. Lobatch, I. R. Rubin, and P. V. Lushnikov, *Phys. Stat. Sol. (b)* **161**, 647 (1990).

[46] M. Kirm, E. Feldbach, R. Kink, A. Lushchik, C. Lushchik, A. Maaroos, and I. Martinson, "Mechanisma of intrinsic and impurity luminescence excitation by synchrotron radiation in wide-gap oxides," *J. Elec. Spec. and Rel. Phenom.* (1995).

[47] P. Bröms, *to be published.*

[48] K. Kareijama, N. Tanigaki, and H. Masuda, *to be published.*

[49] M. Fahlman, J. Birgerson, K. Kaeriyama, and W. R. Salaneck, "Poly(2,5-diheptyl-1,4-phenylene-alt-2,5-thienylene): a new material for blue-light-emitting diodes," *Synth. Met.* 75, 223 (1996).

[50] K. Xing, *to be published.*

[51] M. Fahlman, *to be published.*

Microscopic and Anisotropic Dynamics of Spin Carriers with/without Charge

Kenji Mizoguchi

Dept. of Phys., Tokyo Metropolitan University, Hachi-oji, Tokyo 192-03, JAPAN

Abstract

Electron spin resonance (ESR) studies are reviewed, specifically focusing on an experimental parameter of frequency applied to organic conductive materials. By analyzing the ESR linewidth and/or the spin-lattice relaxation rate measured in a wide frequency range such as several MHz to 24 GHz, one can obtain a frequency spectrum of spin motion that gives characteristic parameters of such motion, anisotropic diffusion rates. The first example is a dynamics of so-called "neutral soliton" in trans-polyacetylene, which is a topological defect having spin half but no charge. In the case of polarons and/or conduction electrons in conducting materials, the dynamics of spin is equivalent to that of the charged carriers responsible for the electrical conduction. Polyaniline (PANI) and polythiophene (PT) are reviewed as examples of the conducting case with a quasi-one-dimensional electronic state. Recent development of this technique suggesting a new relaxation mechanism will be briefly mentioned.

1 Introduction

In most cases the electron spin resonance (ESR) has been studied as a function of temperature, while it has been rarely done over wide frequency range such as from 3 to 24,000 MHz [1]. Major reason to expand a frequency range of ESR has been to resolve origins of resonance shift, linewidth and relaxation and to investigate electronic states. So far, it is not adequately unveiled what kinds of information are embedded in the frequency axis, then it is interesting and important to survey a new world along the frequency axis. For example, in the case of pure Aluminum metal it was reported that the g-shift and the linewidth as functions of temperature and frequency exhibit anomalous behavior [2-4] and a new model for the g-shift and the linewidth in terms of the g-anisotropy in the *k*-space and the electron-electron correlation [3] was proposed to account for such anomalies. However, understanding remains still unclear [4] even in the pure Aluminum metal.

In this paper, ESR studies in the wide range of frequency will be reviewed especially to study the spin dynamics in conductive organic materials as successful examples [1]. Electron spins in these materials can move freely to convey charges. Such spins interact with other electron spins and nuclear spins through dipolar, exchange and hyperfine interactions. Spin motion modulates these interactions and produces a motion spectrum characteristic for the dimensionality of space where the spins are moving; frequency independent in the isotropic three dimension (3D), $\log\omega$ in the two dimension (2D) and $\omega^{-1/2}$ in the one dimension (1D). Then low dimensional systems, in other word, highly anisotropic electronic systems are interesting to apply the spin dynamics technique. Most cases frequently studied are the quasi-one dimensional systems (Q1D) where the characteristic motion spectrum $\phi(\omega)$ is expected to have an expression as follows [5, 6],

$$\phi(\omega)\approx\frac{1}{\sqrt{4D_{//}/\tau_{\perp}}}\sqrt{\frac{1+\sqrt{1+(\omega\tau_{\perp}/2)^2}}{1+(\omega\tau_{\perp}/2)^2}}, \tag{1}$$

Springer Proceedings in Physics, Vol. 81
Materials and Measurements in Molecular Electronics
Editors: K. Kajimura · S. Kuroda

where $D_{//}$ is the diffusion rate along the one-dimensional axis, $\tau_\perp^{-1}$ the cutoff frequency that is, in some cases, equal to the diffusion rate perpendicular to the one-dimensional axis $D_\perp$, in unit of rad/sec for both diffusion rates, and ω the angular frequency. This spectrum can be measured by spin-lattice and spin-spin relaxation rates of ESR or NMR and yields above parameters for the anisotropic spin dynamics that can be reduced to the microscopic electrical conductivity even in polycrystalline materials. Here, only the case of ESR will be reviewed, then see recent reviews for other applications of the spin dynamics with both ESR and NMR [7, 8].

This review is organized as follows. Basic idea for background of the spin dynamics is mentioned in §2. In §3 experimental aspect is reviewed briefly. In §4, as a first example of the spin dynamics study, the neutral soliton dynamics in pristine *trans*-polyacetylene (*t*-PA) by the spin-lattice relaxation rate and the linewidth of ESR is demonstrated. The relaxation mechanism of ESR has been experimentally identified to eq. (1) in this system at first time. In §5, examples in conducting materials, polyaniline, polythiophene and polyheterocyclic polymers are demonstrated. In §6, recent findings that have a possibility to lead us to new information on the electronic states in conducting polymers and TTF-TCNQ will be described briefly and a conclusion in §7.

2 Background of spin dynamics

Idea to study the spin dynamics is based on detection of a motion spectrum that depends on the dimensionality of space where the spins are diffusing. Since the electron spins interact with the other electron spins and nuclear spins via dipolar and scalar couplings, the motion spectrum is generated as a local magnetic field randomly modulated by the spin motion. Such a motion spectrum can be measured by the spin-lattice or spin-spin relaxation rates that are proportional to the spectral density at the Larmor frequency ω_0. A functional form of the motion spectrum can be derived by a diffusion equation [5, 9] or a random walk formalism [6].

2.1 Autocorrelation function

The motion spectrum is described by a Fourier transform of an autocorrelation function $G(t)$ defined by [9]

$$G(t)=\iint p(\boldsymbol{r}_1)\Phi(\boldsymbol{r}_1,\boldsymbol{r}_2,t)F(\boldsymbol{r}_1)F^*(\boldsymbol{r}_2)d\boldsymbol{r}_1 d\boldsymbol{r}_2, \tag{2}$$

where $p(\boldsymbol{r}_1)$ is the probability density to find a spin at $\boldsymbol{r}_1$ and t=0, and $\Phi(\boldsymbol{r}_1, \boldsymbol{r}_2, t)$ the probability density to find such a spin at $\boldsymbol{r}_2$ after t. The $p(\boldsymbol{r}_1)$ is equal to the spin concentration c per unit molecule. $F(\boldsymbol{r})$ is the random function of the implicit parameter t, defined by the interaction Hamiltonian $\mathcal{H}_1=\Sigma_q F^{(q)}A^{(q)}$. In the case of dipolar interaction $F^{(q)}$ and $A^{(q)}$ are given by

$$F^{(0)}=\frac{1-3\cos^2\theta}{r^3}, \qquad A^{(0)}=-\frac{3}{2}\gamma_I\gamma_S\hbar\{-\frac{2}{3}I_zS_z+\frac{1}{6}(I_+S_-+I_-S_+)\}, \tag{3a}$$

$$F^{(\pm1)}=\frac{\sin\theta\cos\theta e^{\mp i\varphi}}{r^3}, \qquad A^{(\pm1)}=-\frac{3}{2}\gamma_I\gamma_S\hbar\{I_zS_\pm+I_\pm S_z\}, \tag{3b}$$

$$F^{(\pm2)}=\frac{\sin^2\theta e^{\mp 2i\varphi}}{r^3}, \qquad A^{(\pm2)}=-\frac{3}{4}\gamma_I\gamma_S\hbar I_\pm S_\pm, \tag{3c}$$

where r is the distance between two spins and θ the angle between the external magnetic field H_0 and the vector $\boldsymbol{r}$. The integration in eq. (2) means ensemble average taken over r in place of t.

2.2 Spectral density and dimensionality

Then the spectral density of the motion spectrum $J(\omega)$ is written by

$$J^{(j)}(\omega)=c\sum_{r_1,r_2}\phi(\boldsymbol{r}_1,\boldsymbol{r}_2,\omega)F^{(j)}(\boldsymbol{r}_1)F^{(j)*}(\boldsymbol{r}_2), \tag{4}$$

where $\phi(\boldsymbol{r}_1,\boldsymbol{r}_2,\omega)$ is a Fourier transform of $\Phi(\boldsymbol{r}_1,\boldsymbol{r}_2,t)$ and the sum is over r_1 and r_2 with all the possible sites. In the case of Q1D electronic systems providing that $\omega \ll D_{//}c_{//}^2/\Delta r_{\rm eff}^2$ where $\Delta r_{\rm eff}$ is the effective range of interaction [7], $\phi(\boldsymbol{r}_1,\boldsymbol{r}_2,\omega)$ is given by eq. (1) as a solution of the 1D diffusion equation $\partial\Phi/\partial t=\mathcal{D}_{//}\Delta\Phi$ modified by an escape probability from the chain, $\exp(-2t/\tau_\perp)$ [10]. Here, $\mathcal{D}_{//}$ (cm^2/s) is the diffusion coefficient, relating to the diffusion rate $D_{//}$ (rad/s) by $D_{//}=\mathcal{D}_{//}/c_{//}^2$, where $c_{//}$ is the intersite distance. Equivalently, anisotropic random walk formalism yielded the same result as the former approach [6]. In the two limits eq. (1) can be rewritten as

$$\phi(\omega)\approx\frac{1}{\sqrt{2D_{//}\omega}},\qquad \text{for } 1/\tau_\perp \ll \omega \ll D_{//}, \tag{5a}$$

$$\phi(\omega)\approx\frac{1}{\sqrt{2D_{//}/\tau_\perp}}=\text{const.},\qquad \text{for } \omega \ll 1/\tau_\perp. \tag{5b}$$

The former is of 1D regime and the latter is of 3D regime. The cutoff frequency $1/\tau_\perp$ is characterized as a crossover frequency between two regimes. $1/\tau_\perp$ is equal to the interchain diffusion rate $D_\perp$, providing that $D_\perp$ is the largest interaction between chains. In the case of chains with finite length, the cutoff frequency is shown to be dominated by the finite chain length effect instead of the interchain diffusion rate $D_\perp$ [7, 11].

2.3 Relaxation rates

The relaxation rates of ESR due to the Q1D spin motion are expressed by the spectral density of the motion spectrum $J^{(j)}(\omega)$ or that of the probability density $\phi(\omega)$ as

$$\begin{aligned} T_1^{-1}&=\frac{3}{2}\gamma_S^4\hbar^2S(S+1)[J^{(1)}(\omega_0)+J^{(2)}(2\omega_0)]\\ &=\gamma_S^4\hbar^2S(S+1)c\Sigma_\ell[0.2\phi(\omega_0)+0.8\phi(2\omega_0)]\\ &=3\gamma_S^2k_BT\chi\Sigma_\ell[0.2\phi(\omega_0)+0.8\phi(2\omega_0)], \end{aligned} \tag{6}$$

$$\begin{aligned} T_2^{-1}&=\frac{3}{8}\gamma_S^4\hbar^2S(S+1)[J^{(0)}(0)+10J^{(1)}(\omega_0)+J^{(2)}(2\omega_0)]\\ &=\gamma_S^4\hbar^2S(S+1)c\Sigma_\ell[0.3\phi(0)+0.5\phi(\omega_0)+0.2\phi(2\omega_0)]\\ &=3\gamma_S^2k_BT\chi\Sigma_\ell[0.3\phi(0)+0.5\phi(\omega_0)+0.2\phi(2\omega_0)] \end{aligned} \tag{7}$$

for the case of the electron-electron dipolar interaction [7, 9, 12], where γ_s is the gyromagnetic ratio of the electron spin and Σ_ℓ the lattice sum $\Sigma P_2(\cos\theta_{12})/(r_1^3r_2^3)$ [5]. The susceptibility χ is in units of emu/unit-molecule. In the second line of eqs. (6) and (7) the spectral density $J^{(j)}(\omega)$ is averaged out for the powder. Quantitative estimation of $D_{//}$ is not easy, since quantitative estimation of the lattice sum is required. On the other hand, quantitative estimation of the cutoff frequency is more accurate, since it depends only on the functional form of eq. (1). The spin-spin relaxation rate is directly related to the linewidth of ESR by a relation, $\Delta H_{pp}=2T_2^{-1}/(\sqrt{3}\,\gamma_S)$ and composed of two parts; frequency-dependent $T_1'^{-1}\propto 0.5\phi(\omega_0)+0.2\phi(2\omega_0)$ (so-called life-time broadening because of the same origin as T_1^{-1}) and frequency-independent $T_2'^{-1}\propto 0.3\phi(0)$ (secular broadening). At the limit of 3D-regime, that is $\omega_0\approx 0$, eq. (6) becomes equal to eq. (7). Such equality was confirmed in pristine trans-polyacetylene [13]. Practically T_1^{-1} measurement is more difficult and tedious than T_2^{-1} and then the linewidth is usually used for the spin dynamics study.

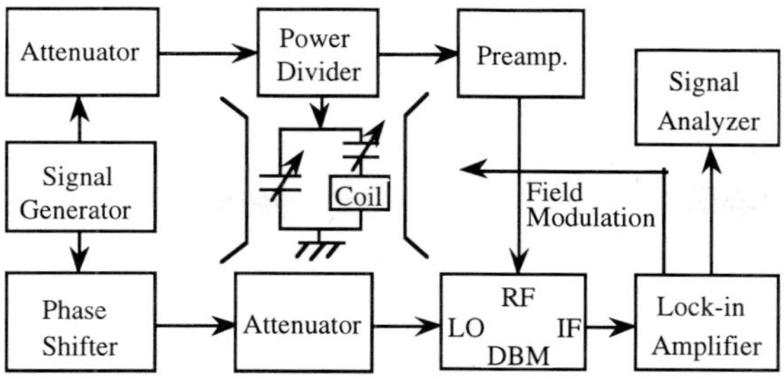

Fig. 1: A block diagram for ESR spectrometer.

3 Experimental

One of the key points of this experiment is to use one unique sample for all experiments over the studied frequency and temperature ranges, since ESR spectrum should be kept free from possible difference in a sample batch caused by any influence due to a presence of impurities such as oxygen and/or humidity, inhomogeneity of doping in different part of sample and etc. Generally samples are sealed into a quartz tube to avoid undesirable ESR signal from glass tube and any influence due to changes of circumstance. To gain sufficient sensitivity in the low frequency range down to several MHz, much more sample quantity than the case of a conventional ESR apparatus at X-band (~9 GHz) is required. We usually use a quartz tube with diameter of 5 mm that can be used up to 24 GHz. A block diagram of a homebuilt ESR spectrometer is shown in Fig. 1. Resonance circuit is composed of a coil and condensers in the lower frequency range than 1 GHz, a loop-gap resonator [14] in the higher frequency than 1 GHz and a cylindrical cavity at 24 GHz are used.

ESR spin-lattice relaxation rate T_1^{-1} is measured by a saturation method instead of a pulse method, since a dead time of signal receiving amplifier after high power *rf* pulses is more than 2 μs at 10 MHz and much longer than T_1 that is typically the order of less than 100 ns. An amplitude of *rf* magnetic field is calibrated by standard free-radical samples, (tri-p-nitrophenyl)methyl radical, $Q(TCNQ)_2$ or $Ad(TCNQ)_2$ where the spin-spin relaxation rate T_2^{-1} satisfies a relation $T_2^{-1}=T_1^{-1}$. When the temperature is changed, the amplitude of *rf*-magnetic field is monitored by a pick up coil located nearby the sample coil to keep accuracy of T_1^{-1} measurement.

ESR linewidth is defined as a peak-to-peak separation of absorption derivative. A least-square fitting with a Lorentzian lineshape is applied to deduce the linewidth from experimental data, which allows to determine the linewidth with resolution of the order of mG. In conductive materials the Lorentzian lineshape is commonly found because of rapid motion of charge carrier with spins which makes satisfy the condition for "the extreme narrowing limit" that assures the observation of the Lorentzian lineshape in ESR [8, 9].

(a)
(b)
(c)

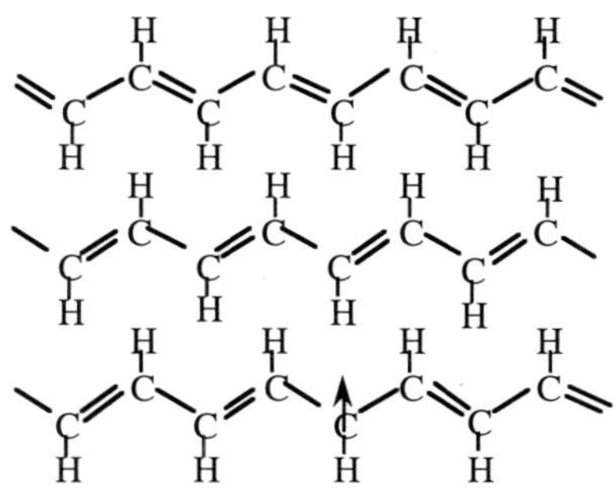

Fig. 2: The chemical structures for *trans*-polyacetylene, (a) phase A, (b) phase B and (c) neutral soliton as a zone boundary of phases A and B.

4 Pristine *trans*-polyacetylene -neutral soliton dynamics- [15-19]

As a good and successful example of the spin dynamics study, ESR investigation as a function of frequency in *trans*-polyacetylene is reviewed. Structure of *trans*-polyacetylene for different bond-alternation phases of A and B is shown in Fig. 2, together with the neutral soliton that is a zone boundary defect of the phases A and B, carrying spin 1/2 but no charge. In principle, the neutral soliton can move freely in the chain, since the ground state energy of the phase A is the same as that of B (degeneracy of ground states in bond alternation), so the energy of polymer chain is not influenced by a position of the neutral soliton. One of the evidences for such free motion of the neutral soliton is an observation of sharp ESR signal (less than 1 G) with Lorentzian lineshape, typical of motionally narrowed ESR spectrum [20]. The other evidence is an observation of the Overhauser effect [21-24], which is an enhancement of nuclear magnetization via saturation of ESR signal at the Larmor frequency of electron spin, $\omega = \omega_e$ [9, 25], providing that a correlation time of motion τ is enough shorter than $1/\omega_0 \approx 10^{-11}$ s. On the other hand, if the correlation time is longer than the inverse of hyperfine coupling frequency ($\approx 10^{-7}$ s), the solid state effect should be observed, which is also the enhancement of the nuclear magnetization, but the irradiation frequency of $\omega = \omega_e \pm \omega_n$ is required [7, 9]. The present study of the spin dynamics enables us to obtain quantitative information on the anisotropic motion of the neutral soliton as follows.

4.1 Frequency dependence of ESR T_1^{-1} and linewidth

Figure 3 shows the frequency dependence of the spin-lattice relaxation rate T_1^{-1} for *t*-$(CH)_x$ and *t*-$(CD)_x$. The concentration dependence of T_1^{-1} shown in Fig. 3 (b) provides important information to identify the relaxation mechanisms. The large intercept for *t*-$(CH)_x$ is reasonably ascribed to the hyperfine interaction with the proton, since the relaxation rate due to the hyperfine coupling with proton nuclei is larger than that with the deuteron by a factor of ≈ 16 in T_1^{-1} [12, 19]. The relaxation rate proportional to the spin concentration is consistent with the prediction of the electron-electron dipolar interaction described by eq. (6) (refer to the second line). The prediction of eq. (6) together with the hyperfine contribution reproduces the data well as shown by the solid curves in Fig. 3. According to eq. (7) it is also expected to find a similar frequency dependence for the ESR linewidth. Actually, the ESR linewidth shows a behavior predicted by eq. (7), as shown by the solid curves in Fig. 4. Characteristic features of this figure are

(1) that the slope of the solid curves gradually increases with decreasing temperature,
(2) that a constant contribution independent of frequency, corresponding to the intercept of the ordinate axis, rapidly increases with decreasing temperature, and

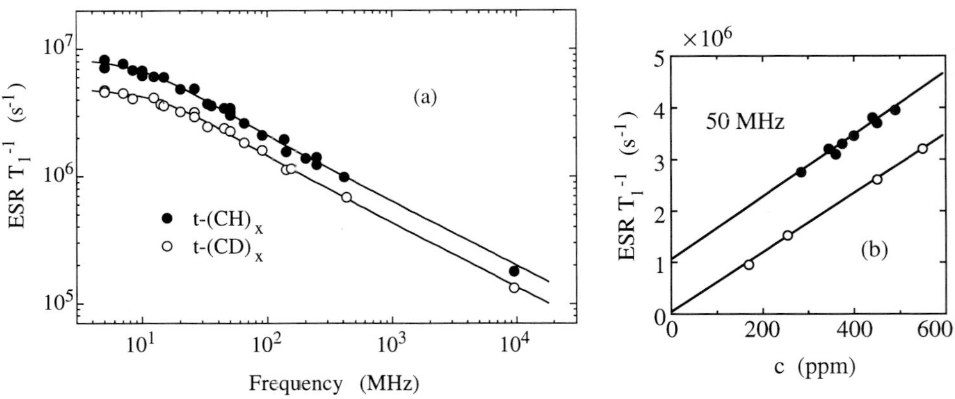

Fig. 3: (a) The frequency dependence of T_1^{-1} for *t*-$(CH)_x$ and *t*-$(CD)_x$. The solid curves indicate predicted behaviors of eq. (6). (b) The concentration dependence of T_1^{-1} for *t*-$(CH)_x$ and *t*-$(CD)_x$. (after K. Mizoguchi, K. Kume and H. Shirakawa, *Solid St. Commun.*, **50** (1984) 213-18)

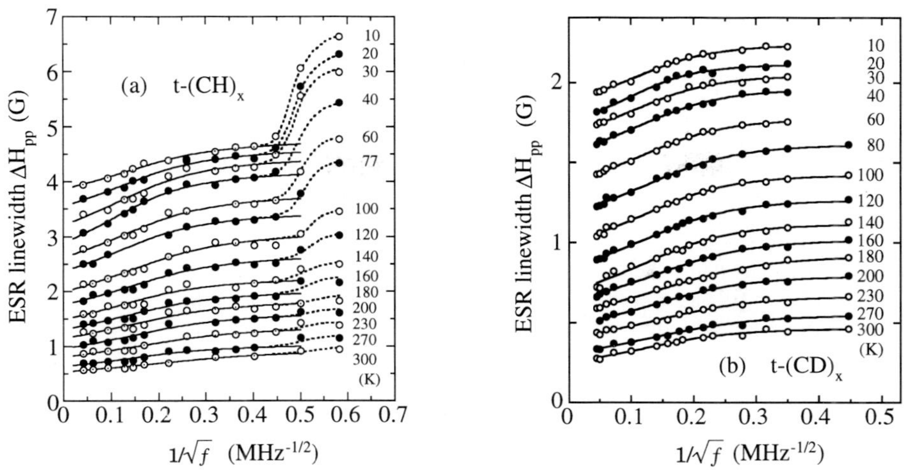

Fig. 4: ESR linewidth in (a) t-$(CH)_x$ and (b) t-$(CD)_x$ versus $1/\sqrt{f}$ with the implicit parameter of temperature. (after (a) K. Mizoguchi, S. Komukai, T. Tsukamoto, K. Kume, M. Suezaki, K. Akagi, and H. Shirakawa, *Synth. Met.*, **28**, 1989, D393-8 and (b) K. Mizoguchi, K. Kume, and H. Shirakawa, Synth. *Met.*, **17**, 1987, 439-45)

(3) that an anomalous steep broadening below 6 MHz (above $1/\sqrt{f}\approx 0.4$) is observed uniquely in t-$(CH)_x$ (refer to Fig. 6 for the lower frequency than $1/\sqrt{f}\approx 0.45$ in t-$(CD)_x$).

The first is interpreted by the Q1D motion of eq. (7), but the second requires to assume a diffuse/trap model for the neutral soliton dynamics proposed by Nechtschein et al. to account for the observed temperature dependence of ESR linewidth at X-band [23]. A schematic explanation for (1) and (2) in terms of the diffuse/trap model is shown in Fig. 5. When the neutral soliton is diffusing in normal sites, the linewidth broadens dynamically by the lifetime described by eq. (7). On the other hand, in a trapping site the linewidth broadens by spatial inhomogeneity of static local field, since the neutral soliton stays for longer duration than in the normal sites by several orders of magnitude. A criterion for the dynamic or static broadening is that the hopping rate of the spin is larger than the Larmor frequency or smaller than the hyperfine interaction frequency, respectively. Then, the diffuse/trap model requires that the linewidth is a sum of two contributions, dynamic and static broadenings, weighted by the ratio of duration being diffusing and trapped as shown in Fig. 5. As the origins of the trapping several possibilities have been proposed [7, 26].

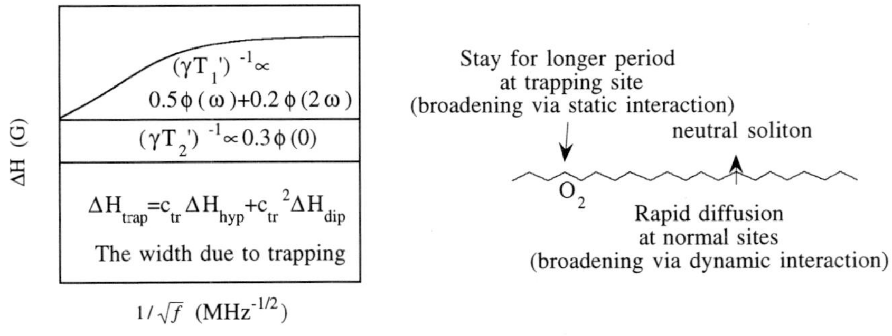

Fig. 5: A schematic figure for the origin of ESR linewidth in terms of the diffuse/trap model. ΔH_{hyp} and ΔH_{dip} are the static broadening due to hyperfine and electron-electron dipolar interactions, respectively. c_{tr} is the ratio of duration stay at the trapping site, $t_{tr}/(t_{tr}+t_{diff})$.

4.2 Anomalous broadening below 6 MHz in t-(CH)$_x$

The above understanding helps us to interpret the anomalous broadening below 6 MHz found uniquely in t-(CH)$_X$. The nature of this anomaly appears in the characteristic pattern of the linewidth against the angle of the external magnetic field to the chain axis in stretch-oriented films of both t-(CH)$_X$ and t-(CD)$_X$ as shown in Fig. 6. Some characteristic indications are found

(1) in the reversed phase of anisotropy in the protonated and deuterated polyacetylenes, and

(2) in the strong enhancement of the anisotropy in the protonated polyacetylene.

The reversed phase is well evidenced that the phase of t-(CD)$_X$ comes from the dynamic origin shown by eq. (7) and that the other phase of t-(CH)$_X$ does from the static one produced during the trapping [16, 27, 28] because of the larger hyperfine width than t-(CD)$_X$. Then, the anomaly in the linewidth is ascribed to the enhancement of the width due to trapping. The reason why such an enhancement appears below 6 MHz solely in t-(CH)$_X$ is due to a crossover from "unlike" spins to "like" spins in the electron and nuclear coupled spin system. Usually the static coupling is caused by $a\boldsymbol{S}_z\bullet\boldsymbol{I}_z$ term that conserves total energy of the spin system even for the "unlike" spins, but not by $b\boldsymbol{S}_\pm\bullet\boldsymbol{I}_\mp$ term due to decoupling by the large Zeeman energy splitting under strong external field. However, under the special condition of Larmor frequency f less than 6 MHz, the spin system becomes "like" spins because of mixing due to the larger hyperfine coupling frequency A ($\mathcal{H}$=$2\pi\hbar A\boldsymbol{S}\bullet\boldsymbol{I}$) than the Larmor frequency. Therefore, $b\boldsymbol{S}_\pm\bullet\boldsymbol{I}_\mp$ term becomes effective to broaden the ESR linewidth below 6 MHz [19]. Using this crossover frequency the maximum spin density ρ of the neutral soliton can be deduced. The effective hyperfine coupling constant A_{eff}=$A\bullet\rho/2$=6 MHz with A=-70 MHz/spin yields ρ to be 0.17 spins/carbon at the center of the neutral soliton extension spread over 18 CH units [29, 30], in good agreement with the ENDOR result in *cis*-polyacetylene [29, 30]. Finally in the case of the deuterated polyacetylene because of the smaller nuclear moment than that of proton, a similar enhancement of the linewidth could be expected if the Larmor frequency were reduced to the comparable magnitude to that of the electron-deuteron coupling.

4.3 Temperature dependence of diffusion rates

With the least square fitting as demonstrated by the solid curves in Figs. 3 and 4, the temperature dependence of the diffusion rate along the chain $D_{//}$ and the cutoff frequency (the diffusion rate across the chains) were derived as shown in Fig. 7, where the correction arising from the trapping was made [19, 23]. The thick solid curve shows $D_{//}$ without correction of the trapping, which indicates importance of the correction below 100 K. Such a correction for the proton NMR entirely

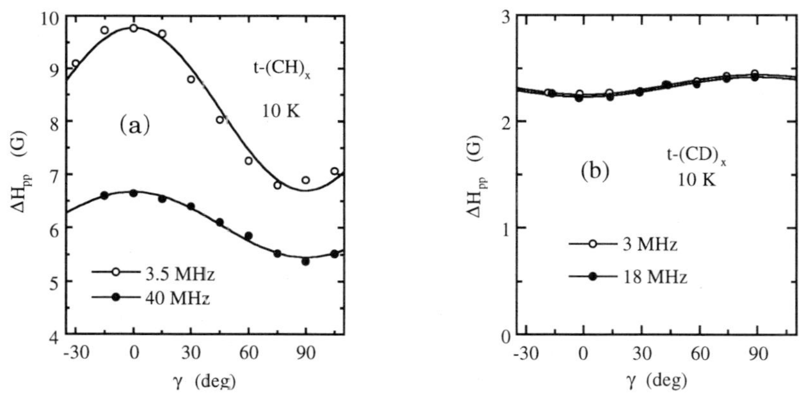

Fig. 6: Angular dependence of the ESR linewidth (a) at 3.5 and 40 MHz for t-(CH)$_X$ and (b) at 3 and 18 MHz for t-(CD)$_X$, at 10 K. Note the reversed phase of anisotropy pattern each other. (after K. Mizoguchi, S. Komukai, T. Tsukamoto, K. Kume, M. Suezaki, K. Akagi, and H. Shirakawa, *Synth. Met.*, **28**, 1989, D393-8)

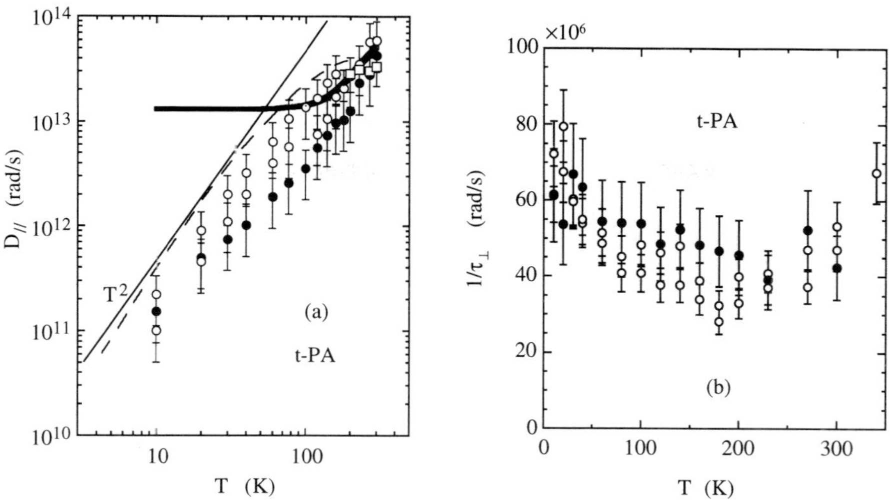

Fig. 7: The temperature dependence of the diffusion rate (a) along the chain and (b) across the chains. ❍ t-$(CD)_x$, ● t-$(CH)_x$ (ESR linewidth), ❑ t-$(CD)_x$, (ESR T_1^{-1}) and --- t-$(CH)_x$ (NMR T_1^{-1}) [23]. (after K. Mizoguchi, S. Masubuchi, K. Kume, K. Akagi, and H. Shirakawa, *Phys. Rev.*, **B51**, 1995, 8864-73)

changes the original behavior to give the corrected result shown in Fig. 7 (a) [23]. The obtained temperature dependence of $D_{//}$ with both ESR and NMR shows good agreement with each other, suggesting rationality of the present experiment and analysis with the diffuse/trap model. Such a temperature variation of $D_{//}$ implies that at absolute zero the neutral soliton is fixed even at the normal sites and that the scattering with phonons activate its diffusion below 300 K. These behaviors can be compared with the theoretical considerations [31-37]. On the cutoff frequency there are several possibilities, hopping of the spins between chains, exchange coupling and spin-spin relaxation itself [38]. The cutoff frequency shows weak variation against the temperature as shown in Fig. 7 (b). The decrease with increasing temperature up to 100 K can be ascribed to motion induced invalidation of exchange coupling among the nearby neutral solitons fixed around the trapping center at absolute zero. Above 200 K it is probable that the hopping of the neutral soliton between the chains could be activated thermally, for example via charged solitons created by unexpected doping as predicted by Kivelson [39], although a serious question of validity of this mechanism has been raised [40]. Here, note that the anisotropy of the neutral soliton dynamics is as large as 10^6, consistent with the topological nature of the neutral soliton.

5 Conducting state of polymers

5.1 Polyaniline (PANI) [41-44]

Polyaniline is a stable conducting polymer in air with chemical structure shown in Fig. 8 and shows the electrical conductivity up to several hundredth S/cm at 300 K by protonating with HCl and camphorsulfonic acid (CSA) [45-47]. It is known that crystalline structure of PANI is not only depends on species of counter ions, but also on sample morphology, powder or film [48-50]. PANI powder protonated with HCl by dipping in aqueous solution of HCl, has ES-I structure, but a four-fold stretch-oriented PANI film cast from N-methylpyrrolidinone (NMP) solution shows ES-II structure [48] with the higher electrical conductivity than in ES-I. In this section the spin dynamics study to clarify microscopic and anisotropic charge conduction in the ES-I powder is reviewed [41-44].

(a)
(b)
(c)

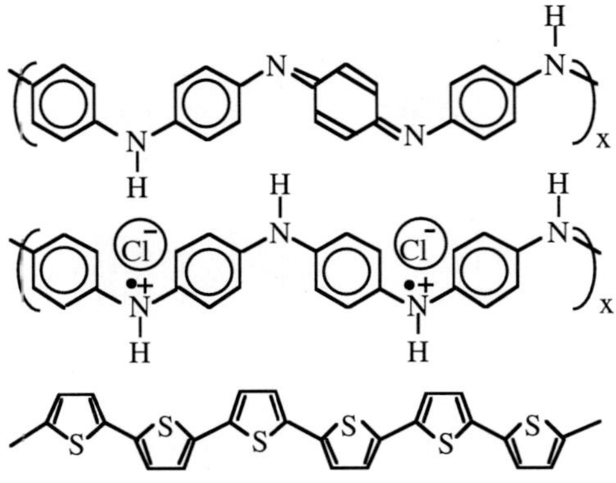

Fig. 8: Chemical structures of (a) neutral (Emeraldine base form, EB) and (b) oxidized (Emeraldine salt form, ES) polyanilines. ES is protonated with HCl. (c) polythiophene. Protons bonded to carbon atoms are abbreviated.

5.1.1 Protonation dependence

Figure 9 (a) shows the ESR linewidth as a function of $1/\sqrt{f}$ [1]. Note that the cutoff frequency $1/\tau_\perp$, crossover frequency from 1D to 3D regime, remarkably increases from $1/\sqrt{f}\approx 0.15$ ($\approx$50 MHz) to 0.01 ($\approx$10 GHz) at y$\approx$02~0.3. On the other hand, $D_{//}$ is almost independent of y as found in Fig. 9 (b). These findings can be understood as a percolative transition due to segregative protonation, in agreement with the conclusion drawn from the Pauli-like susceptibility increasing proportionally with y [52]. ESR observes the spin carriers on the particular chains fully protonated segregatively and other unprotonated chains have no spin carriers, resulting in a single value for $D_{//}$ independent of y. In the same time, the percolative connection of chains strongly enhances the interchain diffu-

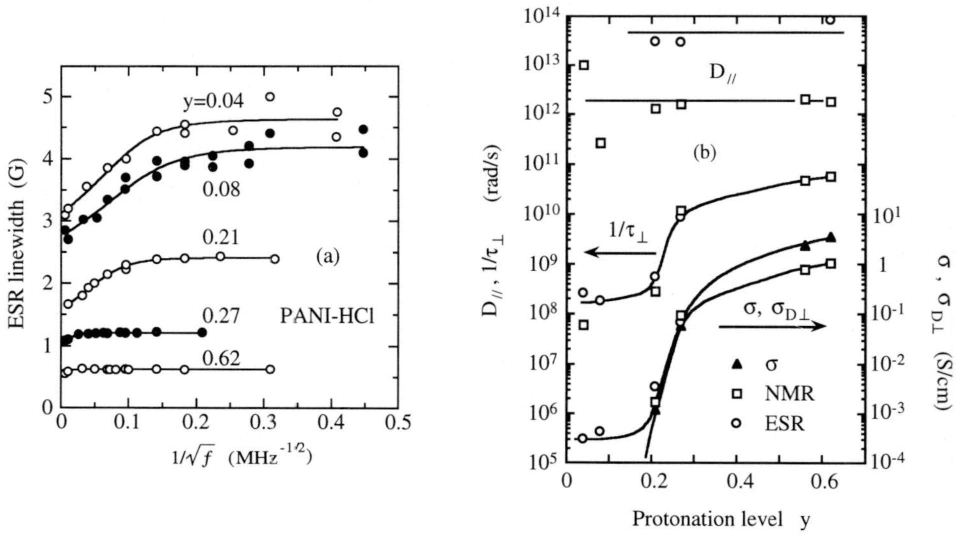

Fig. 9: (a) The frequency dependence of the ESR linewidth with an implicit parameter of protonation level y. (b) The diffusion rate $D_{//}$ and the cutoff frequency $1/\tau_\perp = D_\perp$ with the dc conductivity and the microscopic conductivity deduced from $D_\perp$ against y. Origin of a quantitative difference of $D_{//}$ for ESR from for NMR is probably that each of ESR and NMR requires estimation of similar but different quantities from each other to calculate $D_{//}$. A leveling off of $1/\tau_\perp$ below y$\approx$0.2 is ascribed to the other mechanism than the interchain hopping rate $D_\perp$ for the cutoff frequency $1/\tau_\perp$ [51]. (after K. Mizoguchi, *Jpn. J. Appl. Phys.*, **34**, 1995, 1-19)

sion rate $D_\perp$ ($=1/\tau_\perp$) at y≈0.2~0.3 that corresponds to the percolation threshold p_c≈0.4~0.6. Such a value of p_c is reasonable because the effective dimensionality for the percolation of the interchain hopping is two dimensional triangular lattice that has p_c=0.5 because polyacetylene chain has six neighboring chains at nearly equidistance. The other important feature of Fig. 9 (b) is the protonation dependence of microscopic conductivity $\sigma_{D_\perp}$ derived from $D_\perp$ for the charge carriers with spin, using Einstein relation $\sigma_{D_\perp}=e^2N(E_F)D_\perp c_\perp^2$ for systems with Pauli susceptibility. Here, $N(E_F)$ is the density of states at the Fermi energy and $c_\perp$ the interchain distance. $\sigma_{D_\perp}$ behaves very correspondingly to the dc conductivity measured in the same sample as demonstrated in Fig. 9 (b), which claims that the dc conductivity of PANI-HCl powder is limited by the interchain hopping rate $D_\perp$. Such a correspondence is found also in the temperature dependence of the conductivity [43, 44].

5.1.2 Temperature dependence [44, 53]

The temperature dependence of the ESR linewidth with an implicit parameter of frequency is plotted in Fig. 10 (a) for PANI-HCl with y=0.62. One feature is that the lower the frequency, the larger the temperature dependence, particularly below 100 K. In other words, the lower the temperature, the larger the frequency dependence of the linewidth. This frequency dependence can be ascribed to the quasi-one dimensional motion of the spins [44, 53]. Such an analysis of the frequency dependence yields the residual linewidth corresponding to "the observed linewidth-eq. (7)", which is plotted by the open diamonds in Fig. 10 (a). Origin of the residual linewidth is not clear, but recently new model on the collision with oxygen were proposed [8, 54]. According to this, the line broadening δ_{ox} is proportional to $pC_BD_{//}$, where C_B is the concentration of oxygen and p the efficiency of collision. In the case of "strong collision" via exchange interaction, which forces to lose phase memory of the spin carrier, $\delta_{ox}\propto C_BD_{//}$ since p is constant. On the contrary, if the scattering center works for "weak collision", $\delta_{ox}\propto C_B/D_{//}$ since $p=1/D_{//}^2$. Then, actual line broadening depends on the nature of interactions, "strong" or "weak". Realistically, distance between the charge carrier and the oxygen would distribute over the size of microscopic crystal; the shorter the distance, the stronger the collision, but the longer the distance, the weaker the collision. Therefore, the expected dependence on the diffusion rate $D_{//}$ for the line broadening is dominated by the average "collision strength" and the shape of its distribution. Following such a consideration, we examined the temperature dependence of the residual linewidth to fit with the diffusion rate $D_{//}$ (as indicated by the

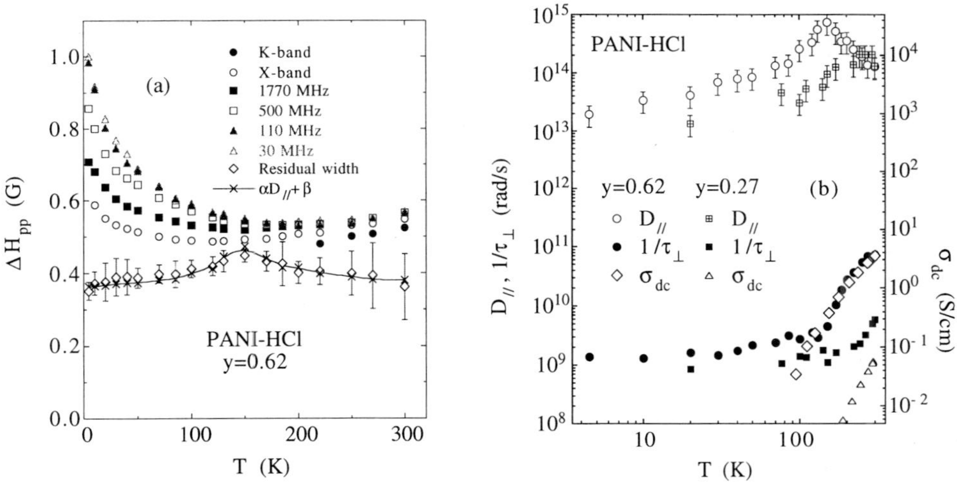

Fig. 10: (a) The temperature dependence of the ESR linewidth with an implicit parameter of frequency. (b) The temperature dependence of the diffusion rate $D_{//}$ and $1/\tau_\perp$. Reason why $1/\tau_\perp$ levels off around 10^9 is the same as the case in Fig. 9(b). (for (b), after K. Mizoguchi and K. Kume, *Synth. Met.*, **69**, 1995, 241-2)

open circles in Fig. 10 (b)) deduced by the Q1D fitting applied to the frequency dependence shown in Fig. 10 (a). The solid curve in Fig. 10 (a) shows a formula $\alpha D_{//} + \beta$ (G) which well reproduces the residual linewidth with $\alpha = 1.5 \times 10^{-16}$ (G•s) and $\beta = 0.36$ (G). This result enables us to conclude that the "strong collision" dominates the residual linewidth, together with the temperature independent width, in other words, the width independent of $D_{//}$. It is possible that the broadening becomes independent of the diffusion rate $D_{//}$ because an average of p becomes proportional to time interacting with oxygen, that is $p = 1/D_{//}$ and then δ_{ox}=const., if one takes into account a distribution of the coupling strength, as in the case of the transition probability in magnetic resonance [9].

Fig. 10 (b) shows some characteristic features that PANI-HCl with y=0.62 shows semiconducting-to-metallic transition around 150 K, but the sample with y=0.27 shows semiconducting behavior up to 200 K. The transition temperature seems to move to higher temperature with decreasing y. Such a microscopic conductivity shows a completely different behavior from the dc conductivity that is governed by the interchain hopping rate $D_\perp$, as clearly demonstrated in Fig. 10 (b) for both y=0.62 and 0.27. Here, note a difference of the scales; the left scale for $1/\tau_\perp$ and the right for σ_{dc} in Fig. 10 (b). Another point is a large anisotropy ratio of $D_{//}/D_\perp$, more than 10^3 for y=0.62 and 10^4 for y=0.27 at 300 K and reaches more than 10^4 at 150 K. Such a large anisotropy yields an estimation of chain length to be $l_{min} = \sqrt{D_{//}/D_\perp}\, c_{//} \approx (3\times10^4)^{1/2} c_{//} = 180 c_{//}$ resulting from a finite chain length effect [1]. This value is consistent with the average molecular weight of 50,000 for PANI [55]. Similar consideration for *trans*-polyacetylene yields $l_{min} \approx 10^3 c_{//}$.

5.2 Polythiophene (PT) [56, 57]

Another example of polythiophene (see, Fig. 8 (c) for chemical structure) applied the spin dynamics study is reviewed [56-58]. PT-ClO_4^- is prepared by electrochemical oxidation in inert atmosphere [58] so that the broadening due to oxygen can not be expected as in the case of PANI-HCl. The frequency dependence of the ESR linewidth is shown in Fig. 11 (a). In this polymer there are three different origin of the linewidth, Q1D (eq. (7)), Elliott mechanism and anisotropic g-shift.

The Elliott mechanism [59] induces a spin-flip scattering via a spin-orbit interaction, which is proportional to the momentum relaxation rate $1/\tau_r$ due to phonon scattering. Since such a relaxation rate $1/\tau_r$ dominates the electrical resistivity ρ, the Elliott broadening ΔH_{Ell} is expected to correspond to the resistivity ρ. Figure 11 (b) demonstrates a good correspondence between the linewidth and the voltage-shorted-compaction (VSC) resistance [60], which suggests the electronic state of this system is metallic consistent with an observation of the Pauli-like susceptibility [56, 57, 61, 62] and

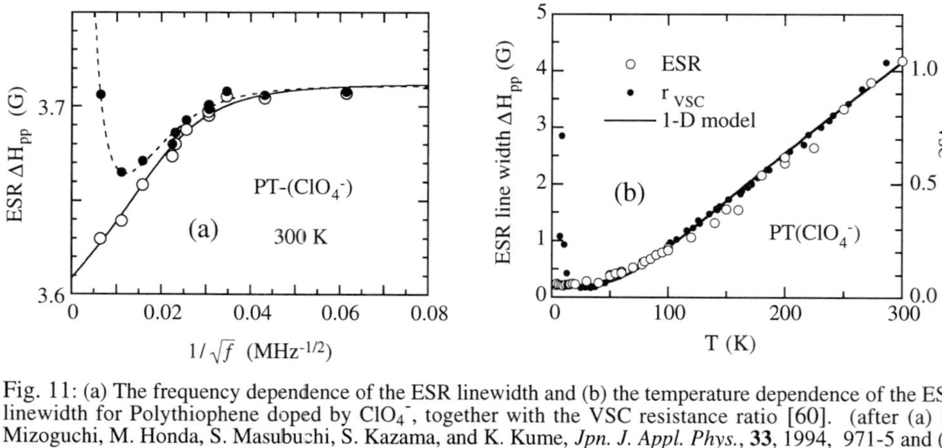

Fig. 11: (a) The frequency dependence of the ESR linewidth and (b) the temperature dependence of the ESR linewidth for Polythiophene doped by ClO_4^-, together with the VSC resistance ratio [60]. (after (a) K. Mizoguchi, M. Honda, S. Masubuchi, S. Kazama, and K. Kume, *Jpn. J. Appl. Phys.*, **33**, 1994, 971-5 and (b) K. Mizoguchi, M. Honda, N. Kachi, F. Shimizu, H. Sakamoto, K. Kume, S. Masubuchi, and S. Kazama, *Solid St. Commun.*, **96**, 1995, 333-7)

Table I. The parameters for polythiophene obtained at 300K from the analysis of ESR broadening due to the diffusive motion and the Elliott mechanism. D is the diffusion rate, τ_r the relaxation time for scattering, V_F the Fermi velocity, ℓ^* the mean free path and σ the electrical conductivity. (after K. Mizoguchi, M. Honda, S. Masubuchi, S. Kazama, and K. Kume, *Jpn. J. Appl. Phys.*, **33**, 1994, 971-5)

	D (rad/s)	τ_r (s)	V_F (cm/s)	ℓ^* ($c_{//}$ or $c_\perp$)	σ (S/cm)
//	1.9×10^{15}	5×10^{-16}	8.0×10^{7}	≈1	1.2×10^{3}
$\perp$	5.5×10^{9}				0.037

linear thermoelectric power [58, 60, 63]. Both the linewidth and the VSC resistance behave as T^2 at lower temperature than 100 K and approach a linear dependence at higher temperatures, which can be understood by a simple 1D metal model [64-66]. Further evidence of the Elliott broadening is found that the linewidth at 300 K shows an expected behavior of the Elliott mechanism, that is, clear proportionality to λ^2, where λ is the spin-orbit coupling constant for the heaviest atoms in not dopant but the polymer backbone of various heterocyclic polymers with five members, such as polythiophene, poly(3-methylthiophene) and polypyrrole doped by a series of dopants [7, 56]. Similar dependence on λ of the alkali metal dopants was reported for other conducting systems, such as polyacetylene [67-69], poly-p-phenylene [70], graphite intercalated compounds (GIC) [71] and fullerides [72]. To validate the Elliott mechanism it requires some deviation from the pure-one-dimensional symmetry in the electronic state; heterocyclic polymers have molecules with asymmetric structure and other alkali-doped systems have sizable contribution of alkali ions.

The broadening due to the g-shift anisotropy is inhomogeneous broadening proportional to the Larmor frequency and arises only in polycrystalline samples, since ESR signal from each crystal with different direction of the crystal axis appears at different magnetic field strength. The most prominent contribution to the ESR linewidth is the Elliott broadening which gives information on the scattering rate $1/\tau_r$ with phonons. The diffusion rates $D_{//}$ and $D_\perp$ deduced from the solid curve in Fig. 11 (a), can be combined with the scattering rate $1/\tau_r$ to give several parameters on the electronic state in PT-ClO_4^- [1, 57], as listed in Table I.

6 Recent findings

As another application of the frequency dependence of ESR, an organic charge transfer salt TTF-TCNQ was investigated, in relation to the NMR result studied as a function of frequency [10]. Although the reported NMR data claimed that the relaxation rate is dominated by Q1D mechanism, the ESR linewidth showed no such a dependence proportional to $1/\sqrt{f}$, instead the linewidth was rather proportional to f~f^2 and leveled off around 50 MHz, keeping constant up to 2,000 MHz as shown in Fig. 12 (a). A similar behavior, but smaller amplitude, has also been found in PANI-HCl film cast from NMP solution (Fig. 12 (b)) and PANI-CSA (camphorsulfonic acid) cast from *m*-cresol solution. In general, the spectral density of the local field approaches zero with the frequency goes to infinite. However, it can occur when the local field strength is proportional to the frequency. One example is the hopping of the spins between TTF and TCNQ stacks with the different g-values, but it is easily shown to be too small in magnitude. Similar mechanism was proposed to explain the anomalous frequency dependence of g-shift in pure aluminum that the g-anisotropy on the Fermi surface combined with the electron-electron correlation could yield the expected frequency dependence [3, 4]. This model claims for the g-shift to show a similar frequency variation to that of the ESR linewidth. An existence of the g-shift change in the same frequency region as for the ESR linewidth has been confirmed in PANI-CSA film, but 30 MHz is too low to measure g-shift in TTF-TCNQ. The quantitative reproduction of the data by this model is not successful, but largely underestimate for TTF-TCNQ. Therefore, in the present status, there is no definite model to explain these data. It still remains as an open question.

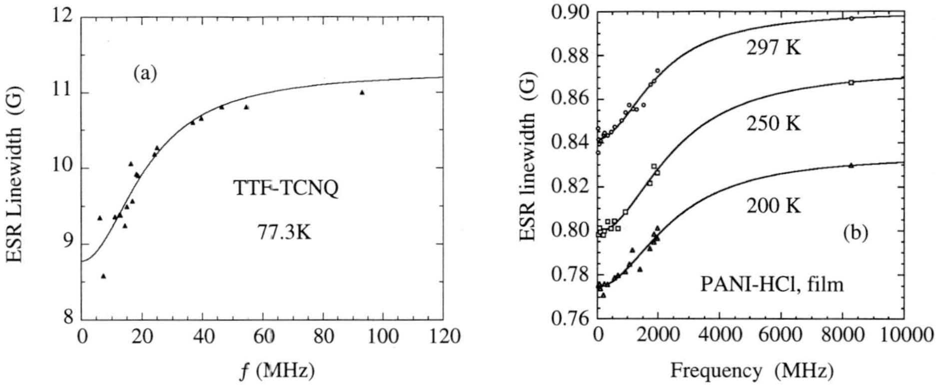

Fig. 12: (a) The frequency dependence of the ESR linewidth in TTF-TCNQ powder. (b) The frequency dependence of the ESR linewidth in PANI-HCl film cast from MNP solution.

7 Conclusion

With the several examples of organic conductive materials it was evidenced that the spin dynamics study is useful and powerful to investigate the dynamics of not only the spin, but also the charge carrier with spin as in the case of conduction electrons and polarons, including its anisotropy even in the polycrystalline materials. Since this field of experiments along the frequency axis in ESR is not fully developed yet, further refinement of experimental technique, analysis and interpretation are required and at the same time it has a possibility to give important information on the dynamics and electronic states of low dimensional conductive systems.

Acknowledgment

The author would like to express his thanks to Profs. H. Shirakawa, K. Akagi and Mr. M. Suezaki (Tsukuba Univ.) for collaboration on polyacetylene, especially preparation of highly oriented polyacetylene, to Drs. M. Nechtschein and J.-P. Travers (CENG, Grenoble) for stimulating discussion and collaboration on polyaniline, to Prof. S. Kazama and Dr. S. Masubuchi (Chuo Univ.) for collaboration on several conducting polymers, to Prof. G. Saito and Mr. M. Kubota for collaboration on charge transfer organic compounds, and to Prof. K. Kume, and present and old members of Kume's laboratory for their kind collaborations. This work was supported in part by Grant-in-Aid for Scientific Research from the Ministry of Education, Science, Sports and Culture (No.03640321 and No.07640487).

References

[1] K. Mizoguchi, *Jpn. J. Appl. Phys.*, vol. **34**, 1995, pp. 1-19.

[2] D. Lubzens, M.R. Shanabarger, and S. Schultz, *Phys. Rev. Lett.*, vol. **29**, 1972, pp. 1387.

[3] D.R. Fredkin and R. Freedman, *Phys. Rev. Lett.*, vol. **29**, 1972, pp. 1390-3.

[4] R. Freedman and D.R. Fredkin, *Phys. Rev.*, vol. **B11**, 1975, pp. 4847-58.

[5] C.A. Sholl, *J. Phys. C*, vol. **14**, 1981, pp. 447.

[6] M.A. Butler, L.R. Walker, and Z.G. Soos, *J. Chem. Phys.*, vol. **64**, 1976, pp. 3592.

[7] K. Mizoguchi and S. Kuroda, Magnetic Properties of Conducting Polymers, in *Handbook of Organic Conductive Molecules and Polymers*, ed. by H.S. Nalwa, John Wiley & Sons, Sussex, 1996, in press.

[8] M. Nechtschein, Electron spin dynamics, in *Handbook of Conducting Polymers*, ed. by T. Skotheim, Marcel Dekker, New York, 1996, in press.

[9] A. Abragam, *The Principles of Nuclear Magnetism*, Oxford Univ. Press, Oxford, 1961.
[10] G. Soda, D. Jerome, M. Weger, J. Alizon, G. Gallice, H. Robert, J.M. Fabre, and L. Giral, *J. de Phys.*, vol. **38**, 1977, pp. 931-48.
[11] W.G. Clark, K. Glover, M.D. Lan, and L.J. Azevedo, *J. Phys. Colloq.*, vol. **44**, 1983, C3 pp. 1493-9.
[12] K. Mizoguchi, *Makromol. Chem., Macromol. Symp.*, vol. **37**, 1990, pp. 53-66.
[13] K. Mizoguchi, K. Kume, and H. Shirakawa, *Mol. Cryst. Liq. Cryst.*, vol. **118**, 1985, pp. 459-62.
[14] J.S. Hyde, W. Froncisz, and T. Oles, *J. Mag. Res.*, vol. **82**, 1989, pp. 223.
[15] K. Mizoguchi, K. Kume, and H. Shirakawa, *Solid St. Commun.*, vol. **50**, 1984, pp. 213-18.
[16] K. Mizoguchi, K. Kume, S. Masubuchi, and H. Shirakawa, *Solid St. Commun.*, vol. **59**, 1986, pp. 465-8.
[17] K. Mizoguchi, K. Kume, and H. Shirakawa, *Synth. Met.*, vol. **17**, 1987, pp. 439-45.
[18] K. Mizoguchi, S. Komukai, T. Tsukamoto, K. Kume, M. Suezaki, K. Akagi, and H. Shirakawa, *Synth. Met.*, vol. **28**, 1989, pp. D393-8.
[19] K. Mizoguchi, S. Masubuchi, K. Kume, K. Akagi, and H. Shirakawa, *Phys. Rev.*, vol. **B51**, 1995, pp. 8864-73.
[20] B.R. Weinberger, E. Ehrenfreund, A. Pron, A.J. Heeger, and A.G. MacDiarmid, *J. Chem. Phys.*, vol. **72**, 1980, pp. 4749-55.
[21] M. Nechtschein, F. Devreux, R.L. Greene, T.C. Clarke, and G.B. Street, *Phys. Rev. Lett.*, vol. **44**, 1980, pp. 356-9.
[22] K. Holczer, J.P. Boucher, F. Devreux, and M. Nechtschein, *Phys. Rev.*, vol. **B23**, 1981, pp. 1051-63.
[23] M. Nechtschein, F. Devreux, F. Genoud, M. Guglielmi, and K. Holczer, *Phys. Rev.*, vol. **B27**, 1983, pp. 61-78.
[24] W.G. Clark, K. Glover, G. Mozurkewich, S. Etemad, and M. Maxfield, *Mol. Cryst. Liq. Cryst.*, vol. **117**, 1985, pp. 447.
[25] C.P. Slichter, *Principles of Magnetic Resonance*, vol. **1**, 3rd ed. Springer, Berlin, Heidelberg, 1989.
[26] K. Mizoguchi, H. Sakurai, F. Shimizu, S. Masubuchi, and K. Kume, *Synth. Met.*, vol. **68**, 1995, pp. 239-42.
[27] K. Mizoguchi, K. Kume, S. Masubuchi, and H. Shirakawa, *Synth. Met.*, vol. **17**, 1987, pp. 405-11.
[28] S. Kuroda, M. Tokumoto, N. Kinoshita, and H. Shirakawa, *J. Phys. Soc. Jpn.*, vol. **51**, 1982, pp. 693-4.
[29] S. Kuroda and H. Shirakawa, *Phys. Rev.*, vol. **B35**, 1987, pp. 9380-2.
[30] S. Kuroda, *Int. J. Mod. Phys. B*, vol. **9**, 1995, pp. 221-60.
[31] Y. Wada and J.R. Schrieffer, *Phys. Rev.*, vol. **B18**, 1978, pp. 3897-912.
[32] M. Ogata and Y. Wada, *Prog. Theor. Phys. Suppl.*, 1988, pp. 115-27.
[33] K. Maki, *Phys. Rev.*, vol. **B26**, 1982, pp. 4539-42.
[34] K. Maki, *Phys. Rev.*, vol. **B26**, 1982, pp. 2187-91.
[35] K. Maki, *Phys. Rev.*, vol. **B26**, 1982, pp. 2181-6.
[36] S. Jeyadev and E.M. Conwell, *Phys. Rev.*, vol. **B36**, 1987, pp. 3284-93.
[37] Y. Wada, *Progr. Theor. Phys. Suppl.*, vol. **113**, 1993, pp. 1-23.
[38] F. Devreux, *Phys. Rev.*, vol. **B25**, 1982, pp. 6609.
[39] S. Kivelson, *Phys. Rev.*, vol. **B25**, 1982, pp. 3798-21.
[40] D. Baeriswyl, D.K. Campbell, and S. Mazumdar, An overview of the theory of π-conjugated polymers, in *Conjugated Conducting Polymers*, vol. **102**, ed. by H.G. Kiess, Springer-Verlag, Berlin, Heidelberg, 1992, pp. 107.
[41] K. Mizoguchi, M. Nechtschein, J.-P. Travers, and C. Menardo, *Phys. Rev. Lett.*, vol. **63**, 1989, pp. 66-9.

[42] K. Mizoguchi, M. Nechtschein, J.P. Travers, and C. Menardo, *Synth. Met.*, vol. **29**, 1989, pp. E417-24.
[43] K. Mizoguchi, M. Nechtschein, and J.-P. Travers, *Synth. Met.*, vol. **41**, 1991, pp. 113-16.
[44] K. Mizoguchi and K. Kume, *Solid St. Commun.*, vol. **89**, 1994, pp. 971-5.
[45] A.G. MacDiarmid, J.C. Chiang, M. Halpern, W.S. Huang, S.L. Mu, N.L.D. Somasiri, W.Q. Wu, and S.I. Yaniger, *Mol. Cryst. Liq. Cryst.*, vol. **121**, 1985, pp. 173.
[46] M. Reghu, Y. Cao, D. Moses, and A.J. Heeger, *Synth. Met.*, vol. **57**, 1993, pp. 5020-5.
[47] A.P. Monkman and P.N. Adams, *Synth. Met.*, vol. **41-43**, 1991, pp. 627.
[48] J.P. Pouget, M. Laridjani, M.E. Jozefowicz, A.J. Epstein, E.M. Scherr, and A.G. MacDiarmid, *Synth. Met.*, vol. **51**, 1992, pp. 95-101.
[49] J.P. Pouget, Z. Oblakowski, Y. Nogami, P.A. Albouy, M. Laridjani, O. E.J., Y. Min, A.G. MacDiarmid, J. Tsukamoto, T. Ishiguro, and A.J. Epstein, *Synth. Met.*, vol. **65**, 1994, pp. 131-40.
[50] J.P. Pouget, C.-H. Hsu, A.G. MacDiarmid, and A.J. Epstein, *Synth. Met.*, vol. **69**, 1995, pp. 119-20.
[51] C. Jeandey, J.P. Boucher, F. Ferrieu, and M. Nechtschein, *Solid St. Commun.*, vol. **23**, 1977, pp. 673.
[52] A.J. Epstein, J.M. Ginder, F. Zuo, R.W. Bigelow, H.-S. Woo, D.B. Tanner, A.F. Richter, W.-S. Huang, and A.G. MacDiarmid, *Synth. Met.*, vol. **18**, 1987, pp. 303-9.
[53] K. Mizoguchi and K. Kume, *Synth. Met.*, vol. **69**, 1995, pp. 241-2.
[54] E. Houzé and M. Nechtschein, submitted to Phys. Rev. **B**.
[55] E.J. Oh, Y. Min, J.M. Wiesinger, S.K. Manohar, E.M. Scherr, P.J. Prest, A.J. MacDiarmid, and A.J. Epstein, *Synth. Met.*, vol. **55-57**, 1993, pp. 977.
[56] K. Mizoguchi, M. Honda, N. Kachi, F. Shimizu, H. Sakamoto, K. Kume, S. Masubuchi, and S. Kazama, *Solid St. Commun.*, vol. **96**, 1995, pp. 333-7.
[57] K. Mizoguchi, M. Honda, S. Masubuchi, S. Kazama, and K. Kume, *Jpn. J. Appl. Phys.*, vol. **33**, 1994, pp. 971-5.
[58] S. Masubuchi, S. Kazama, K. Mizoguchi, H. Honda, K. Kume, R. Matsushita, and T. Matsuyama, *Synth. Met.*, vol. **57**, 1993, pp. 4962-7.
[59] R.J. Elliott, *Phys. Rev.*, vol. **96**, 1954, pp. 266-79.
[60] S. Masubuchi and S. Kazama, *Synth. Met.*, vol. **74**, 1995, pp. 151-8.
[61] K. Mizoguchi, K. Misoo, K. Kume, K. Kaneto, T. Shiraishi, and K. Yoshino, *Synth. Met.*, vol. **18**, 1987, pp. 195-8.
[62] F. Moraes, D. Davidov, M. Kobayashi, T.C. Chung, J. Chen, A.J. Heeger, and F. Wudl, *Synth. Met.*, vol. **10**, 1985, pp. 169-79.
[63] S. Masubuchi and S. Kazama, *Synth. Met.*, vol. **69**, 1995, pp. 315-16.
[64] S. Kivelson and A.J. Heeger, *Synth. Met.*, vol. **22**, 1988, pp. 371-84.
[65] F. Shimizu, K. Mizoguchi, S. Masubuchi, and K. Kume, *Synth. Met.*, vol. **69**, 1995, pp. 43-4.
[66] F. Shimizu, Thesis: Tokyo Metropolitan Univ., 1994.
[67] D. Billaud, J. Ghanbaja, J.F. Marêché, E. McRae, and C. Goulon, *Synth. Met.*, vol. **28**, 1989, pp. D147-54.
[68] P. Bernier, C. Fite, A. El-Khodary, F. Rachdi, K. Zniber, H. Bleier, and N. Coustel, *Synth. Met.*, vol. **37**, 1990, pp. 41.
[69] F. Rachdi and P. Bernier, *Phys. Rev.*, vol. **B33**, 1986, pp. 7817-19.
[70] L. Kispert, J. Joseph, G.G. Miller, and R.H. Baughman, *Mol. Cryst. Liq. Cryst.*, vol. **106**, 1984, pp. 418.
[71] P. Lauginie, H. Estrade, J. Conard, D. Guerard, P. Lagrange, and M. El Makrini, *Physica B & C*, vol. **99**, 1980, pp. 514-20.
[72] K. Tanigaki, M. Kosaka, T. Manako, Y. Kubo, I. Hirosawa, K. Uchida, and K. Prassides, *Chem. Phys. Lett.*, vol. **240**, 1995, pp. 627.

Optical and Dynamical Properties of One-Dimensional Excitons in Conjugated Polymers

Shuji Abe

Electrotechnical Laboratory, 1-1-4 Umezono, Tsukuba 305, Japan

Abstract

The nature of excitons in conjugated polymers is discussed from a theoretical point of view. An intrachain exciton can be essentially described as a Wannier exciton. However, its one-dimensional character leads to an unusually large binding energy and an unusually small exciton size, for which the electronic correlation length turns out to be a decisive factor. These extraordinary features have important implications on the linear and nonlinear optical properties of conjugated polymers. In addition to these intrachain characteristics of excitons, we discuss the influence of interchain excitation transfer. It is demonstrated that the excitation transfer depends upon the interchain distance and the chain length in a complicated way due to the large molecular size. The limitation of the point dipole approximation as well as that of the extended dipole approximation are elucidated.

1 Introduction

Excitons are primary optical excitations in insulating and semiconducting solids. In organic molecular crystals Frenkel excitons are dominant excitations, whereas Wannier excitons are important in covalent semiconductors. In the case of conjugated polymers, π electrons are delocalized on each chain, so that one can employ the description of a Wannier-type exciton. However, the confinement of an exciton in a one-dimensional structure imposes a peculiar feature to the exciton: the size of the lowest-energy exciton is much smaller than the ordinary effective Bohr radius. This implies that the exciton has a character close to the Frenkel type. The nature of this one-dimensional exciton was discussed in detail in Ref. 1, and is briefly surveyed here in Sec. 2. Its relevance to various nonlinear optical properties is discussed in Sec. 3, including the effect of exciton-exciton interactions. In Sec. 4 we consider the intermolecular motion of an exciton. Hopping of electrons among chains is usually very small, but the exciton can transfer to nearby molecules as a Frenkel exciton using a matrix element of Coulomb interaction. In this sense the exciton has both Frenkel and Wannier character in a condensed state of polymers.

2 Exciton on a Single Chain

In conjugated polymers π electrons are well delocalized along the chain, forming wide valence and conduction bands separated by a Peierls gap due to bond alternation. It is natural to think that Coulomb attraction between an electron and a hole results in the formation of a Wannier exciton as in ordinary semiconductors. In fact, the light masses of electrons and holes allow us to work on the Wannier exciton model with the effective mass approximation [2]. However, there is an intrinsic difficulty of this approach in the present case. The eigenenergies of one-dimensional exciton is given by

Springer Proceedings in Physics, Vol. 81
Materials and Measurements in Molecular Electronics
Editors: K. Kajimura · S. Kuroda

$$E_n^{(1D)} = -\frac{Ry^*}{n^2}, \; n=0,1,2,... \tag{1}$$

where Ry^* is the effective Rydberg energy [3]. Therefore, the energy of the lowest exciton (n=0) turns out to be negative infinity. This stems from the singular nature of the Coulomb attractive potential in one dimension. To avoid this divergence, a lower cutoff length must be introduced for the potential. In other words, the binding energy of the lowest exciton state is finite for a potential with a cutoff, but it tends to infinity in the limit of the vanishing cutoff length. Now the question is what is the appropriate cutoff length in conjugated polymers. It was postulated in Ref. 2 that the cutoff length should be given by the electronic correlation length (or the delocalization length) associated with the Peierls gap. A scaling property of exciton energy levels was obtained on the basis of this postulate [2].

However, this assumption was introduced rather intuitively, and there are some uncertainties. The correlation length is clearly defined for the case of non-interacting electrons, but not in the presence of Coulomb interactions. In this sense it is much better to work with a tight-binding model, where we do not need to worry about the cutoff length. The use of a tight-binding model is suitable also for the calculation of nonlinear optical response. Therefore, we switch to a tight-binding model in the following.

We consider a model of π electrons in the presence of bond alternation and Coulomb interactions. The former is taken into account by transfer modulation δt, and the latter, by the Pariser-Parr-Pople model: the total Hamiltonian reads [4]

$$H = H_0 + H_{\mathrm{int}}, \tag{2}$$

$$H_0 = -[\, t + (-1)^n \delta t \,]\, (\, C_{n+1,s}^{\dagger} C_{n,s} + \mathrm{H.c.}\,), \tag{3}$$

$$H_{\mathrm{int}} = \sum_n U \, \rho_{n\uparrow} \rho_{n\downarrow} + \frac{1}{2} \sum_{n,s} \sum_{m(\neq n),s'} V_{n-m}\, \rho_{n,s}\, \rho_{m,s'}, \tag{4}$$

where $C_{n,s}^{\dagger}$ creates an exciton at site n and spin s, and $\rho_{n,s} = C_{n,s}^{\dagger} C_{n,s}$. H_0 and H_{int} describe the transfer energy modulation (bond alternation) and Coulomb interactions, respectively. The model is illustrated in Fig. 1. For the long range interaction we use the form $V_p = V/|p|$.

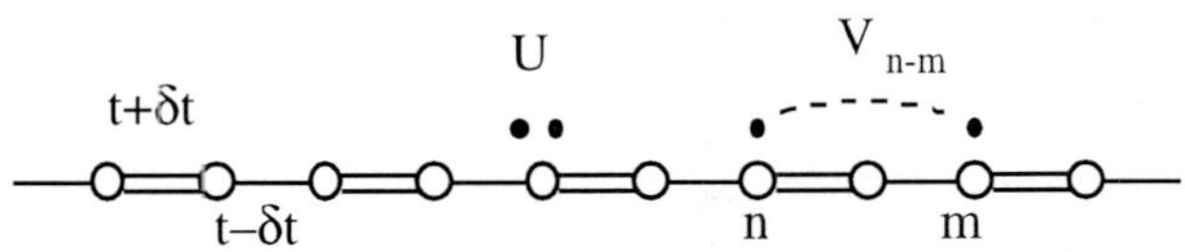

Figure 1 Model of conjugated polymer with bond alternation and long-range electron-electron interaction

To calculate exciton states, we need to take into account configuration interaction (CI) among one-electron excited states. This is the intermediate exciton approach, or the single-excitation CI method. In this method, we first obtain the Hartree-Fock one-electron states and then calculate Hamiltonian matrix elements among single-electron excited states and diagonalize the resultant matrix. Although electron correlation in the ground state is not taken into account at the level of single CI, the electron-hole correlation in excited states is properly treated by this method.

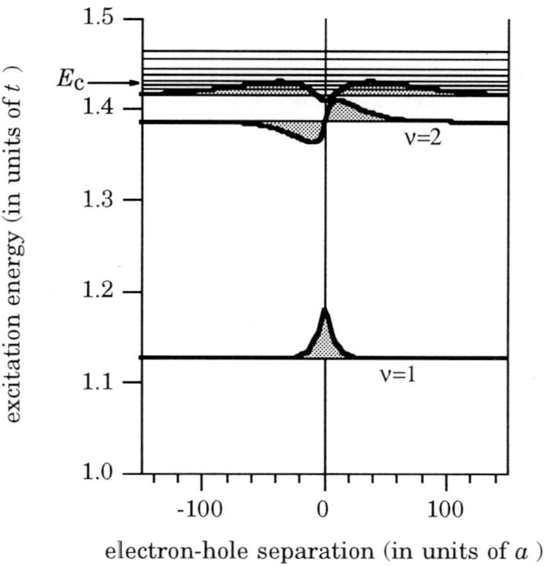

Figure 2 Exciton energies and envelope functions of the three lowest K=0 states calculated for a model polymer chain with U=2t, V=t and δt=0.2t. E_c indicates the lower edge of the electron-hole continuum, and a is the interatomic spacing.

The results of such calculations in fact indicate that the exciton is a Wannier type: the distance between the constituent electron and hole is much larger than the interatomic spacing. Figure 2 displays an example of calculated exciton wave functions. However, the exciton is different from the ordinary Wannier exciton in the sense that the size of the lowest exciton is always much smaller than the effective Bohr radius. Figure 3 shows the dependence of the size of the calculated exciton on the parameter δt. The corresponding Bohr radius is shown with a dotted curve. Clearly, the size of the second exciton matches the Bohr radius, but that of the lowest exciton is much smaller. It turns out that the latter scales with the so-called correlation length (or the delocalization length) defined by $\xi = (t/\delta t)a$, where a is the bond length. (This definition of ξ does not take electron correlation into account.) This implies that the "cutoff length" for the Coulomb attractive potential is approximately given by ξ. This is reasonable, because ξ governs the spatial extension of electron (hole) Wannier orbital in Peierls insulators [2]. Actually, ξ determines the length scales of almost all elementary excitations, including solitons and polarons, in conjugated polymers [5].

The discussion above has been based on single-CI calculations. Although the essential features of excitons are described in this approach, we need to take into account the excitation of two excitons to discuss optical nonlinearity near resonance. For this purpose it is necessary to perform CI among double excitations. Such calculations [6] indicate that two excitons are strongly bound to each other to form a biexciton (see below).

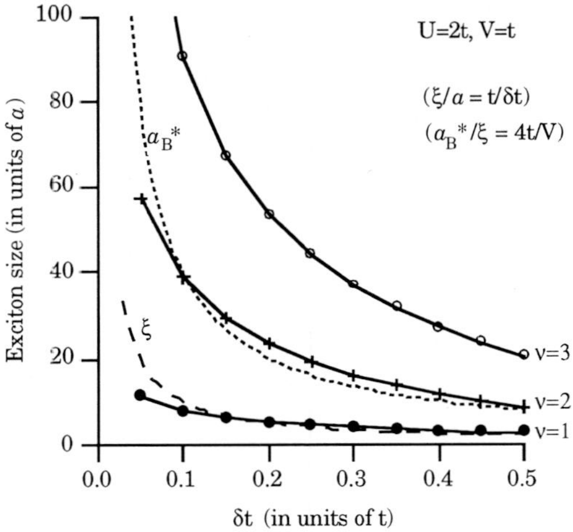

Figure 3 Calculated exciton size against δt, in comparison with the correlation length ξ and the effective Bohr radius a_B*.

3 Nonlinear Optical Response

An important consequence of the one-dimensional exciton effect is the concentration of oscillator strength into the lowest exciton state from interband transitions [2,7]. The oscillator strength of the entire interband continuum is negligibly small compared with that of the exciton peak, as shown in Fig. 4. In actual conjugated polymers, the absorption peak is usually quite broad except in the case of polydiacetylene crystal. The broad absorption band has been sometimes interpreted as interband transitions, but it is more likely to be an exciton peak broadened by various mechanisms such as electron-lattice coupling and disorder.

To clarify the nature of the observed absorption peak, it is quite useful to study third-order nonlinear optical spectra $\chi^{(3)}(\omega)$,. It turns out that each nonlinear spectrum exhibits each characteristic structure, as summarized in Fig. 4: (b) two-photon absorption (TPA), (c) electroabsorption, and (d) third-harmonic generation [7]. For example, the electroabsorption spectrum is characterized by a Stark shift at the exciton energy and a oscillatory structure around the interband edge E_c. These structures have been observed in polydiacetylene crystals [8,9]. The most interesting spectrum is the third-harmonic generation, which exhibits three peaks — two three-photon resonance peaks and a two-photon resonance peak at an uppermost frequency. The former have been observed in polydiacetylene [10] and polythiophene [11].

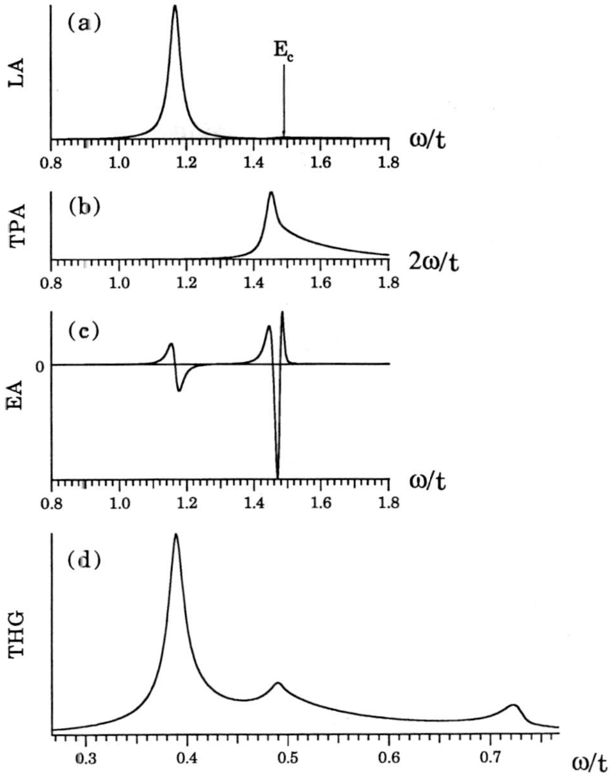

Figure 4 Characteristic features of (a) linear absorption, (b) two-photon absorption, (c) electroabsorption, and (d) third-harmonic generation, calculated with single-CI for a ring of N=1000 sites with U=2t, V=t and δt=0.2t [12].

The spectra shown in Fig. 4 were obtained in the single-CI calculations. Therefore, for example, the TPA spectrum in Fig. 4(b) exhibits a peak at the second exciton state, but the results at much higher energies are not very reliable, where two-exciton states become important. For this purpose we turn to the results of the double-CI calculations. Figure 5 displays an example of calculated two-photon absorption spectra for N=40 sites [6]. (The spectrum contains many structures compared with Fig. 4(b), but this is due to the smaller system size used in the double-CI calculations.) In the lower part of the figure, the component of double excitations to each two-photon excited state is plotted. The low energy peaks originate mainly from single excitations, whereas the high energy intense peaks essentially come from double excitations. The peak of the TPA is located far below the one-photon resonance energy, indicating that the state is a biexciton state with a large binding energy.

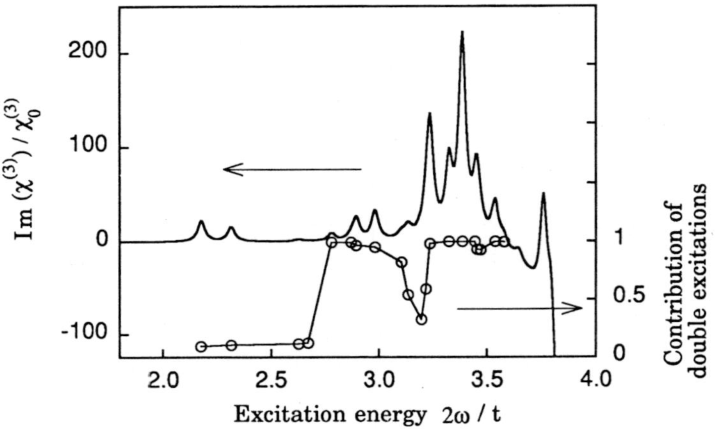

Figure 5 The two photon absorption spectrum and the contribution of double excitations (open circles) to the two-photon excited states, calculated with double-CI for a ring of N=40 sites with U=2t, V=t and δt=0.2t [6].

These spectra were obtained by use of the standard Orr-Ward formula [13] of the third-order nonlinear susceptibility. Now we would like to discuss the most interesting region of resonant nonlinearity just at the exciton absorption peak. In this case the Orr-Ward approach is not appropriate, and we should use a formula obtained with a density matrix approach [14], where diagonal as well as off-diagonal damping rates are incorporated. Such a calculation has been carried out recently [15] with the double-CI description of the PPP model.

Figure 6(a) displays the calculated degenerate four wave mixing spectrum, i.e., the absolute value of $\chi^{(3)}(-\omega;\omega,-\omega,\omega)$, together with the linear absorption, i.e., the imaginary part of $\chi^{(1)}(\omega)$. We would like to discuss the relationship between the two quantities, because there have been an experimental report [16] on a "scaling" relation between them. In Fig. 6(b), the relationship of the two quantities is displayed. The solid line corresponds to the low energy side of the absorption peak, while the broken curve to the high energy side. Clearly there is a power law dependence:

$$\left|\chi^{(3)}(-\omega;\omega,-\omega,\omega)\right| \approx c\left\{\text{Im}\left[\chi^{(1)}(\omega)\right]\right\}^{p}. \tag{5}$$

The power p is about 1.2 on the low energy side and about 1.7 on the high energy side. The low-energy behavior is in agreement with the empirical value p~1 obtained by Bubeck *et al.* [16].

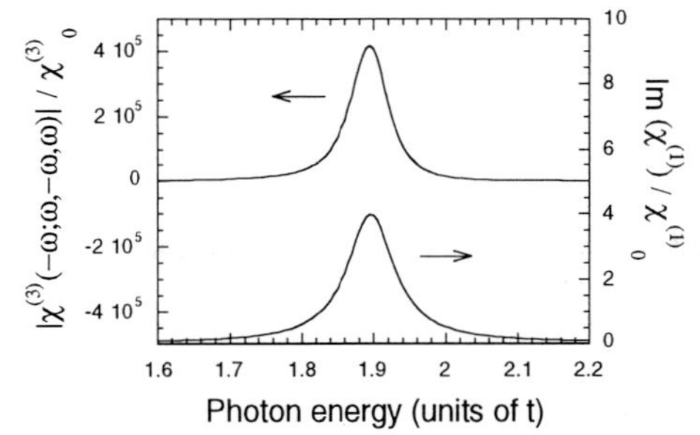

(a)

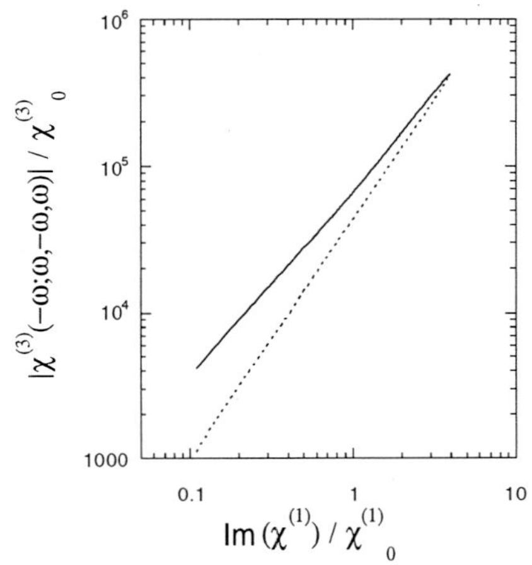

(b)

Figure 6 (a) An example of calculated dispersion of the third-order nonlinear susceptibility (upper curve) and linear absorption (lower curve) around the exciton resonance, and (b) the scaling relationship between the two quantities in logarithmic scales, with solid and broken curves corresponding to the low and high energy side of the absorption peak. The used model is a polymer ring of N=20 sites with U=2t, V=t and δt=0.2t. Here $\chi^{(3)}{}_0$ and $\chi^{(1)}{}_0$ are normalization factors defined in Ref. 15.

Another interesting result obtained in this study is the dephasing-induced extra resonance [15]. In the two-photon absorption region of $\chi^{(3)}$ spectra, the calculated spectrum exhibits quite a few additional peaks, for which the doubled peak energy does not correspond to an energy of two-photon allowed state. The photon energy of an extra resonance corresponds to an energy difference between two excited states. The height of an extra peak grows with Γ_2/Γ_1, where Γ_1 and Γ_2 are longitudinal and transverse relaxation rates, respectively. The effect is similar to the extra resonance discussed by Bloembergen *et al.* in a different context [17]. An important point in the present case is that the extra resonance is not a weak effect. Some of the extra peaks can become higher than ordinary two-photon absorption peaks. This theoretical result is relevant for experiments, because the ratio Γ_2/Γ_1 in real conjugated polymers are usually quite large.

4 Interchain Transfer of an Exciton

We turn to the problem of *interchain* transfer of excited states. If we consider two chains, the electron-electron interaction part $H_{\rm int}$ of eq.(4) contains an interchain part, say $H_{1\text{-}2}$. Suppose we first obtain excited states $|\,\phi_{i\nu}>$ of each chain (i = 1, 2) by the single CI method. Then we calculate the interchain matrix elements

$$J_{\nu\mu} = <\phi_{1\nu}\,|\,H_{1\text{-}2}\,|\,\phi_{2u}>. \tag{6}$$

Let us consider two identical chains of N sites facing each other in parallel with a distance R, as illustrated in Fig. 7(a). In this configuration the coupling between two in-phase dipoles is repulsive, so that J is expected to be positive.

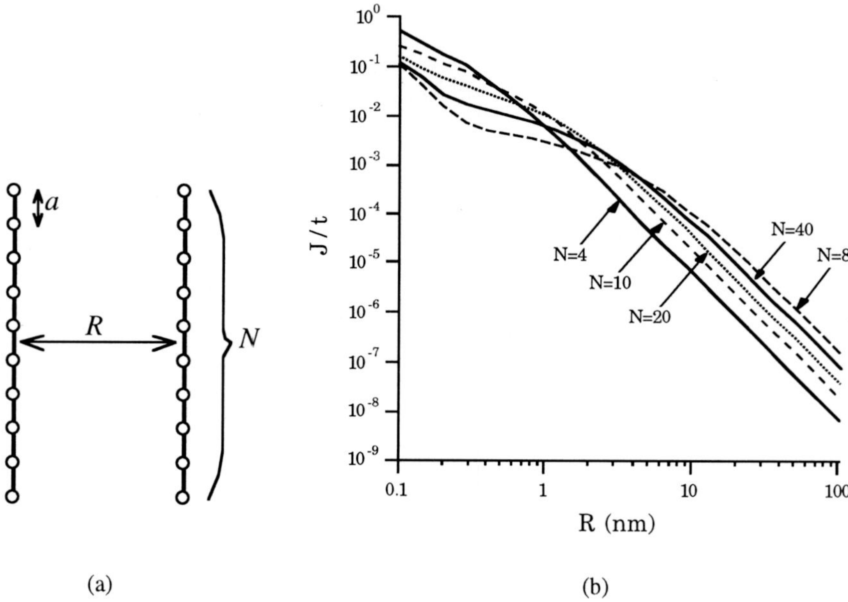

Figure 7 (a) Model of two chains; (b) dependence of interchain transfer J on interchain distance R for various chain size N, The used paremeters are U=4t, V=2t, δt=0.2t, a=0.14 nm.

Since the lowest exciton state (ν=1) has the largest transition dipole moment, we first examine J_{11}. Figure 7(b) displays the calculated R-dependence for several fixed chain lengths. For a short chain, the overall dependence is R^{-3}, corresponding to the behavior of usual dipole-dipole coupling. However, for a longer chain, the dependence deviates from R^{-3} at short distances. Another way of looking at the dependence is plotting as a function of the chain length Na for a fixed distance R. This is depicted in Fig. 8. For a short chain, J is approximately proportional to N. This is nothing but the behavior expected from the dipole-dipole coupling, because the transition dipole moment of each chain is proportional to $N^{-1/2}$ (in other words, the oscillator strength, which is proportional to the square of the dipole, scales as N). However, with increasing N, J tends to saturate and then starts to decrease again.

The dipole approximation breaks down when Na becomes larger than the interchain distance R. One can introduce a simple interpretation of the dependence. This is the so-called *extended dipole* model [18] to simulate the behavior beyond the point dipole approximation. In this scheme, instead of a point dipole we attribute two opposite charges $+q$ and $-q$ at the two ends of each chain, with $q=D/L_{\text{eff}}$, where D is the calculated transition dipole (see Fig. 8(b)). J is calculated as the sum of Coulomb energy among the four charges on two chains. The broken lines in Fig. 8(a) corresponds to the prediction of this model. We see that the extended dipole model simulates well the behavior of the calculated J.

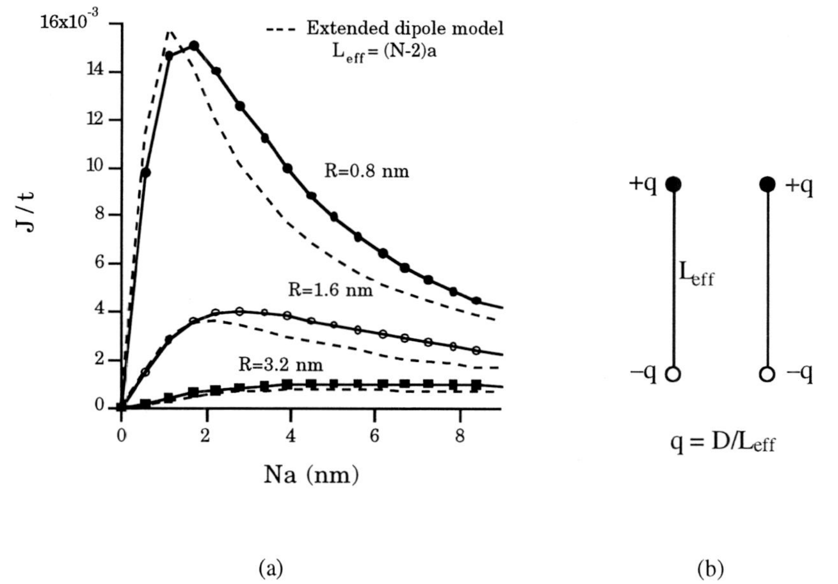

Figure 8 (a) Dependence of intermolecular transfer J on chain size N for the same parameters as in Fig. 7, compared with the extended dipole model (broken line) with $L_{\text{eff}} = (N\text{-}2)a$. (b) Extended dipole model.

But this is not the whole story. So far we have seen the behavior of J among the lowest excited states of the two chains. There are other matrix elements among arbitrary excited states ν and μ. Actually the "diagonal" elements ($\nu=\mu$) turns out to be dominant. In Fig. 9 we display an example of the calculated $J_{\nu\nu}$ and the dipole moment D_ν as functions of the excitation energy E_ν. First of all, we see that J is not proportional to D^2. The dipole is largest for the lowest excitation, but J has a maximum at a higher energy.

This is quite an usual situation. Although the lowest exciton state has the largest transition dipole moment, intermolecular transfer is not effective for this state. This implies that the optical response of an exciton is predominantly determined by its intrachain one-dimensional character. On the other hand, the exciton can be more easily transferred to neighboring molecules if excited by molecular vibrations or by some other scattering mechanism to a higher state with larger J.

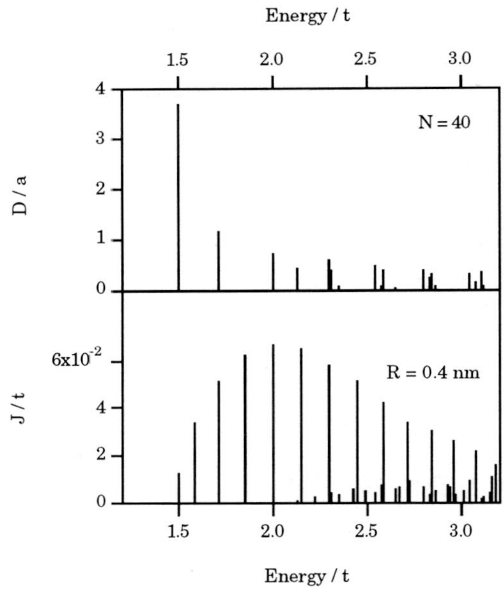

Figure 9 Transition dipole D (upper panel) and interchain transfer J (lower panel) of each state plotted against its excitation energy for the same paramters as in Figs. 7 and 8.

5 Concluding Remarks

Characteristic features of excitons in conjugated polymers have been discussed from a theoretical point of view. It has been demonstrated that the intrachain one-dimensional exciton has drastic effects on the linear and nonlinear optical spectra. Overall agreement between theory and experiment are obtained for moderately strong electron-electron interaction. The behavior of interchain excitation transfer is also peculiar in the sense that the transfer energy J is not necessarily determined by the transition dipole moments of the exciton state, due to the one-dimensional delocalization of the exciton. Thus the optical properties are mainly determined by the intrachain character, but the interchain effects are important for the dynamics of the exciton.

There are important effects which have not been discussed in this chapter. For example, phenomenological damping constants were used here for the broadening of optical spectra, but for the discussion of actual spectra we must explicitly consider the effect of electron-lattice coupling and disorder. These effects can be taken into account in part by numerical experiments for a classical model of lattice fluctuations [19]. Such calculations in fact yield a broad, asymmetric line shape of the exciton absorption peak in overall agreement with experimental spectra. Another interesting topic is the formation of an exciton-polaron, which is relevant for the explanation of photo-induced absorption [20].

Acknowledgments

I wish gratefully to acknowledge the contribution of V. A. Shakin in calculating nonlinear optical spectra with the double-CI method. I am grateful to many colleagues, especially Y. Shimoi, K. Harigaya, W.-P. Su, and M. Schreiber, for illuminating discussions.

References

[1] S. Abe, in *Relaxation in Polymers* edited by T. Kobayashi (World Scientific, Singapore, 1993) pp. 215.
[2] S. Abe, J. Phys. Soc. Jpn. **58** (1989) 62.
[3] R. Loudon, Amer. J. Phys. **27** (1959) 649.
[4] S. Abe, J. Yu and W. P. Su, Phys. Rev. B **45** (1992) 8264.
[5] A. J. Heeger, S. Kivelson, J. R. Schrieffer, and W.-P. Su, Rev. Mod. Phys. **60** (1988) 781.
[6] V. A. Shakin and S. Abe, Phys. Rev. B **50** (1994) 4306.
[7] S. Abe, M. Schreiber, W.P. Su and J. Yu, Phys. Rev. B **45** (1992) 9432.
[8] L. Sebastian and G. Weiser, Chem. Phys. Lett. **64** (1979) 396.
[9] Y. Tokura, Y. Oowaki, T. Koda, and R. H. Baughman, Chem. Phys. **88** (1984) 437.
[10] T. Kanetake, K. Ishikawa, T. Hasegawa, T. Koda, K. Takeda, M. Hasegawa, K. Kubodera, and H. Kobayashi, Appl. Phys. Lett. **54** (1989) 2287.
[11] W. E. Torruellas, D. Neher, R. Zanoni, G. I. Stegeman, F. Kajzar, and M. Leclerc, Chem. Phys. Lett. **175** (1990) 11.
[12] S. Abe, Y. Shimoi, V. A. Shakin, and K. Harigaya, Mol. Cryst. Liq. Cryst. **256** (1994) 97.
[13] B.J. Orr and J.F. Ward, Mol. Phys. **20** (1971) 513.
[14] R.W. Boyd, *Nonlinear optics* (Academic, London, 1992), Chap.3.
[15] V. A. Shakin, S. Abe, and T. Kobayashi, Phys. Rev. B **53** (1996), in print.
[16] C. Bubeck, A. Kaltbeitzel, A. Grund and M. LeClerc, Chem. Phys. **154** (1991) 343.
[17] N. Bloembergen, H. Lotem, and R. T. Lynch, Indian J. Pure Appl. Phys. **16** (1978) 151.
[18] V. Czikklely, H. D. Forsterling, and H. Kuhn, Chem. Phys. Lett. **6** (1970) 207.
[19] M. Schreiber and S. Abe, Synth. Metals **55-57** (1993) 50.
[20] Y. Shimoi and S. Abe, Phys. Rev. **49** (1994) 14113.

Part III

Langmuir–Blodgett Films

LANGMUIR-BLODGETT MICRO-SANDWICHES

Siegmar Roth, Steffi Blumentritt, Marko Burghard, Claudius M. Fischer, Carsten Müller-Schwanneke, Günther Philipp

Max-Planck-Institut für Festkörperforschung, Heisenbergstr. 1, D-70569 Stuttgart, Germany

Abstract

This paper reports on LB films in microstructured gold sandwiches. The active area of these sandwiches are so small that there is a fair chance of finding junctions without pinholes, or of seeing the effects of individual conductive paths. Selected current-voltage characteristics will be shown and discussed in terms of Coulomb staircases, resonant tunneling and molecular rectification.

1. Introduction and Historical Survey

The concept of molecular rectification is illustrated in Fig. 1, in which we relate the molecular orbitals of a π-conjugated molecule to the band structure of a classical semiconductor. The valence band corresponds to the bonding (π) orbitals and the conduction band to the antibonding (π*) orbitals. The valence band edge is analogous to the highest occupied molecular orbital (HOMO) and the conduction band edge to the lowest unoccupied molecular orbital (LUMO). A semiconductor is doped by inserting atoms with a different number of electrons as compared to the host atoms.

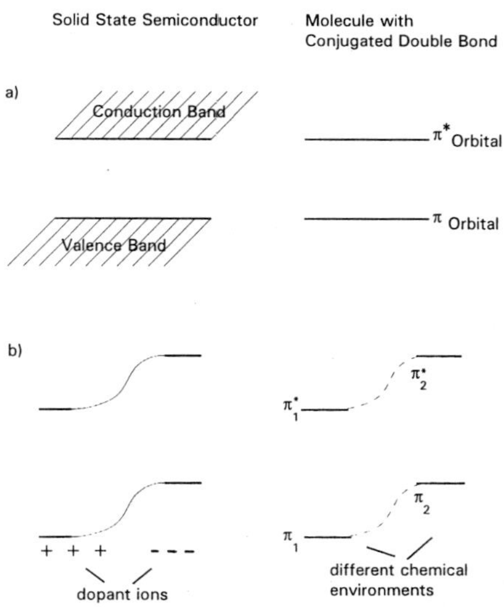

Figure 1: Comparison of semiconductor band structure and molecular levels of a π conjugated molecule: a) undoped case, b) band and orbital shifts after doping or modification of the chemical environment

Springer Proceedings in Physics, Vol. 81
Materials and Measurements in Molecular Electronics
Editors: K. Kajimura · S. Kuroda

These atoms ionize, and the space charges of the ions push the bands either upward or downward. In a p-n junction, p- and n-doped regions meet, and the bands are displaced as shown in Fig. 1b. Single molecules cannot be doped in the same way as semiconductors, since different chemical environments (electron donating and withdrawing groups) will modify the electron affinity and lead to analogous shifts of the molecular levels. More than 20 years ago Aviram and Ratner [1] proposed the A-σ-D molecule shown in Fig. 2. The symbol A stands for acceptor (electron withdrawing group), D for donor (electron donating group) and σ for a saturated group, which has no electroactive function but acts as a spacer (tunnel barrier) and mechanically keeps the D and A parts together. The D and A moieties are well known in one-dimensional physics [2] as TTF (tetrathiafulvalene) and TCNQ (tetracyanoquinodimethane) and are the basis of the organic charge transfer salts. Aviram and Ratner showed that the molecule of Fig. 2 would act as a rectifier if it could be connected to an electric circuit which would allow the measurement of current-voltage characteristics.

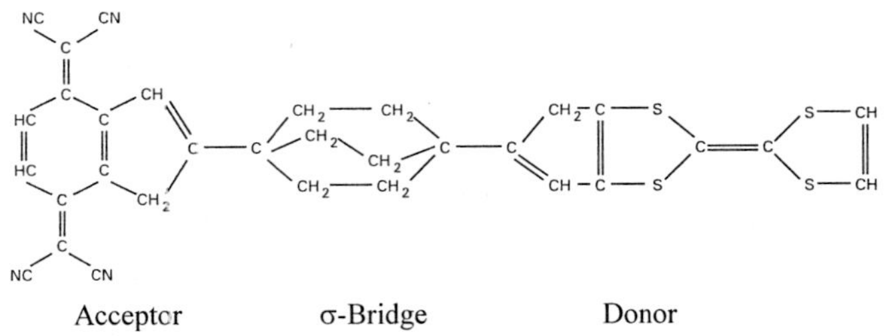

Figure 2: "D-σ-A molecule" proposed by Aviram and Ratner in 1974 [1]

Three different ways of contacting individual molecules are shown in Fig. 3. The most direct way is to use a split electrode. Itoua, amongst others, has followed this route [3]. A thin metallic strip is produced by electron beam lithography and by the method of "pixel jumping" the strip is interrupted to form a small gap. A chain molecule is then laid across the gap. If the molecule has sulfur groups at its ends, covalent bonds might form between the metal of the electrode and these sulfur groups. At the moment, the gaps which can presently be fabricated by the physicists are not quite small enough for the length of the molecules which can be synthesized by the chemists.

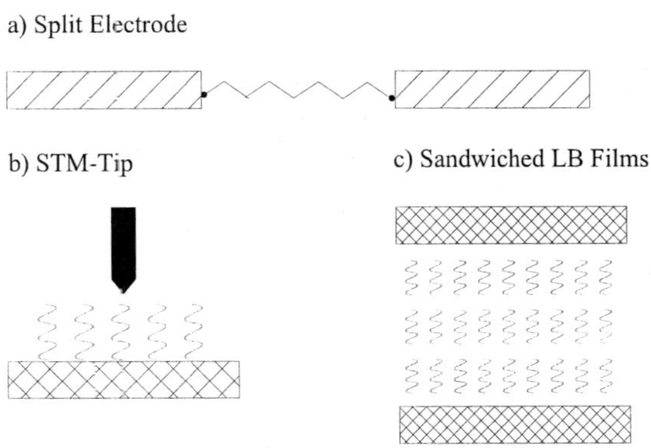

Figure 3: Three principal ways of contacting rectifying molecules (or other functionalized molecules of interest for molecular electronics)

After the development of the scanning tunneling microscope it was tempting to employ this technique to rectifying molecules. One of the first report on such an experiment was given by Aviram et al. [4]. The results are reproduced in Fig. 4, which shows the current and voltage traces on an oscilloscope screen. In reverse bias the current rises to about 3 nA if the voltage increases from 0 to 0.5 V. In forward bias a much larger current is observed, running out of scale at about 8 nA. Undoubtedly the experiment shows rectification, but it turned out to be difficult to unambiguously assign this behaviour to the properties of the molecules used (a situation still quite general in all experiments in the field of molecular electronics).

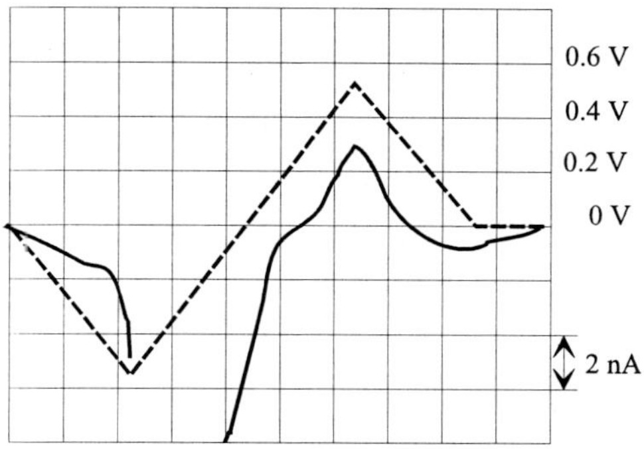

Figure 4: Molecular rectification in a STM experiment. Current (——) and voltage (----) traces on the oscilloscope [4]

Stabel et al. [5] looked with the STM at alkylated hexabenzocoronene adsorbed on highly oriented pyrolytic graphite. The STM image shows black and white features, which are attributed to the electrically inert alkyl chains and to the π conjugated system in the coronene moiety. When the tip is over the inert parts, symmetric tunnel characteristics are observed (trace (a) in Fig. 5), and when the tunnel current passes through the π-conjugated electrons of hexabenzocoronene, rectifying behaviour is displayed. Rectification is explained by the spatially asymmetric position of the π system with respect to the electrodes (STM tip and graphite substrate).

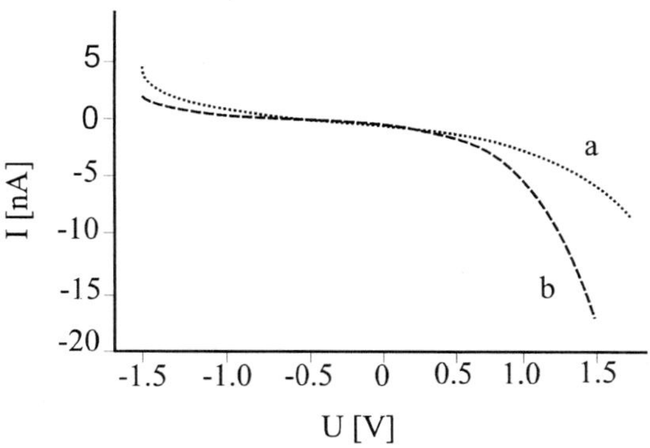

Figure 5: Tunnel characteristics through alkylated hexabenzocoronene adsorbed on pyrrolytic graphite. Trace a): symmetric characteristic when tip is over alkylated part. Trace b): rectifying characteristic when tip is over aromatic coronene part [5]

The third way of contacting molecules indicated in Fig. 3 is by incorporation into sandwiched LB films. Kuhn [6] has already pointed out that in a LB film electroactive molecules could be diluted by inert molecules and if the contact area is small enough, "single" electrons could be investigated. The Exeter group has pioneered in investigating the rectifying behaviour of functionalized LB films [7]. In Fig. 6 current-voltage characteristics of LB films containing zwitterionic molecules are presented. The zwitterionic molecules are shown in Fig. 7. They are similar to the D-σ-A molecules of Fig. 2, except that the bridge is not a saturated σ section, but also contains π conjugated electrons. This leads to a stronger coupling between donor and acceptor and to a strong permanent dielectric moment. The rectifying behaviour in Fig. 6 is clearly seen. To demonstrate that rectification is related to the zwitterionic molecules Martin et al. have performed two blind experiments. They have separated the zwitterionic molecules from the metallic electrodes by inert layers to avoid chemical reactions at the organic-metal interface and have even measured with inert layers alone, in which case symmetric characteristics were observed. Furthermore, they have bleached the zwitterions by chemical treatment, and again restored the symmetric characteristics. A review on the "quest for molecular rectification" has been given by Metzger [8].

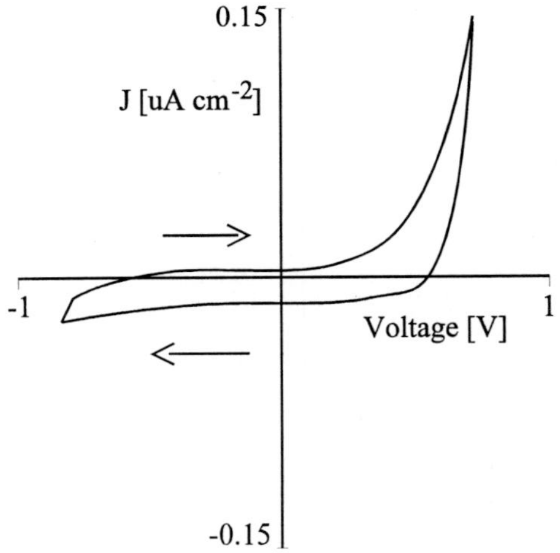

Figure 6: Current-voltage characteristics of LB films of the zwitterionic molecules shown in Fig. 7 (after Martin et al. [7])

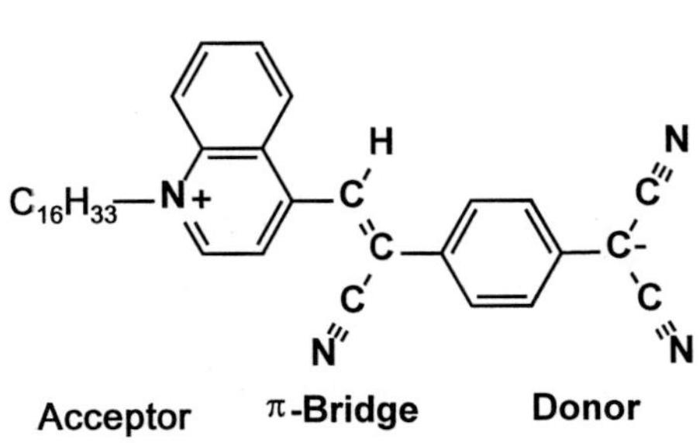

Figure 7: Zwitterionic molecule used by Martin et al. to prepare rectifying LB films [7]

2. Micro-Sandwiches

One of the main problems in measuring transport through thin organic layers is that of pinholes in these layers. If electrodes are evaporated on these layers, the electrode metal will fill the pinholes and this will lead to short circuits. However, if only very small sections of a film are contacted, there is a fair chance that pinhole-free junctions can be prepared. In addition, individual pinholes, "almost-pinholes" and other defects could be detected and perhaps new phenomena would be observed. This idea follows the lines of the gigaseal technique, developed by Neher and Sakmann to study ionic transport through biological membranes. In such membranes there are channels which open and close at random and if a capillary of only a few micrometer in diameter is closed by a piece of such a membrane, individual channels can be investigated. Figure 8 is taken from Neher's Nobel lecture [9]. Trace (a) shows a voltage pulse of 50 ms and trace (b) the current response integrated over some hundred pulses. Traces (c) correspond to one pulse each and show single channel openings. The resistance of a Neher-Sakmann junction is larger than 1 GΩ, hence the name gigaseal.

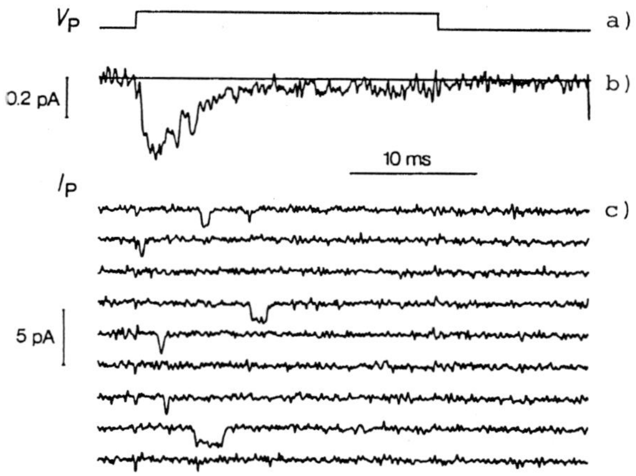

Figure 8: Ion transport through biological micromembranes (after E. Neher [9]). a) voltage pulse; b) current response integrated over some hundred pulses; c) current response to individual pulses. The gigaseal technique developed by Neher and Sakmann to study intracellular communication has inspired the LB micro-sandwich technique described in this paper

The cross section of the LB junctions investigated in this work were made so small, that the room temperature resistance of many of the junctions was larger than 1 GΩ. Sandwiches with smaller resistances were regarded as short circuits and were not investigated further. In addition to microstructuring we also used the technique of multiplexing, i.e. we put up to 50 micro-sandwiches on one glass plate or silicon chip, so that we at least obtained "some" working devices on a chip. Finally it should be noted that most measurements were carried out at liquid helium temperature. At low temperatures pronounced structures are seen in many current-voltage characteristics. At high temperature these structures are masked by "leakage" currents, which we attribute to variable range hopping between point defects.

In Fig. 9 we have plotted the percentage of junctions with short circuit (i.e. with $R<1G\Omega$) as a function of the cross section for 4 and 9 monolayer micro-sandwiches of the polymer PolC5F (a poly-(maleicmonoester-co-1-octadecene)-derivative with fluorinated ester groups). We see that the yield of short-circuit free junctions increases with the number of monolayers and with decreasing junction area.

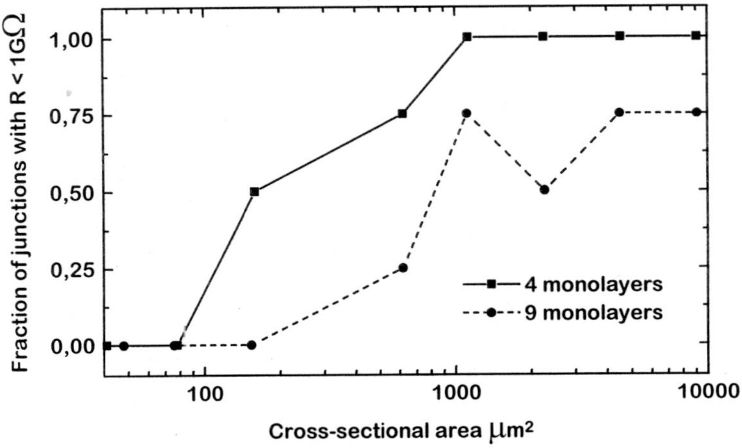

Figure 9: Dependence of the number of quasi short circuits on the cross sectional area studied on two sets of micro-sandwiches of nominally 9 respective 4 monolayers of PolC5F

Several electrode configurations have been used in our devices. One example is depicted in Fig. 10. Narrow gold fingers are evaporated as bottom electrodes onto a glass plate. Several monolayers of organic molecules are then transferred by the LB technique and one gold top electrode is then evaporated crosswise. Another electrode configuration is shown in Fig. 11. Here the junctions have different cross sections so that the area dependence of the current through the junctions can be studied. In addition the junctions are less interdependent than in Fig. 10.

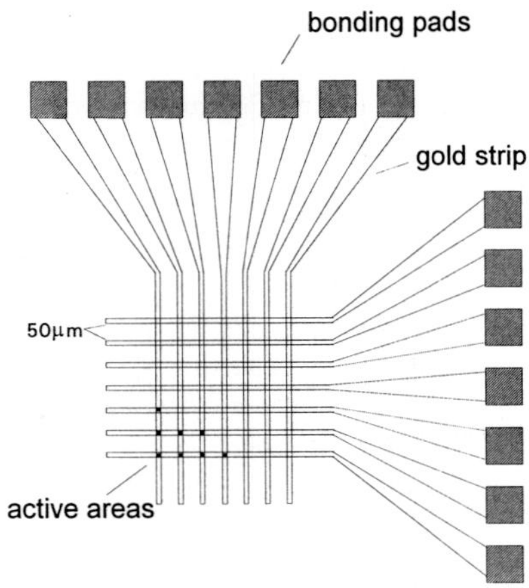

Figure 10: Example of an electrode configuration to study an array of LB micro-sandwiches

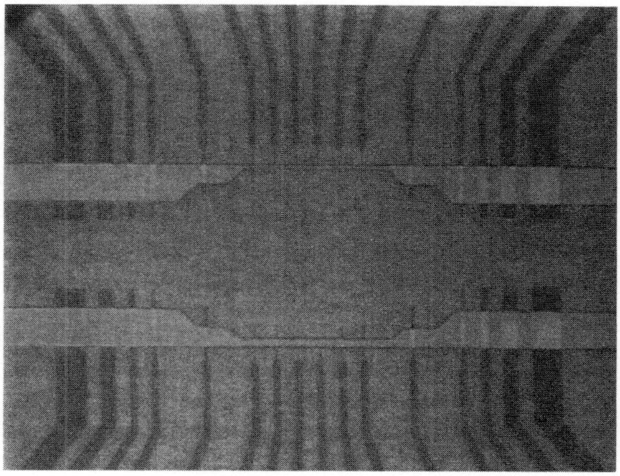

Figure 11: Electrode configuration for studying the dependence of the current on the junction cross section

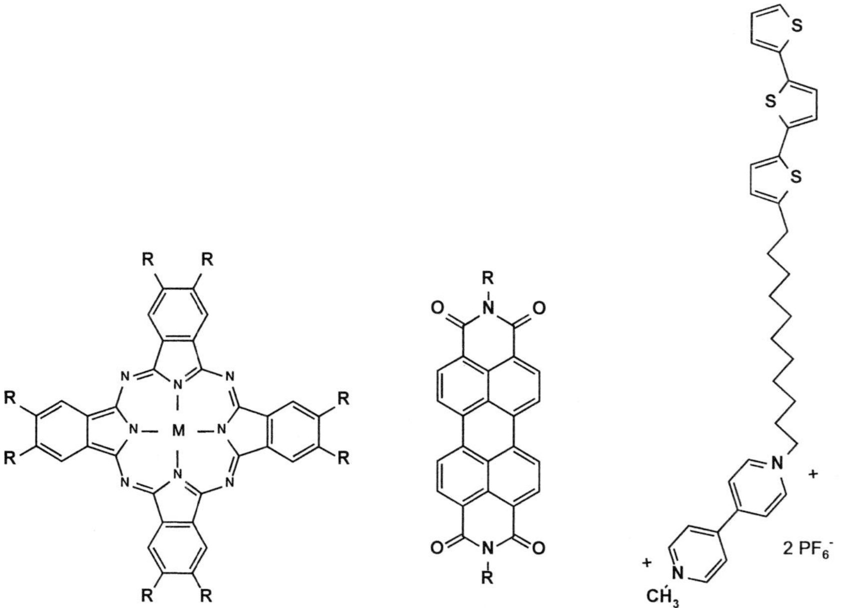

PTCDI-OEt: R=CH_2-CH$(OEt)_2$

PcPd: M=Pd, R=OC_5H_{11} | PTCDI-SPent: R=CH_2-CH(SPent) | T_3-V^{2+}

(a) (b) (c)

Figure 12: Example of molecules used in these investigations

Examples of the molecules used in these investigations are shown in Fig. 12. The palladiumphthalocyanine- (Fig. 12a) and perylenetetracarboxyldiimide-derivatives (Fig. 12b) are molecules with fairly large π conjugated electronic systems, the restgroups R being chains which have been added to enable film formation. The molecule in Fig. 12c is a D-σ-A molecule, but with D, A, and σ parts different from those originally proposed by Aviram and Ratner. A common feature of these molecules is that the π orbitals (HOMO's) are energetically close to the Fermi level of gold, so that in a gold-LB-gold sandwich these orbitals could participate in the charge transport across the junctions and perhaps lead to distinct structures (resonances) in the current-voltage characteristics. The level scheme of a Au/PcPd/PTCDI-R_2/Au sandwich is shown in Fig. 13. The data are compiled from published literature. We have characterized the LB films by surface-pressure versus molecular area diagrams, transfer ratios, UV/VIS and IR optical spectroscopy, grazing incident X-ray reflection, and several other physicochemical methods and hence we know quite well how the molecules are oriented in the films. The details have been published in Refs. [10-12]. Unfortunately these methods give little information on individual and localized defects and therefore we have to speculate on these from transport measurements.

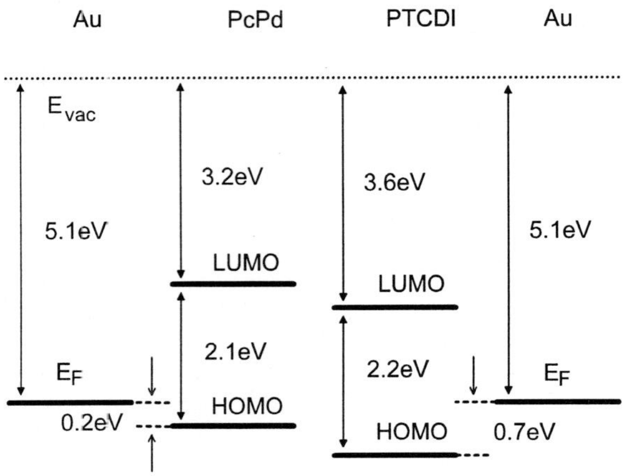

Figure 13: Level scheme of a gold/phthalocyanine/perylene/gold sandwich. The data has been approximated by published literature

3. Selected Current-Voltage Characteristics

Apart from short circuits (room temperature resistance <1GΩ) we observe two types of current voltage characteristics: smooth characteristics and characteristics with distinctive structures (kinks, steps, thresholds). A set of smooth characteristics taken on a gold/10 monolayers of phthalocyanine/gold micro-sandwich at various temperatures is shown in Fig. 14. The freeze-out of the current at low temperatures is clearly observed, indicative of either variable range hopping or some similar transport mechanism. We believe that we would get smooth characteristics when our films are 10 monolayers or thicker and if all layers are intact. Such films are thicker than 200 Å and direct tunneling from electrode to electrode is negligibly small.

We have good reasons to assume that usually not all layers are intact. If the layers were intact the current through the junction should increase linearly with the cross-sectional area. This is not observed and we take this lack of linear scaling as an evidence that even in those junctions which we have classified as "free of short circuits" transport is often dominated by such imperfections as partially short-circuited layers and other local defects.

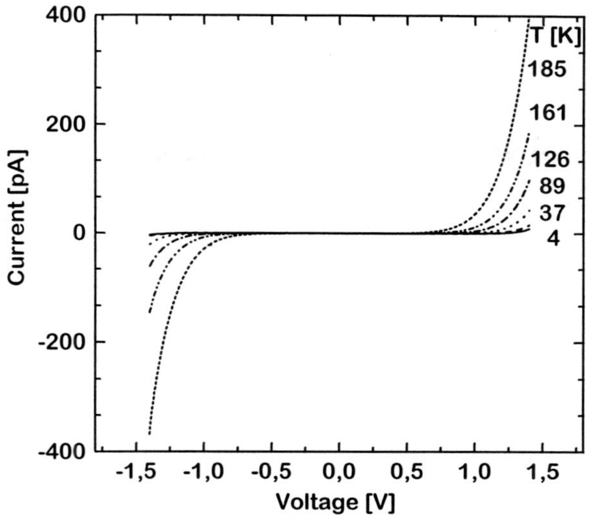

Figure 14: "Smooth" current-voltage characteristics at various temperatures, taken on a 10 monolayer gold/PcPd/gold micro-sandwich

Current voltage characteristics with pronounced structures are usually observed on thin films (4 monolayers). If they are encountered in nominally thick films we suspect that some monolayers are interrupted or short circuited. The LB junction might then look as depicted in Fig. 15, where a very beautiful staircase is seen. This curve was measured on a LB micro-sandwich of 4 monolayers

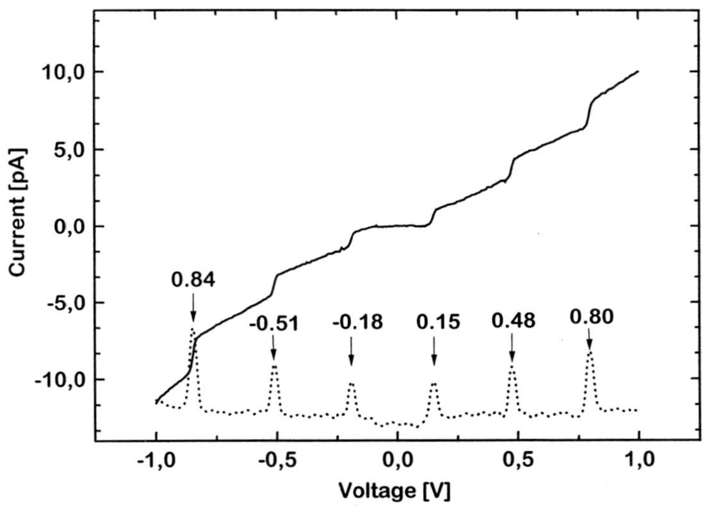

Figure 15: Current-voltage characteristic of a LB micro-sandwich with 4 monolayers of a saturated polymer. The structure is interpreted as Coulomb staircase when a gold cluster is charged by single electrons. The gold cluster might have separated from one of the electrodes and migrated into the film

of the polymer PolC5F which contains only saturated bonds, so that the structure cannot be caused by resonant tunneling into π orbitals. We see a series of equidistant steps, separated by voltage intervals of 0.33 V (to show this equal spacing more clearly we have also plotted the derivative of the current, as dotted line at the bottom of the figure).

We interpret the observed structure as Coulomb staircase of a gold cluster which might have separated from one of the gold electrodes and migrated into the film. If an electron tunnels to such a metallic island it charges and the potential therefore changes by $\Delta V=e^2/C$, where e is the electronic charge and C the capacitance. The next electron cannot follow (Coulomb blockade) unless the voltage at the electrode is raised so far that it compensates for the charging energy of the island. The Coulomb staircase comes from single electrons having tunneled to the gold cluster. Similar effects have been observed in semiconductor quantum dots [13] and in metal particles adsorbed on solid surfaces, which have been approached by STM tips [14-17]. The geometry of a gold cluster in a LB micro-sandwich is schematized in Fig. 16 and the corresponding energy level scheme in Fig. 17. Here the Fermi level of the cluster is marked with a prime, E'_F, since it need not necessarily coincide with the Fermi level E_F of the electrode (in gold single crystals the work function is known to strongly depend on the crystallographic orientation of the surface on which it is measured). The level shifts by charging the island are also indicated in the figure. If $E_F \neq E'_F$ and if the cluster is not centered between the electrodes the Coulomb staircase will also be shifted to higher or lower voltages and the current-voltage characteristic will be "rectifying".

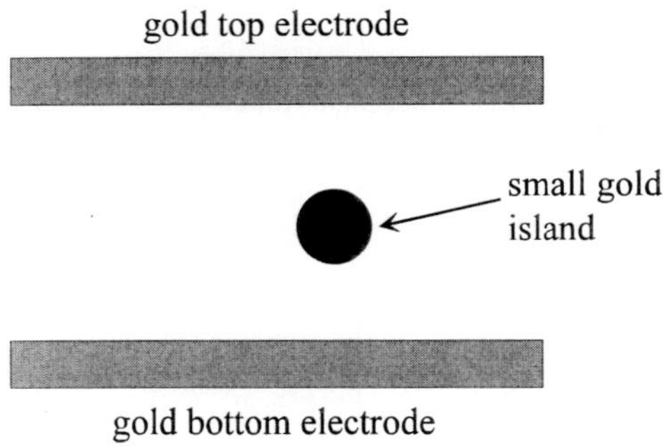

Figure 16: Gold cluster in an LB micro-sandwich

In the lower part of Fig. 17 the gold cluster is replaced by a phthalocyanine molecule. There are two differences with respect to gold: i) the relevant molecular levels might be even further away from E_F than was E'_F. ii) There is no continuum of states on the molecule and the next state (HOMO-1 or LUMO+1) will be out of reach (strictly speaking, there is no continuum on the small gold cluster either, because of electron confinement leading to level splitting, but this splitting is of the same order or even smaller than the charging energy). To accommodate several electrons in phthalocyanine we would have to assume that the molecules are stacked as in a single crystal and that the interaction within the stack leads to band formation. In Fig. 18 we reproduce current-voltage characteristics with pronounced thresholds, recorded on LB micro-sandwiches containing various π-conjugated molecules [12]. Since these thresholds seemed to coincide with the energetic positions of the HOMO (respectively LUMO) of the molecules the features had been assigned to resonant tunneling through the π-molecular orbitals.

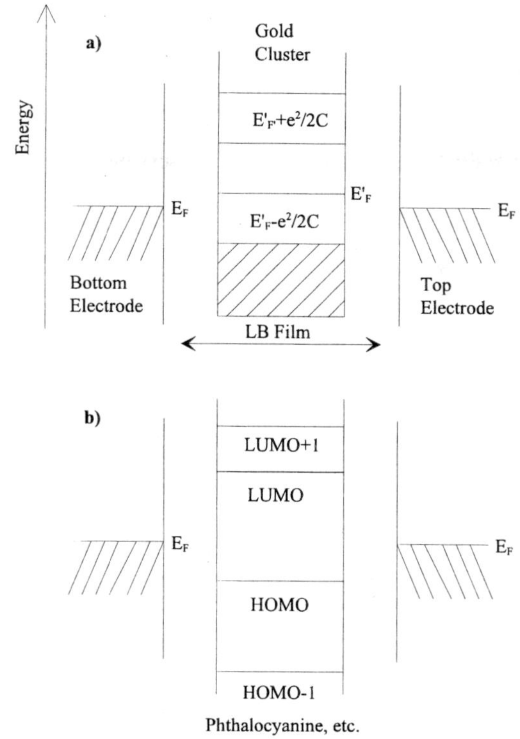

Figure 17: Level scheme (schematically) of a gold cluster (a) and a π conjugated molecule (b) in a LB micro-sandwich between gold electrodes to explain resonant tunneling and single electron effects

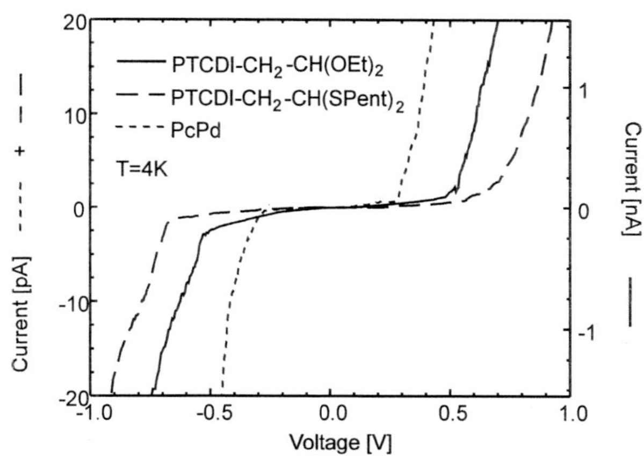

Figure 18: Current-voltage characteristics of LB micro-sandwiches [12] of various π conjugated molecules with kinks coinciding with the HOMO positions as derived in Fig. 11

The voltage plotted in the abscissa of the current-voltage diagrams is the potential difference between the gold electrodes. If the phthalocyanine molecule is in the middle between the electrodes one can easily rationalize that resonance should occur when $V=\pm 2\,(E_F-E_{HOMO})$. Figure 19 shows the current-voltage characteristics of LB micro-sandwich composed of 6 monolayers of the phthalocyanine- and 6 monolayers of a PTCDI-derivative. We have already shown the corresponding energy level scheme of this junction in Fig. 13 and in view of our above argumentation we should not be surprised that this asymmetric junction acts as a rectifier [18].

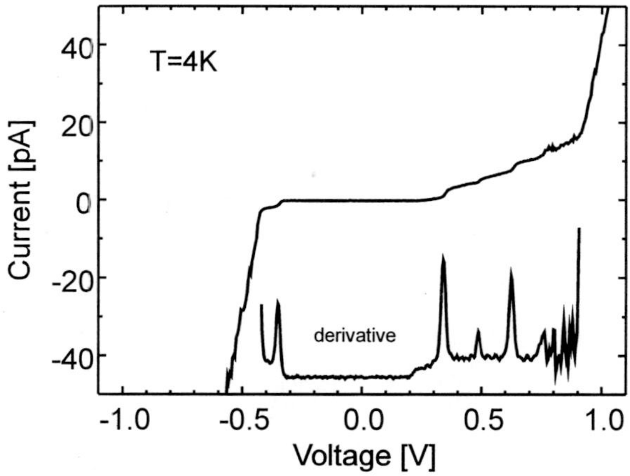

Figure 19: Rectifying current-voltage characteristics of a gold/phthalocyanine/perylene/gold LB micro-sandwich [18]

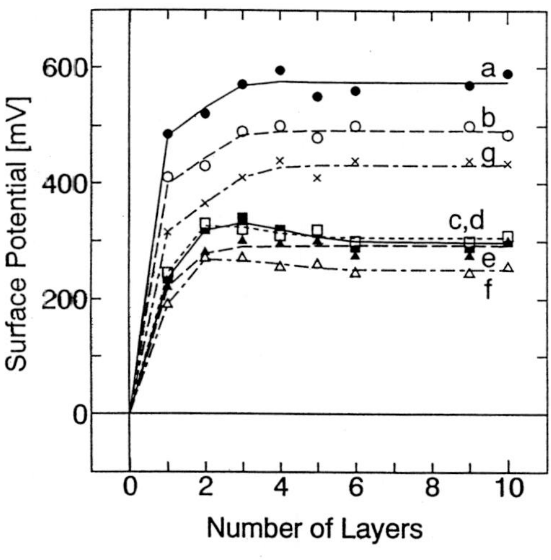

Figure 20: Surface potential measurements of PcPd LB layers on gold (measured at room temperature) demonstrating a band offset at the metal-organic interface. The curves a) to f) correspond to different environments (air, vacuum, etc.) and to different annealing temperatures

It must be stressed, however, that this assignment is only tentative and that the quantitative agreement found could just be fortuitous. So far, in our discussion we have neglected the possibility of space charges. Interactions at the metal-organic interface or even between different LB layers could lead to defect states which might capture electrons and thus cause space charges to accumulate. Figure 20 [19] shows the surface potential as measured by a Kelvin probe on one of our samples which we have sent to the Tokyo Institute of Technology. We see that on the interface between phthalocyanine and gold space charge effects can lead to a band offset as large as 0.5 V. Therefore more data and more systematic studies are needed before a clear view of charge transport in LB micro-sandwiches can be obtained.

References

[1] A. Aviram, M.A. Ratner: *Chem. Phys. Lett.* 29, 277 (1974).

[2] For an introduction see: S. Roth, *One-Dimensional Physics*, VCH Verlag, Weinheim, 1995.

[3] S. Itoua, PhD Thesis, Toulouse 1995.

[4] A. Aviram, C. Joachim, M. Pomerantz: *Chem. Phys. Lett.* 146, 490 (1988).

[5] A. Stabel, P. Herwig, K. Müllen, J. Rabe: *Angew. Chem.* 107, 1768 (1995).

[6] H. Kuhn: *Thin Solid Films* 178, 1 (1989).

[7] S. Martin, J.R. Sambles, G.J. Ashwell: *Phys. Rev. Lett.* 70, 218 (1993).

[8] R.M. Metzger: *Biomolecular Electronics*, ed. by R.R. Birge, American Chemical Society Advances in Chemistry Series 1994, Vol. 240, p. 81.

[9] E. Neher: *Angew. Chem.* 104, 837 (1992).

[10] M. Schmelzer, M. Burghard, C.M. Fischer, S. Roth, W. Göpel: *Synth. Metals* 71, 2087 (1995).

[11] M. Burghard, M. Schmelzer, S. Roth, P. Haisch, M. Hanack: *Langmuir* 10, 4265 (1994).

[12] M. Burghard, C.M. Fischer, M. Schmelzer, S. Roth, M. Hanack, W. Göpel: *Chem. Mater.* 7, 2104 (1995).

[13] B. Su, V.J. Goldman, J.E. Cunningham: *Bull. Am. Phys. Soc.* 36, 400 (1991).

[14] J.B. Barner, S.T. Ruggiero: *Phys. Rev. Lett.* 59, 807 (1987).

[15] E. Bar-Sadeh et al.: *Phys. Rev. B* 50, 3961 (1994).

[16] H. van Kempen, J.G.A. Dubois, J.W. Gerritsen, G. Schmid: *Physica B* 204, 51 (1995).

[17] M. Amman, R. Wilkins, E. Ben-Jacob, P.D. Maker, R.C. Jaklevic: *Phys. Rev. B* 43, 1146 (1991).

[18] C.M. Fischer, M. Burghard, S. Roth, K. von Klitzing: *Europhys. Lett.* 28, 129 (1994).

[19] Iwamoto et al.: to be published.

Control of Electrical and Optical Properties of Langmuir-Blodgett Films Using Photoisomerization of Azobenzene

Mutsuyoshi Matsumoto and Hiroaki Tachibana

National Institute of Materials and Chemical Research (NIMC), Tsukuba, Ibaraki 305, JAPAN

Abstract

Electrical and optical properties of Langmuir-Blodgett (LB) films were controlled using photoisomerization of supermolecules having switching, transmission, and working units. Conductivity switching was observed in single-component LB films of the supermolecules on alternate photoirradiation with UV and VIS light. In the mixed LB films with cyanine dye, J-aggregate formation was induced by photoisomerization of the supermolecule. It was suggested that the signal produced by the molecular deformation could be conveyed to the surrounding molecules, causing the cyanine molecules to form J-aggregate.

1. INTRODUCTION

The Langmuir-Blodgett (LB) technique has been attracting considerable interest since it enables the fabrication of molecular assemblies with well-defined structures[1-4]. Potential applications of LB films include conductive and photochromic materials, biosensors, and nonlinear optics. The materials employed cover a wide range of molecules such as low molecular weight amphiphiles, fullerenes, polymers, and biological compounds. With the repopularization of the term *molecular electronics*, many researches have concentrated on the construction of molecular devices for more than ten years now[5,6]. Despite considerable efforts, however, molecular devices have not been used practically.

In general, there are two ways by which molecular electronics may be realized. The first one utilizes the assembly of supermolecules, multi-functional molecules, which are each endowed with two or more functional units. The second one relies on the assembly of molecular components with different functions. In both of the *supramolecular* and *molecular component* systems, great emphasis is laid on how to design, synthesize molecules and assemble them as desired.

We have been doing research on the construction of functionalized molecular materials using LB films[7,8] and found that the photoisomerization of molecules can be used as a trigger to control the structures and functions of LB films. This paper focuses on controlling the electrical and optical properties of LB films using the photoisomerization of azobenzene. Electrical conductivity switching is realized using *supramolecular* systems and optical properties are controlled using *molecular component* systems.

2. Photoisomerization of Azobenzene

Figure 1 illustrate the photoisomerization of azobenzene. Azobenzene is the best-known molecule that undergoes photoisomerization. Two isomers exist and the trans isomer is thermodynamically

Springer Proceedings in Physics, Vol. 81
Materials and Measurements in Molecular Electronics
Editors: K. Kajimura · S. Kuroda

more stable than the cis one. By the irradiation with UV light, the trans isomer isomerizes to the cis one, which isomerizes back to the trans isomer by the irradiation with visible light. The cis isomer also isomerizes thermally to the trans one.

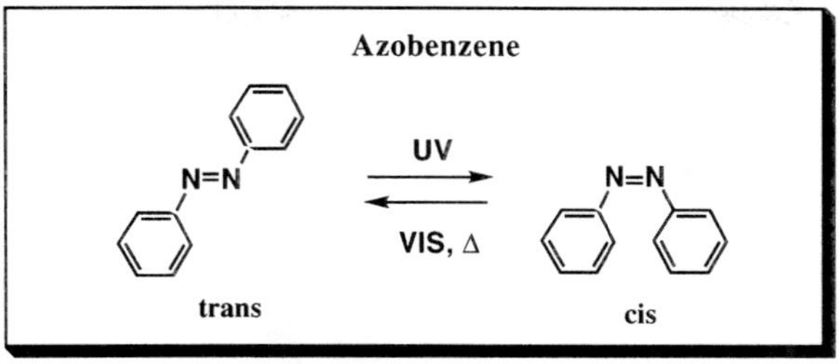

Figure 1: Photoisomerization of azobenzene

Photoisomerization gives rise to a change in absorption spectrum which is know as photochromism. But, photoisomerization also causes changes in molecular conformation, *i.e.*, molecular deformation, cross-sectional area, and dipole moment. Actually, the latter two increase with trans-to-cis isomerization. What we are interested in is that the molecular deformation caused by photoisomerization serves as a trigger to initiate changes in the structures and functions of LB films.

3. Electrical Conductivity Switching Using Supramolecular LB Films

Supermolecules have been attracting interest for the investigation of physical processes such as photoinduced electron transfer by modeling the photosynthetic centers. Different functional units are combined in order to stabilize the charge separation. It is essential to immobilize the supramolecular systems if they are to be used practically and the LB technique provides an easy way of doing this. The usage of supermolecules has another advantage that the appropriate combination of functional units spatially may give rise to a novel function which is not easily realized by mixing different molecules with different functional units. A typical example is given below[9,10].

Figure 2 shows a schematic representation of a switching device based on the LB film of a supermolecule consisting of a switching, a transmission, and a working unit. When the molecule of this kind is assembled properly in molecular materials, the external stimulus received by the switching unit gives rise to a kind of signal which is conveyed through the transmission unit to the working unit. This signal changes the structure of the working unit and hence the function associated with the working unit. Out of a number of possible external stimuli such as heat, electric and magnetic field, and chemical substances, light seems to serve this purpose best because of the compatibility with the existing information processing devices. Along this line, electrical conductivity seems to be one of the best candidates as the function of the materials.

Figure 3 shows the structure of the supermolecules of this type. More specifically, azobenzene or phenylazonaphthalene is selected as a switching unit to take advantage of the photoisomerization as a trigger to control the function, and a pyridinium$(TCNQ)_2$ complex is employed as a working unit which is associated with the function, *i.e.*, electrical conductivity, of the LB films. An alkyl chain serves as a transmission unit which connects the above two functional units. With this arrangement, a photon received by the switching unit causes the azobenzene unit to isomerize from a trans to a cis isomer. The signal produced by this molecular deformation is considered to reach the working unit and change the electrical conductivity of the film.

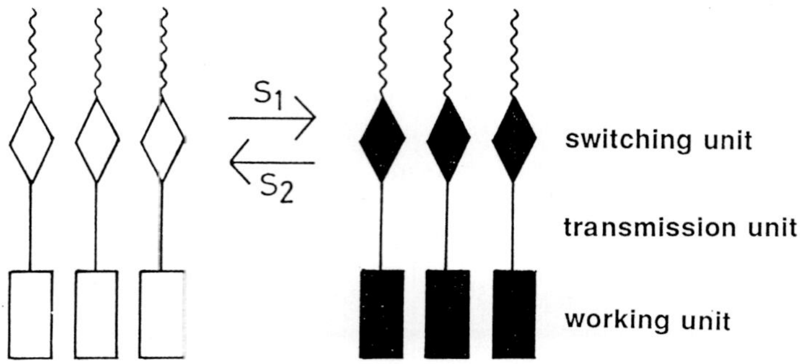

Figure 2: Schematic representation of a switching device based on the LB film of a supermolecule (S_1, S_2: external stimuli), reproduced from Ref. [10]

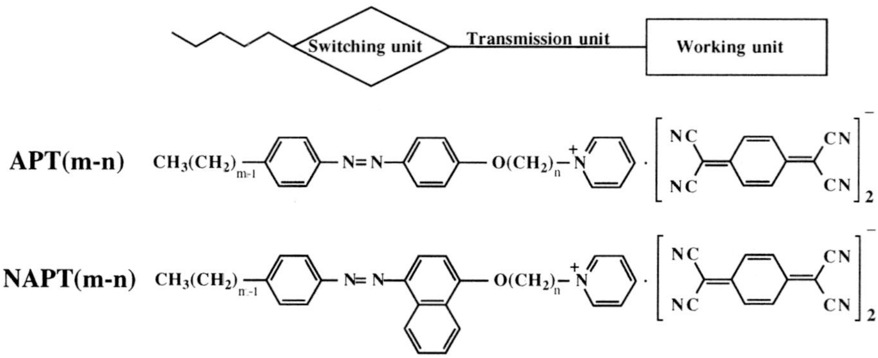

Figure 3: Chemical structure of APT(m-n) and NAPT(m-n)

Figure 4 shows the absorption spectra of the LB film of APT(8-12) before and after photoirradiation with UV (365 nm) and VIS (436 nm) light. The curve (a) is an absorption spectrum of the as-deposited LB film, and (b) and (c) correspond to the ones after irradiation with UV and VIS light, respectively. Intense monomeric absorption of trans-azobenzene decreases on irradiation with UV light while the absorption at 450 nm due to cis-azobenzene increases, indicating trans-to-cis isomerization. Conversion to the cis isomer in the photostationary state was estimated to be ca. 25% from a calculation based on the difference spectra. By the irradiation with VIS light, trans-azobenzene is regenerated. The spectra are reproducible for the further irradiation cycles ((b) $\leftrightarrow$ (c)).

Figure 5 shows the change in the absorbance at 356 nm due to trans-azobenzene and the lateral conductivity of the LB film of APT(8-12) upon alternate irradiation with UV and VIS light. It is clearly seen that the conductivity increases simultaneously with trans-to-cis photoisomerization. The increase in conductivity in the photostationary state is ca. 20%. Then the conductivity reverts with cis-to-trans photoisomerization on irradiation with VIS light. Thus the conductivity associated with the TCNQ moiety is controlled by the photoisomerization of the switching unit. The reversible cycle of the conductivity switching can be repeated more than 100 times. The apparent low switching speed is due to the low intensity of the light used. The actual switching speed is less than 100 μs when a laser pulse is used and this value 100 μs is imposed by the instrumental limit.

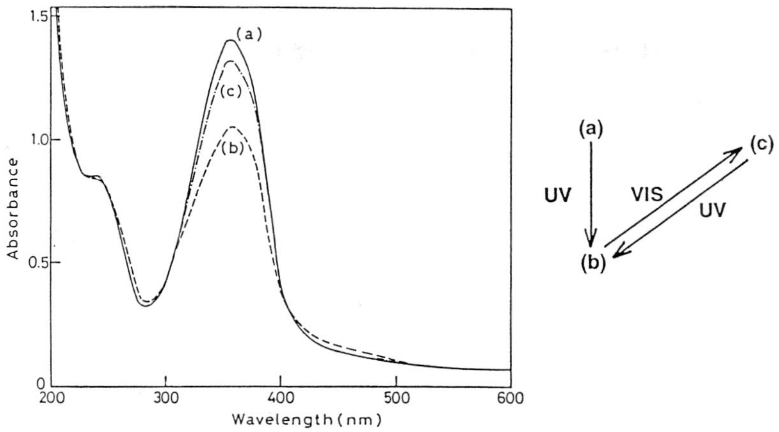

Figure 4: Absorption spectra of APT(8-12) LB film; (a) before irradiation, (b) after irradiation with UV light for 5 min, (c) after irradiation with VIS light for 3 min, reproduced from Ref. [10]

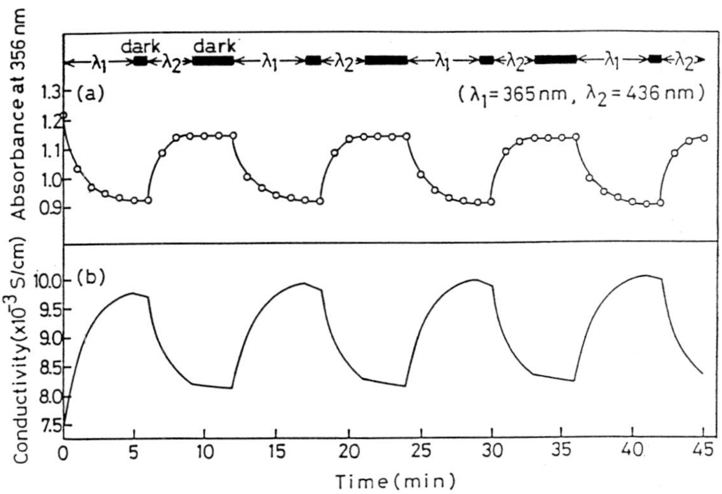

Figure 5: Change in (a) absorbance at 356 nm and (b) conductivity of APT(8-12) LB film on alternate irradiation with UV and VIS light, reproduced from Ref. [10]

IR spectroscopy gives information on a change in orientation of the TCNQ moiety with the conductivity switching in the LB films. This change in orientation can be monitored by measuring the intensities of specific two vibration bands in reflection-absorption spectra: the one located at 840 cm^{-1} whose transition moment is perpendicular to the TCNQ molecular plane, and the other at 1504 cm^{-1} whose transition moment is parallel to the long axis of TCNQ. With the irradiation of UV light, trans-to-cis photoisomerization of azobenzene proceeds and the orientation of TCNQ changes in such a manner that the long axis of TCNQ moves toward the surface normal. The orientation of TCNQ reverts to the original state with the irradiation of VIS light which causes cis-to-trans photoisomerization. The conductivity switching seems to be due to this reversible change in orientation of TCNQ since the conductivity of the LB film is associated with the orientation and the columnar structure of TCNQ moiety although it should be kept in mind that the analyses of IR spectra give the average tilt angles of specific vibrations with no information on the distributions of the tilt angles.

Three types of photochemical switching phenomena are observed by controlling the length of the transmission unit[11,12]. Changing the length of the transmission unit means changing the distance between the switching and the working units. Figure 6 shows the change in the absorbance at 356 nm and the lateral conductivity of APT(8-14) LB film on alternate irradiation with UV and VIS light. A reversible change in conductivity induced by cis-trans isomerization is also observed but the direction of the conductivity switching is the opposite compared with APT(8-12).

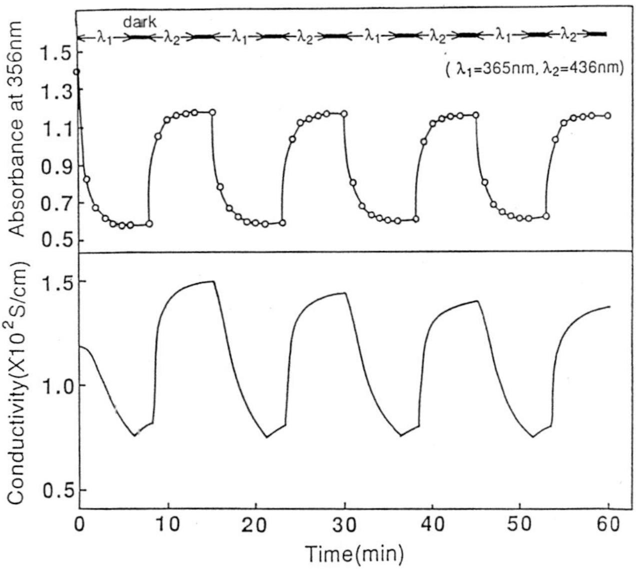

Figure 6: Change in (a) absorbance at 356 nm and (b) conductivity of APT(8-14) LB film on alternate irradiation with UV and VIS light

Another type of conductivity switching is observed for APT(8-6) LB film as shown in Fig. 7. Photoisomerization is reversible like other LB films but the conductivity change is stepwise: the conductivity increases with trans-to-cis isomerization while the conductivity stays unchanged with cis-to-trans isomerization. Another irradiation of UV light causes another increase in conductivity. The increment per cycle, however, decreases with increasing number of cycle and the conductivity tends to become constant at the cycle repeated more than 30 times. This behavior can be compared to the learning process since the LB film acquires an improved skill (higher conductivity) with repeated practice (alternate irradiation of UV and VIS light). In other words, electrons can pass through the LB films more easily as the stimulation (photoirradiation) is repeated. The mechanism involved in the plasticity in this phenomena is unknown at present. A tentative explanation is as follows: as-deposited, the LB film is in a quasi-stable state, but the activation energy to be overcome is not small.

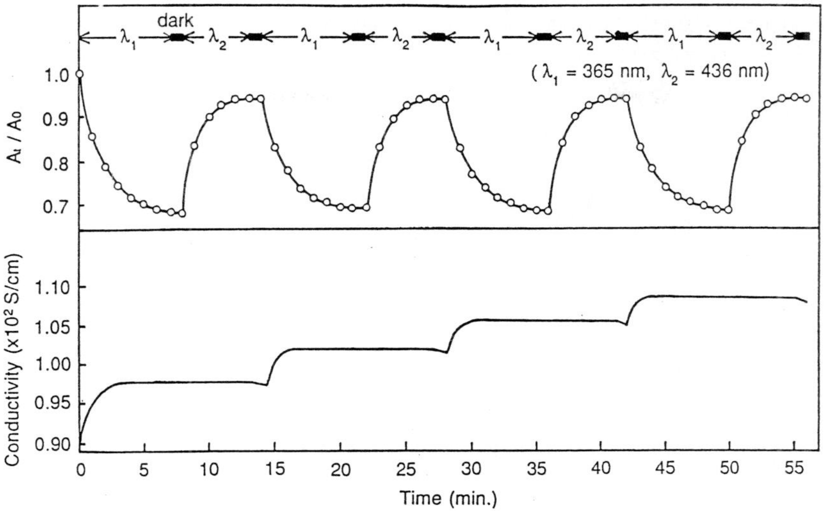

Figure 7: Change in (a) absorbance at 356 nm and (b) conductivity of APT(8-6) LB film on alternate irradiation with UV and VIS light

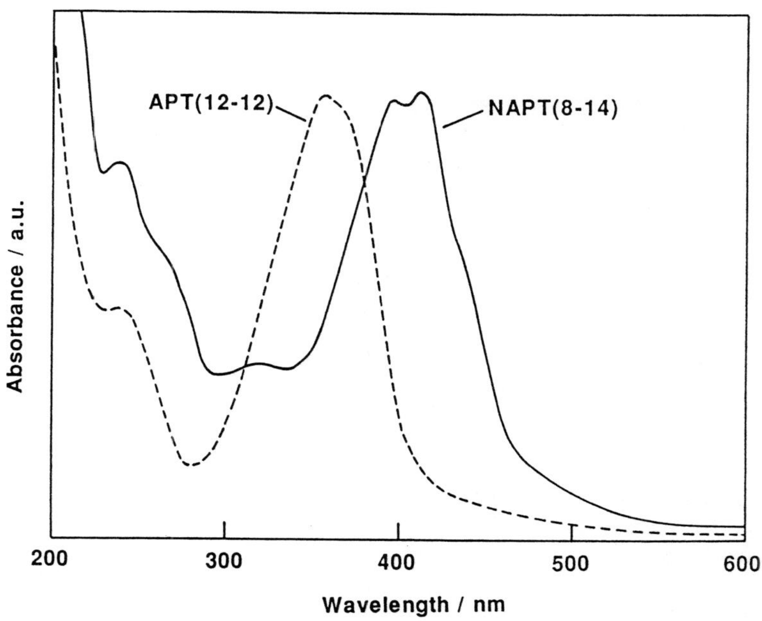

Figure 8: Absorption spectra of the LB films of APT(12-12) and NAPT(8-14), reproduce from Ref. [13]

In this case, the film will be relaxed to a more stable form when a stimulus with sufficient energy is imposed. In our system, the switching phenomena having plasticity would be understood if we assume that the stimulus produced by the trans-to-cis isomerization of azobenzene has sufficient energy but the one by the cis-to-trans isomerization does not have. This is supported by the fact that trans-to-cis isomerization is generally more difficult than the cis-to-trans isomerization in monolayers or LB films since the former includes the process in which the area per molecule increases.

The fabrication of a multiple switching device can be done by combining two different photochemical switching LB films with different switching units[13,14]. For this purpose, APT(12-12) and NAPT(8-14) are used. APT(12-12) LB film behaves like APT(8-12) LB film except that the conductivity change of the former LB film is larger than that of the latter. NAPT(8-14) is designed as follows: the wavelength of the control light is determined by the chemical structure of the switching unit. The easiest way to shift the control light toward longer wavelength region would be to extend the π-conjugated system of the switching unit. This is done by substituting a naphthalene ring for one of the benzene rings of APT molecules. The absorption spectra of the LB films of these compounds are shown in Fig. 8. It is clearly seen that the absorption band is really shifted toward longer wavelength region by this chemical modification.

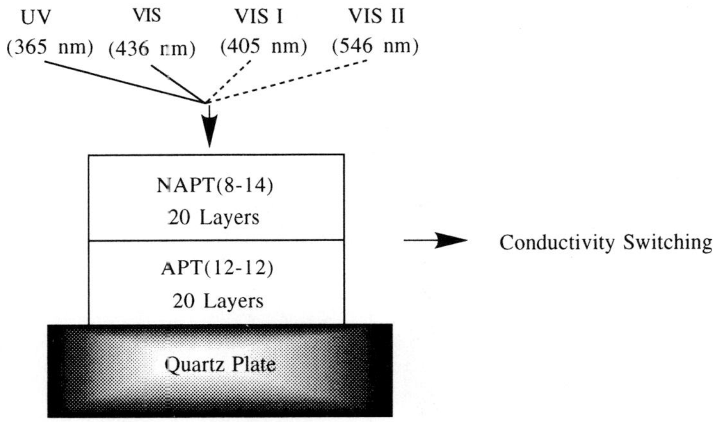

Figure 9: Structure of a prototype of a multiple switching device based on LB films

Figure 9 shows schematically the fabrication of a prototype of the multiple switching device consisting of a hybrid LB film by transferring sequentially the monolayers of APT(12-12) and NAPT(8-14) on the same quartz plate. UV and VIS are control light for APT(12-12) and VIS I (405 nm) and VIS II (546 nm) are control light for NAPT(8-14). The conductivity change of this hybrid LB film, when irradiated with the above four types of light, is shown in Fig. 10. On irradiation with VIS I light, trans-to-cis isomerization of phenylazonaphthalene of NAPT(8-14) proceeds and consequently the conductivity increases. The conductivity increases further with trans-to-cis isomerization of azobenzene of APT(12-12) on irradiation with UV light. The conductivity reverts to the initial value by irradiating sequentially with VIS and VIS II light, which are associated with cis-to-trans isomerization of APT(12-12) and NAPT(8-14), respectively. These features rely on the fact that there is no significant electronic interaction between the LB films of APT(12-12) and NAPT(8-14) and that each LB film responds to the irradiated light almost independently from each other. The results indicate that this hybrid LB film can serve as a prototype of a multiple switching device since the conductivity of the film is controlled by irradiation with four types of light.

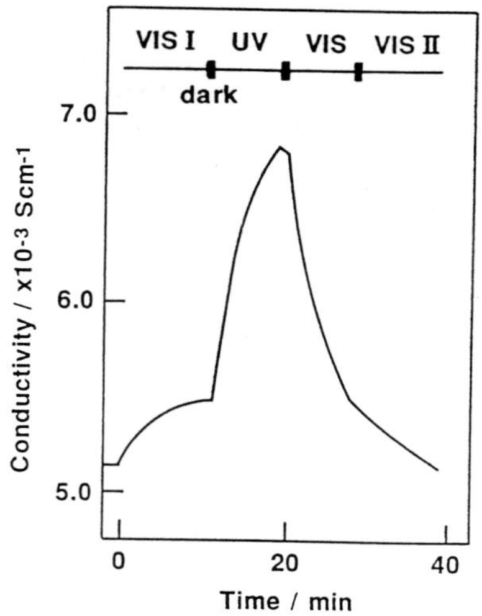

Figure 10: Change in conductivity of the multiple switching device on sequential irradiation with VIS I, UV, VIS and VIS II light, reproduced from Ref. [13]

4. Control of Optical Properties Using Mixed LB Films

In the previous part, photoisomerization of APT and NAPT molecules are used to control the electric conductivity of the LB films: the conductivity switching is observed in single-component LB films. A kind of signal produced by the isomerization is conveyed to a different part within the molecule. The signal produced by the photoisomerization can be transmitted to the surrounding molecules when a mixed LB film is used. This signal transfer is clearly seen in a mixed LB film of APT(12-12) and S120[15].

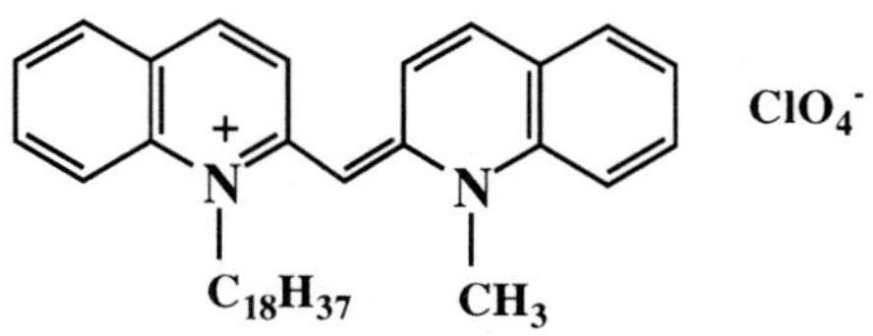

Figure 11: Chemical structure of S120

Figure 12 shows the surface pressure-area isotherms of APT(12-12), S120, and S120/APT(12-12)=1/1. The abscissa of the isotherm of the mixed monolayer is area per APT(12-12) molecule. It is seen that the area of the mixed monolayer is not the sum of those of the component monolayers at the same surface pressure, especially at lower surface pressures. This indicates that the two components should be molecularly mixed and not phase-separated. This feature should be very important when the signal is to be transmitted to different component molecules. Figure 13 shows the absorption spectra of a solution and an LB film of S120. A broad band in the solution spectrum is due to the absorption of a monomer and a dimer.

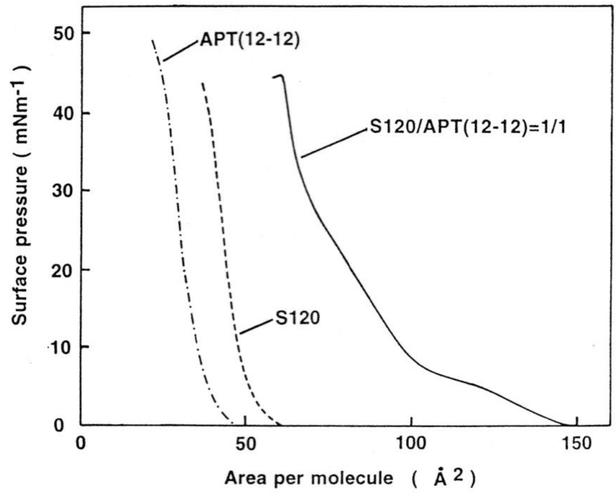

Figure 12: Surface pressure-area isotherms of APT(12-12), S120, and S120/APT(12-12)=1/1

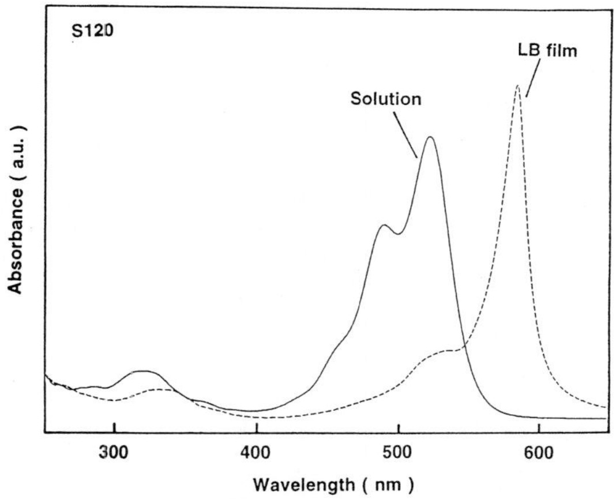

Figure 13: Absorption spectra of a solution and an LB film of S120

In the LB film spectrum, a relatively narrow band is located at longer wavelengths which is assignable to J aggregate. This means that molecular aggregation is apparent in LB film of pure S120. Figure 14 shows the change of the absorption spectrum of a S120/APT(12-12)=1/1 mixed LB film on alternate photoirradiation with UV and VIS light. In the absorption spectrum of the as-deposited LB film, J aggregate formation is barely seen and most of S120 molecules are considered to be in a monomeric or a dimeric state. On irradiation with UV light, trans-to-cis photoisomerization of APT(12-12) proceeds. Simultaneously with this isomerization, the absorption due to J aggregate of S120 increases. This absorption also increases with cis-to-trans photoisomerization of APT(12-12) on irradiation with VIS light. J aggregation proceeds further with alternate photoirradiation of UV and VIS light though the photoisomerization of APT(12-12) seems almost reversible. The saturation of J aggregate formation is seen after a certain number of the irradiation cycle: in the saturated state, the absorption due to J aggregate is almost unchanged by the reversible photoisomerization of APT(12-12). The existence of APT(12-12) is critical since the development of J aggregate is not seen when an amphiphilic inert matrix is used instead of APT(12-12). The results indicate that J aggregation of S120 is induced by the photoisomerization of APT(12-12) in the same LB film. In this LB film, two components are considered to be molecularly mixed and the signal produced by the molecular deformation of APT(12-12) could be efficiently conveyed to the surrounding S120 molecules, causing S120 to form J aggregate although the exact number of molecules necessary for one J aggregate is unknown at present.

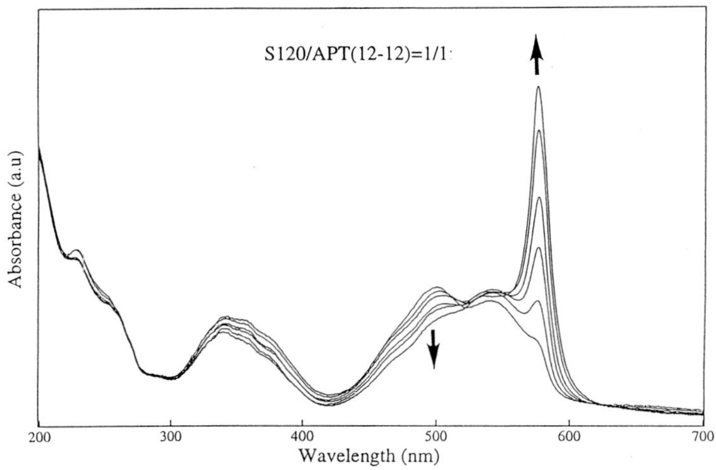

Figure 14: Change of the absorption spectrum of a S120/APT(12-12)=1/1 mixed LB film on alternate photoirradiation with UV and VIS light

5. Conclusion

Photoisomerization of azobenzene can be used as a trigger to control the electrical conductivity of pure APT(m-n) LB film and J aggregation of S120 in a mixed LB film. In both of the systems, the signal conveyed seems to be produced by the molecular deformation of APT molecules. In particular, this signal can be used to control the aggregation structure of the co-existing molecules in the mixed LB film. The results suggest future applications of these phenomena to switching or memory devices using hybrid LB films systems.

References

[1] H. Kuhn, D. Mobius, H. Bucher, *Physical Methods of Chemistry*, ed by A. Weisserberger, B.W. Rossiter, Wiley Interscience, New York, Vol. 1, Part IIIB, 1972, pp.577-702.

[2] *Langmuir-Blodgett Films*, ed by G. G. Roberts, Plenum Press, New York, 1990.

[3] A. Ulman, *An Introduction to Ultrathin Organic Films: From Langmuir-Blodgett to Self-Assembly*, Academic Press, Boston, 1991.

[4] *Proceedings of the Sixth International Conference on Organized Molecular Films*, ed by R. M. Leblanc, C. Salesse, Trois-Revieres, Quebec, Canada, July 4-9, Elsevier, Lausanne, 1993: *Thin Solid Films*, Vol. 242-244, 1994.

[5] *Molecular Electronic Devices*, ed by F. L. Carter, Marcel Dekker, New York, 1982.

[6] R. M. Metzger, C. A. Pannetta, The Quest for Unimolecular Devices, *New Journal of Chemistry,* Vol. 15, 1991, pp.209-221.
[7] H. Tachibana, M. Matsumoto, Functionalized Langmuir-Blodgett Films -Toward the Construction of Molecular Devices, *Advanced Materials*, Vol. 5, 1993, pp.796-803.
[8] M. Matsumoto, H. Tachibana, T. Nakamura, *Organic Conductors: Fundamentals and Applications*, ed by J.-P. Farges, Marcel Dekker, New York, 1994.
[9] H. Tachibana, T. Nakamura, M. Matsumoto, H. Komizu, E. Manda, H. Niino, A. Yabe, Y. Kawabata, Photochemical Switching in Conductive Langmuir-Blodgett Films, *J. Am. Chem. Soc.*, Vol. 111, 1989, pp.3080-3081.
[10] H. Tachibana, A. Goto, T. Nakmura, M. Matsumoto, E. Manda, H. Niino, A. Yabe, Y. Kawabata, Photoresponsive Conductivity in Langmuir-Blodgett Films, *Thin Solid Films*, Vol. 179, 1989, pp.207-213.
[11] H. Tachibana, R. Azumi, T. Nakamura, M. Matsumoto, Y. Kawabata, New Types of Photochemical Switching Phenomena in Langmuir-Blodgett Films, *Chemistry Letters*, 1992, pp.173-176.
[12] H. Tachibana, Y. Nishio, T. Nakamura. M. Matsumoto, E. Manda, H. Niino, A. Yabe, Y. Kawabata, Control of Photochemical Switching Phenomena by Chemical Modification, *Thin Solid Films*, Vol. 210/211, 1992, pp.293-295.
[13] H. Tachibana, E. Manda, R. Azumi, T. Nakamura, M. Matsumoto, Y. Kawabata, Multiple Photochemical Switching Device Based on Langmuir-Blodgett Films, *Applied Physics Letters*, Vol. 61, 1992, pp.2420-2421.
[14] H. Tachibana, M. Matsumoto, E. Manda, Conductivity Switching of Langmuir-Blodgett Films Using Photoisomerization of Phenylazonaphthalene, *Molecular Crystals and Liquid Crystals*, Vol. 267, 1995, pp.341-346.
[15] H. Tachibana, F. Sato, M. Matsumoto, in preparation.

Comparative Study on Langmuir-Blodgett Films and Crystals based on TTF Derivatives with Long Alkyl Chains

Mitsuru IZUMI and Hitoshi OHNUKI

Laboratory of Applied Physics, Tokyo University of Mercantile Marine,
Etchu-jima, Koto-ku 135, Tokyo, JAPAN
e-mail: izumi@ipc.tosho-u.ac.jp ; Fax:+81-3-5245-7462

Abstract

In the basic research of Langmuir-Blodgett (LB) films, most of the difficulties to deduce the inherent structural and physical properties come from the existence of many small regions with differing orientation as two dimensional (2D) domain plates in each monolayer. To overcome such difficulties we emphasize a strategy of studying the single crystal and the LB films which have the identical molecular organization. As an example, we show a series of comparative studies on a stable conducting LB films of $[EDT\text{-}TTF(SC_{18})_2]_2I_3$ and its single crystal. The success of growth of the single crystals leads to a proposal of possible 3D model of molecular organization and clarification of the electronic states responsible for the conductivity.

1. Introduction

In thin organized molecular films such as multilayered ultra-thin films, the molecular assemblies with the well-defined structure are essential for the realization of the potential applications [1]. The term "Langmuir-Blodgett (LB) Films" is used to denote monolayer and multilayers deposited from the liquid-gas interface onto a solid substrate with ordered structure. It is a historical trend to try to fabricate highly conducting LB films based on charge-transfer (CT) complex or CT salt by using electroactive donor and/or acceptor molecules among fruitful functional molecules from the viewpoint of molecular electronics[2]. Most of the studies following the above trend have been encouraged by the rapid progress in the study of single crystals of CT compounds based on a donor molecule TTF (tetrathiafulvalene, Fig. 1-**1**) or an acceptor molecule TCNQ (tetracyanoquinodimethane) and their derivatives [2]. The LB technique allows one to produce films of a precisely determined number of layers in principle, which ordinary crystallization methods do not and LB films is much easier to incorporate into practical device structure [1]. On the other hand, as the disadvantage of LB method we note that proper long-range order in the film plane has not yet been achieved and it prevents to clarify physical properties [3]. In spite of the importance to the in-plane molecular packing structure responsible for conductivity as described later, the existence of many small regions with differing orientation in each monolayer makes difficulties in the study of intrinsic structural and physical properties in the study on conducting LB films. Although x-ray diffraction, dc conductivity, infrared absorption and electron spin resonance measurements are

Springer Proceedings in Physics, Vol. 81
Materials and Measurements in Molecular Electronics
Editors: K. Kajimura · S. Kuroda

crucial ones to obtain the informations on the molecular organization and electronic transport properties in the case of the crystals, these experimental techniques are strongly influenced by the effect of the existence of the above mentioned domains in the case of LB films. Hence, the results of such experiments show the averaged properties of randomly distributed domains. Therefore, it is impossible to obtain the inherent physical properties on a single domain that should be essential to conductivity. In this paper, we show that one of the strategies to deduce the inherent information on structural and physical properties from the LB films is to study comparatively on the single crystal which is composed of the same molecules and assembly with that of the LB films. Such approach would provide the mechanism to realize higher conductivity and show the ways improve the conductivity. As an example of the study following the present strategy, we exhibit what is going on the comparative study on conducting LB films of cation radical salt based on EDT-TTF(SC_{18})2 ((ethylenedithio)-bis(octadecylthio)tetrathiafulvalene Fig. 1-**4**) and its single crystal.

In the present organization, firstly we give a brief summary of the progress in the study of organic CT crystals, which has given rise to the motivation of the study of conducting LB films (Sec. 2). Secondly, we deal with the electronic state to obtain conducting organic CT molecular assembly and the difficulties to achieve a well-defined electronic transport in LB films (Sec. 3). Subsequently, in Sec. 4 we describe a basic strategy as the comparative study. In Sec. 5, we exhibit a general preview on the conducting LB films of cation radical salt $[EDT\text{-}TTF(SC_{18})_2]_2I_3$ which we have applied the present strategy. In Sec. 6, we report that the success in the crystal growth leads to a proposal of the model of molecular assembly and precise clarification of inherent physical properties.

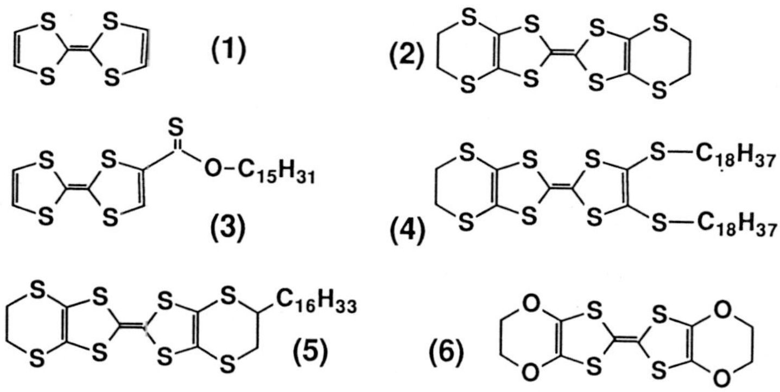

Fig. 1 TTF and its derivatives with long alkyl chain which provides highly conducting Langmuir-Blodgett films. TTF (**1**), BEDT-TTF (**2**), HDT-TTF (**3**) [16], EDT-TTF(SC_{18})2 (**4**), C_{16}-BEDT-TTF (**5**) [17] and BEDO-TTF(BO) (**6**) [18].

2. Historical background

Since the discovery of metallic conductivity peak in the crystal of CT complex, TTF-TCNQ in 1973

[4], the research field of crystals of organic low-dimensional conductor has been drastically extended via the synthesis of a number of superconducting CT salts with the critical temperature up to 12.8 K in BEDT-TTFCu[N(CN)$_2$]Cl [5]. Thanks to the low ionization potentials of the constituent donor molecules and the high electron affinities of the acceptor molecules, the several classes of binary materials assigned as CT complex or CT salt (ion-radical salt) have been studied as the ionic or mixed valence molecular species in the ground state. In these materials, a CT donor-acceptor intermolecular interaction after Mulliken [6] is essential to the appearance of the electrical conductivity, where its magnitude and the difference in the orbital energies of adjacent molecules are important [7]. Crystals formed by both open-shell molecular ions and closed-shell counter-ions exhibit numerous fruitful physical properties. Cation-radical salts of 2-1 stoichiometry (we note it as D_2X, where D is a donor TTF derivatives and X is an anion, respectively) issued from the TTF (Fig. 1- **1**) backbones present a series of organic conductors with a variety of ground states [8-9]. They are characterized by segregated stacks of TTF-type molecules with a regular packing of dimers, where mixed valence dimers D_2^+ are formed with molecules mostly associated with a monovalent diamagnetic anion X^-. In that case, carriers are able to move by hopping from dimer to dimer, which conducts to metallic or semiconducting electronic properties. In physical interpretation of such electronic system, during the last two decades, an extended Hubbard electronic and spin Hamiltonian have described well those physical properties in most of quasi-one dimensional (1D) organic conductors and during the last decade one finally achieved Fermi surface formed by two dimensional (2D) molecular overlap in the electronic system, which has provided stable metallic conductors and/or superconductors. Among the TTF derivatives, many compounds based on the donor BEDT-TTF (bis(ethylenedithio)-tetrathiafulvalene, ET : Fig. 1-**2**) molecule have been synthesized [10]. In this system, it is shown that the anion size and anion -CH_2 hydrogen interactions are critical factors that control the packing of the ET donor molecule and determine the transport properties [11]. It can be easily seen that a series of quite fruitful electronic and magnetic properties originate from different packing schemes of donor in cation-radical salts of ET molecule [12]. The discovery of the superconductivity of CT systems seems to have been accepted with great technological relevance.

To submit such interesting donor-type TTF derivatives to the fabrication of the multilayered thin films by LB techniques, it is necessary to obtain the stable monolayer on the water surface [2,13]. In contrast to the case of organic crystals, donor-type TTF molecules generally oblige to have long aliphatic chains to constitute a stable packed molecular assembly on the water surface, since the molecule should be amphiphilic. Therefore, if we use the same TTF-type donor molecule and anion to fabricate LB films as used in the CT crystal, the obtained molecular assembly is not generally identical in both LB films and crystals. Historically, Saclay group [14] in France succeeded to build conducting LB films of charge-transfer complex based on an acceptor-type TCNQ for the first time and subsequently the fabrication of LB films of TTF/TCNQ has been done by Nakamura et al [15]. Figure 1 exhibits several TTF-family molecules which have given highly electrical conductivity in the form of LB films. As summarized by Roberts [1], mostly ohmic behavior is observed with room temperature dc conductivities in the range 10^{-4} - 10^2 S/m. The conductivities are thermally activated as in doped inorganic semiconductors. However, higher conductivities than those above are inferred from optical conductivities, which has been interpreted that "grain boundaries" in the film plane may well be limiting the dc values [1].

3 . What controls the electronic properties in LB films?

In the following, we will focus on the conducting LB films of 2-1 cation radical salts based on long chain alkyl TTF derivatives.Several key factors have been proposed to obtain a well-defined conducting LB films [2,13]. We can summarize them as: (1) stability and transferability of monomolecular layer on the water surface, (2) stability of multilayered films after the deposition, (3) achievement of the dimer with mixed-valence state as cation-radical salts which is also essential in the case of crystals. However, we think that the above mentioned points are not sufficient and the following important point should compensate : (4) formation of 2D in-plane molecular packing to achieve sufficient values of transfer integrals intra- and/or interdimers. The last point (4) is quite important for higher conductivity, since the development of 2D lateral overlap of the molecular orbital gives a metallic state and dominates electronic properties. That is, the origin of electronic properties is attributed to the dimer packing arrangement in the film plane. As described in the preceding section, we can easily see in the case of crystals that a series of quite fruitful electronic and magnetic properties originate from 2D different packing schemes of donor in cation-radical salts of ET molecule [12].

Let's us consider more detailed electronic state in dimers to provide electrical conductivity. Supposing the mixed valence isolated dimer as D_2^+ or D^+D^0, the electronic state has two energy levels separated by an energy gap with 2t between bonding and antibonding orbitals from HOMO, where t is the electron transfer interaction (transfer integral) between two equivalent monomers as shown in Fig. 2 [19]. If relatively strong dimerization exists, the interdimer overlap integral t' can be defined. With increasing t', half-filled upper band becomes broad as shown in Fig. 2(b) and subsequently in Fig. 2(c). Consequently, as a function of 2t-t' in the electronic structure we shall obtain the electronic properties responsible for metallic conductivity. In the case of LB films based on CT salt, in principle, the electrical conductivity should be realized by following the same mechanism as described above. However, it is reminded as described in Sec. 2 that donor molecules are generally obliged to have long aliphatic chains to constitute a stable amphiphilic structure on the water surface. To prepare a conducting LB films of CT salts, several methods have been developed. In the first way, one obtained LB films of donor molecules with long alkyl chain and final conducting LB films as D_2X were fabricated after the oxidation process under the exposure to halogen gas. In the second way, one could reach an idea of autodoping (homodoping strategy) by mixing, for example, D^0 and D^{+1} molecules (but not realized for TTF derivatives) [20]. Many interesting TTF type donor molecules with long alkyl chains do not have desirable amphiphilic character but rather hydrophobic. In this situation, mixed LB films of the interesting cation-radical salt with fatty acid appeared and it has sometimes given better electrical conductivity [21]. Among

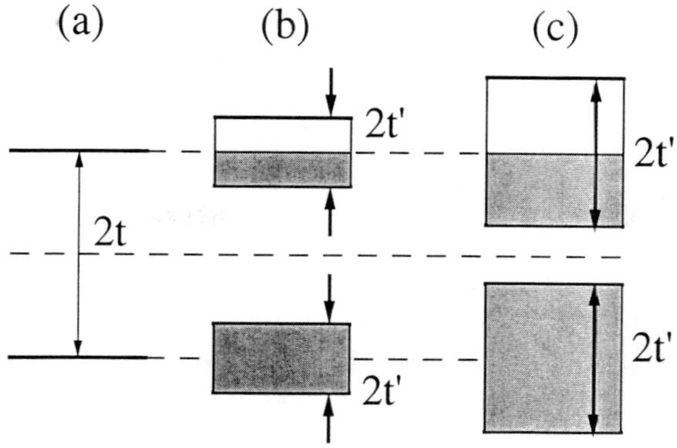

Fig. 2 Schematic energy spectra for (a) isolated donor dimer, (b) chain of quasi-isolated donor dimers (t' << t), and (c) chain of strongly interacting donor dimers (t' << t).

conducting LB films based on such CT salts, most of them are semiconductors even with the existence of the mixed valence state dimer and quite a few ones exhibit metallic electrical conduction [1,22]. Then, how can we obtain the reason why they exhibit semiconducting properties? Is it only due to the existence of "grain boundaries"? Is it intrinsically semiconductor or metal? How can we achieve any information to improve the electrical conductivity? The first step to further breakthrough is to clarify the structure of the molecular organization in the segregated layer stacks which remain not to be clarified well in most of the conducting LB films. In particular, packing structure of donor molecules in the film plane is suffered by the lateral packing scheme of long alkyl chains. Therefore, we easily imagine that both of them are deeply correlated to the control of both t and t' values cited above. Possibly, the electronic state of the molecular condensates such as LB films shall be attributed to the packed structure determined by a competition between the charge-transfer interaction between donor-counter ion molecules and intermolecular interaction in long alkyl chains. Thanks to the knowledge of the in-plane structure of molecular assembly, we shall be able to interprete the physical properties and reconsider the molecular conformation to improve the conductivity.

4. Basic strategy to study physical and structural properties of conducting LB films

As described before, the existence of the ensemble of 2D domains in each mono-molecular layer gives a limited information on structural and physical properties and makes difficulties to deduce intrinsic properties. When we consider that the starting point of material research is to get the knowledge of precise structure of the material, one of the strategies to deduce the inherent structural and physical properties is to study comparatively on the single crystal which is composed of the same molecules and assembly with that of the LB film. Single crystal X-ray diffraction will exhibit the precise crystal structure and compensate deficient structural parameters in the case of X-ray diffraction on the LB films. There, we will be able to get precise packing structure of donor TTF-type molecules which is crucial information to take account of the conducting properties. In the next step, one would study the inherent physical properties induced from the determined structure. For example, polarized infrared-absorbance spectroscopy will give the anisotropy of the overlap integral of donor-molecular orbitals by the analysis of the charge-transfer absorption band. This result may provide the quantitative estimation of intra- and interdimer transfer integral. By measuring the electrical conductivity in single crystals we shall overcome the effect of the domains on the electrical conduction in the case of LB films and deduce the precise properties of the conductivity. In contrast to the case in the LB films, we will be able to clarify any in-plane anisotropy in conductivity, dielectric constant, magnetic susceptibility by the experiments on single crystals. As the candidate following the present strategy, we focused on the conducting LB film of cation radical salt based on EDT-TTF$(SC_{18})_2$ ((ethylenedithio)-bis(octadecylthio)tetrathiafulvalene Fig. 1-**4**). Properties of the LB film formed by the mixture with behenic acid have been extensively studied [21,23,24]. As it becomes a stable semiconducting film after the oxidation process with iodine vapor and shows dc conductivity up to ~1 S/cm. Moreover, its semiconducting property has been submitted to the test for the FET device [25]. To understand the mechanism of the electronic transport, firstly one have to clarify whether the observed semiconducting properties are originated from the effect of "grain boundaries" or intrinsic properties. Secondly, we have to understand what kind of structural assembly of EDT-TTF brings about the electronic states (t or t') responsible for the conductivity.

5. General description of conducting LB films of [EDT-TTF$(SC_{18})_2]_2I_3$

Mixed LB films of [EDT-TTF$(SC_{18})_2]_2I_3$ and behenic acid has been investigated intensively due to stability of the conducting state [21,23,24]. The technique to fabricate LB films is a vertical dipping-type classical method as described in earlier paper [26]. After the Y-type deposition of the 1:1 mixed LB films of EDT-TTF$(SC_{18})_2$ and behenic acid, the oxidation process by iodine vapor is performed and finally we obtain purple colored conducting LB films by annealing process at 50°C for 2 hours. Such annealing process is necessary to achieve a stable film, since the iodine deintercalation process to the neutral phase occurs after the doping process at room temperature without annealing. This is easily detected by X-ray diffraction. Figure 3 exhibits the X-ray diffraction profiles from LB films oxidized by iodine vapor. Each profiles was observed at 0 hours (a), 24 hours (b), 72 hours (c), 5 days (d), 3 weeks (e) and 6 weeks (f) after the oxidation without annealing process. In Fig. 3 four stable diffraction peaks (*) are assigned as those from behenic acid clusters. It is clearly observed that both the angles and the intensity of diffraction peaks with 55 Å periodicity from EDT-TTF$(SC_{18})_2$ (a) are modified to those with 45 Å periodicity from [EDT-TTF$(SC_{18})_2]_2I_3$ as in (b), (c), (d) and (e) due to the iodine intercalation. However, after 6 weeks, the similar profile with (a) can be seen in (f) as the result of the deintercalation process of iodine. Figure 4 shows the x-ray diffraction profile of the stabilized LB films after the annealing process. The obtained stacking periodicity is 45 Å from the stabilized [EDT-TTF$(SC_{18})_2]_2I_3$ [21]. The maximum conductivity reached 1 S/cm in the film plane for the 50 layered films (25 bilayers

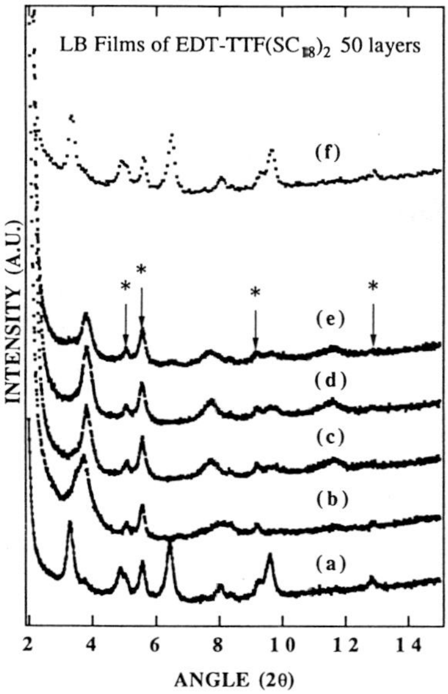

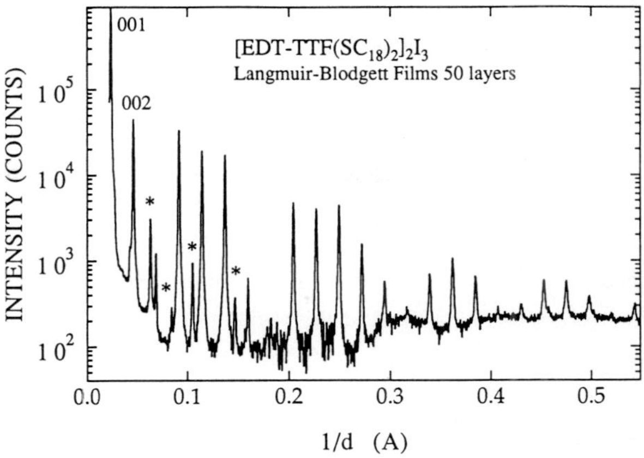

Fig. 4 X-ray diffraction profiles of the stabilized LB films after the annealing process. Four diffraction peaks (*) are assigned as those from behenic acid cluster.

Fig. 3 X-ray diffraction profiles of LB films as a function of time after the iodine oxidation at room temperature : (a) 0 h, (b) 24 h, (c) 72 h, (d) 5 days, (e) 3 weeks, (f) 6 weeks after the oxidation. Sample was placed in the atmosphere. CuKα line was used. The diffraction peaks (*) are assigned as those from behenic acid cluster.

followed by the definition of Y-type films). By the plot of room temperature conductivity vs. number of layers or mixed ratio, it has been concluded that the conduction follows more or less a 3D percolation model and the conductivity 1 S/cm of films with 1:1 mixed ratio has been assigned as one at percolation limit [21].

In the study of organic superconductors, the great interest to the infrared absorbance has been given due to the proposed mechanism of superconductivity mediated the electron-molecular vibration (e-mv) coupling by Yamaji [27]. One can deduce the e-mv coupling constants from the optical spectra of these compounds [7]. In the mixed LB films of [EDT-TTF$(SC_{18})_2]_2I_3$ and behenic acid, the vibronic mode based on the e-mv coupling was observed associated with the charge-transfer absorption as shown in Fig. 10 of ref. 21 (Torrance's "A" band [28]). From the absorption profile, Dourthe et al. [21] have analyzed by using the optical dielectric function of charge-transfer excitation associated with the e-mv coupling under the assumption of the existence of the ensemble of single charged isolated dimers in the first approximation with a Drude-Lorentz model (in presence of a narrow continuum). The above mentioned progress has not yet reached the goal to clarify the origin of the electrical conductivity. The definitive reason for this is that we have not yet clarified the packed structure composed of the donor molecules in the film plane, which shall be submitted to some comparison with the FT-IR data.

Thus, as in most works on conducting LB films, the research for individual materials is obliged to stop at the equivalent stage on which they stay at the point of success of the determination of the molecular orientation. As described in the preceding sections it is more important to get the exact information on the molecular organization in the film plane. To obtain any advancement, it is clear to need the information on not only orientation correlation but also spatial correlation, in other words, the column structure and/or the packing of TTF dimers in the film plane. In summary, when one would try to definitively clarify the mechanism of the electrical conductivity, the following tasks always remain. Firstly, one have to study molecular organization of EDT-TTF in the film plane which gives us the opportunity to estimate t and t' and intrinsic mechanism of electrical conduction.

6. An introduction to the comparative study on structural and electronic properties in both LB films and crystals of $[EDT\text{-}TTF(SC_{18})_2]_2I_3$

Single crystals were obtained by electrochemical crystallization and the detailed description was given in ref. 24. X-ray diffraction study exhibited that the crystal symmetry is orthorhombic, a= 10.5 Å, b= 5.4 Å and c= 45 Å [24]. Even by using both rotating anode X-ray source and imaging plate, it is difficult to collect enough number of intensified diffraction spots. Thus, the determination of the crystal structure by the conventional crystal structure analysis has not been succeeded. The length of the c-axis of the crystal coincides with that of the stacking periodicity in the LB films. The intensity distribution of the present diffraction series in crystals and LB films is similar each other, which suggests that linear electron density distribution along the c axis (the direction of the layer stacking in LB films) is identical in both LB films and crystals. Then, the emphasis is placed on bilayer packing arrangement in the film plane. From the viewpoint of the lateral packing of alkyl chains, the subcell type of alkyl chains which forms the orthorhombic one is O type [29]. Taking into account of the space filling of the alkyl chain and assuming the O[±3,0] subcell structure, the cell dimensions are calculated to be a=10.62 Å and b=4.96 Å which reproduce well the observed dimension of a-b plane. Under the substantial coincidence of inclination angle of alkyl chain of LB films and crystals with the result of FT-IR on LB films [21], molecular assembly of in-plane structure was proposed. Subsequently, a full unit cell structural model including bilayer and occupation of triiodide has been reported as shown in Fig. 5 [24], which exhibits a possible model of the molecular organization of the observed orthorhombic cell. In this model, the interdigitation between upper and lower layer of $EDT\text{-}TTFS_2$ parts takes in place the bilayered Y-type structure, which leads to the appearance of shorter intermolecular distance between adjacent EDT-TTF parts and brings about an enhancement of overlapping integral along the b axis. It may be worthwhile to note that the presently proposed herringbone packing of EDT-TTF skeleton head parts resembles the crystal structure of $\alpha\text{-}(BEDT\text{-}TTF)_2I_3$ which is one of the typical packing structure for ET molecules [30].

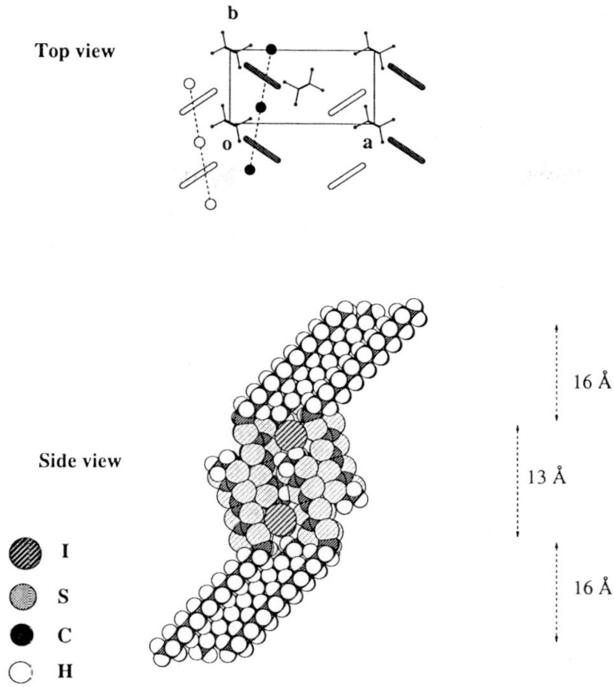

Fig. 5 Structural model proposed based on X-ray diffraction and XANES study.

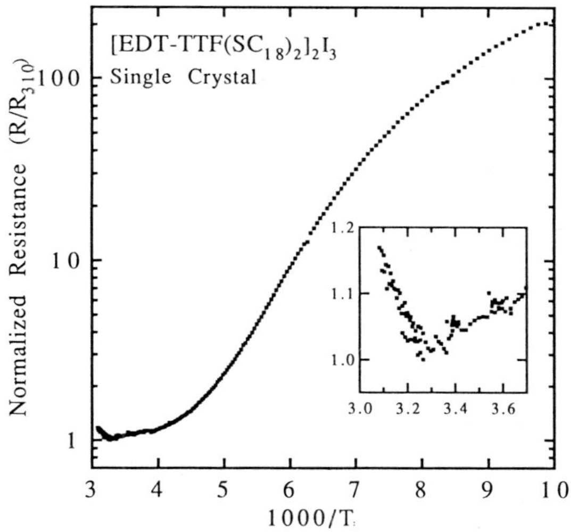

Fig. 6 Temperature dependence of the DC electrical resistance of single crystal of [EDT-TTF(SC$_{18}$)$_2$]$_2$I$_3$. It clearly exhibits a sign of the onset of a metal-semiconductor transition at 310 K.

The dc electrical conductivity along the b axis is 1.1 S/cm (T=300K). It is in good agreement with 1.0 S/cm in LB films. In single crystal, both dc and microwave conductivity under 50 GHz as a function of temperature exhibit semiconducting behavior below room temperature [24,31], which confirms that the observed semiconducting characteristics in LB films shows the intrinsic properties of cation-radical salt $[EDT\text{-}TTF(SC_{18})_2]_2I_3$. Thus, the obtained result suggests that the LB films is not a semiconductor originated from the existence of possible "grain boundaries" but an intrinsic semiconductor. Metal-Insulator transition was observed at 310 K as shown in Fig. 6, which suggests the onset of delocalization of molecular-orbital holes of the $EDT\text{-}TTF(SC_{18})_2$ above 310 K in the present material. No evidence of the structural phase transition has been given by X-ray diffraction.

Figure 7 shows the absorbance spectra of crystals and LB films in 800 to 6000 cm^{-1} range. As assigned in ref. 21, broad electronic absorption band around 2200 cm^{-1} and asymmetric resonance peak around 1230 cm^{-1} are originated from charge-transfer and vibronic excitation based on e-mv coupling, respectively. The former excitation corresponds to dimer charge oscillations and is characteristic of the existence of single charged dimer state. The anti-resonance dip at 1450 cm^{-1} between the former and latter peaks is observed together with CH_2 bending absorption peak. There, in the presence of some symmetry breaking, the e-mv interactions induce oscillations of the conduction electron density along the direction of conducting axis at frequencies close to those of the unperturbed totally symmetric intramolecular modes [32]. On the other hand, the charge-transfer excitaion energy ($h\omega_{CT}$) is attributed to the excitaion from the bonding state to anti-bonding state in zero-order approximation (see Fig. 2(a)). The dielectric function along the direction joining the molecular entities in the dimer, in the presence of the e-mv coupling, may be written for the assembly of isolated dimers in the following way[7,32]:

$$\tilde{\varepsilon}(\omega) = \varepsilon_\infty + \frac{f_0}{\omega_{CT}^2(1 - D(\omega)) - \omega^2 - i\omega\Gamma}$$

with

$$f_0 = 2\omega_{CT}\frac{e^2d^2}{4}|\langle\beta|n_1 - n_2|0\rangle|^2, \quad n_i = \Sigma_\sigma a_{i\sigma}^+ a_{i\sigma} \quad \text{and} \quad D(\omega) = \Sigma_\alpha \frac{\lambda_\alpha \omega_\alpha^2}{\omega_\alpha^2 - \omega^2 - i\omega\gamma_\alpha},$$

where d is the intermolecular spacing, n_i is the electron number operator for molecule i=1,2. $D(\omega)$ is the propagator which presents the α vibronic modes defined by bare frequency ω_α and damping constant γ_α . It couples linearly to the electrons with the e-mv coupling constant λ_α. Figure 5 exhibits the result of the model calculation with $\omega_{CT} = 2t = 1500\ cm^{-1} = 0.18$ eV, $\omega_\alpha = 1480\ cm^{-1}$, $\lambda_\alpha=0.2$ and $\Gamma=2000\ cm^{-1}$. On the vibronic feature, two central C= C stretching a_g vibrations locate around 1400-1500 cm^{-1} in ET family molecules [33]. Then it is natural to conclude that this vibration is strongly coupled to electrons. As shown in Fig. 2 a qualitative description of the electronic states of such narrow-band conduction materials gives a variety of the conducting

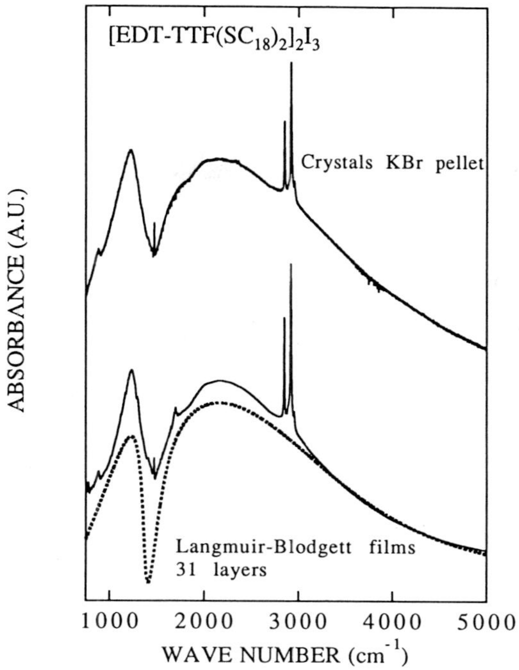

Fig. 7 Infrared absorbance spectra of crystals and LB films of $[EDT\text{-}TTF(SC_{18})_2]_2I_3$.The broken curve exhibits the calculated absorbance on the basis of the microscopic theory.

properties. For example, the half-filled upper band is narrow in the case of poor interdimer overlap, the dc conductivity is lower. On the other hand, supposing a slight dimerization, half-filled conduction band has a larger width, hence we may expect higher conductivity. In the strongly dimerized stacks, within the atomic limit one can expect simply two excitations 2t and 2t'. If there exists strongly dimerized stacks it is probable to observe two CT bands around "A" band region associated with the intradimer and interdimer charge excitations as $(DIMET)_2SbF_6$ [34].

As shown in Fig. 7, single broad band CT excitation was observed in the present single crystals and LB films. The charge-transfer excitation energy, ω_{CT} is almost identical in both crystal and LB films. These results suggest that the electronic state in crystals is close enough to that in LB films. Its excitation energy value suggests the existence of relatively narrow band width associated probable slightly dimerized stack. The composition and the precise origin of the broad band CT absorption cannot be clarified in the present stage. However, the success of larger single crystals would conducts to the study of the infrared absorption under the polarization configurations.

7. Concluding Remarks

There exist two subjects to improve in the basic research of conducting Langmuir-Blodgett (LB) films based on TTF derivative with long alkyl chains. Firstly, it is essential to achieve the knowledge of packing structure of molecular assembly in the film plane, since the crystals of ET salts are composed of 2D arrangement of dimerized molecular units with various polymorphisms and the donor molecular arrangement in the film plane may be different from that in the case of crystals without long alkyl chains. Secondly, we must deduce not only the inherent structural properties as described above but also intrinsic electronic properties to characterize the present LB films. The difficulties to realize both of them come from the existence of many small regions with differing orientation in the film plane. To overcome such difficulties we emphasize a strategy in which one studies comparatively the single crystal and the LB films which have the identical molecular organization. A series of studies on a stable conducting LB films of $[EDT\text{-}TTF(SC_{18})_2]_2I_3$ and its single crystal was introduced. The success of the growth of single crystals leads to a proposal of possible 3D model of molecular organization and clarification of the electronic states responsible for the conductivity. From presently proposed 3D model, 2D S-S network seems to be the key to achieve highly conducting LB films. Comparative FT-IR study shows that the electronic system is characterized by the dimerized stack with narrow bandwidth. Therefore, it is stressed that the increase of the interdimer interaction t' is necessary to realize higher conductivity in the present system. A series of comparative studies on LB films and crystals give us fruitful and intrinsic information about structural and electronic properties.

Acknowledgments

The authors wish gratefully to acknowledge the contribution of Reizo Kato for his help in the preparation of the samples. They are grateful to V. M. Yartsev for valuable discussions. Many thanks are due to Nicolet Japan Co. Ltd for their technical assistance.

References

[1] G. Roberts, *Langmuir-Blodgett Films*, Plenum Press, New York and London, 1990.

[2] P. Delhaes, *Lower Dimensional Systems and Molecular Electronics*, ed. by R. M. Metzger, P. Day and G. C. Papavassiliou, NATO ASI Series B, Vol. 248, Plenum Press, New York, 1991 43.

[3] R. H. Tredgold, The Physics of Langmuir-Blodgett Films, *Rep. Prog. Phys.*, Vol. 50 1987, 1609.

[4] L. B. Coleman, M. J. Cohen, D. J. Sandman, F. F. Yamagishi, A. F. Garito and A. J. Heeger, *Solid State Commun.*, Vol. 12 (1973) 1125.

[5] J . M. Williams, A. M. Kini, H. H. Wang, K. D. Carlson, U. Geiser, L. K. Montgomery, G. J. Pyrka, D. M. Watkins, J. M. Kommers, S. J. Boryshuk, A. V. Strievy Crouch, W. K. Kwok, J. E. Schirber, D. L. Overmyer, D. Jung and M.-H. Whangbo, *Inorg. Chem*., Vol. 29 1990, 3272.

[6] R. S. Mulliken ad W. B. Person, *Molecular Complexes* (John Wiley & Sons, Inc.) 1969.

[7] R. Bozio and C. Pecile, Charge Transfer Crystals and Molecular Conductors, *Spectroscopy of Advanced Materials*, ed. by R. J. H. Clark and R. E. Hester, Advances in Spectroscopy Vol. 19, John Wiley & Sons, 1991, 1.

[8] P. Delhaes, *Organic and Inorganic Low-Dimensional Crystalline Materials*, ed. by P. Delhaes and M. Drillon, NATO, ASI Series B, Physics, Vol. 168, Plenum, New York, 1987

[9] T. Ishiguro and K. Yamaji, *Organic Superconductors*, Springer Series in Solid State Sciences Vol. 88, Springer Verlag, Berlin, 1990.

[10] P. Delhaes and L. Ducasse, Localized and Itinerant Molecular Magnetism, to be published in NATO, ASI Series, Plenum, New York, 1996.

[11] M-H. Whangbo, J. M. Williams, A. J. Schultz, T. J. Emge and M. A. Beno, *J. Am. Chem. Soc.*, Vol. 109, 1987, 90.

[12] A. Kini, *The Physics and Chemistry of Organic Superconductors*, Vol. 51, (Springer-Verlag, Berlin) 1990 334.

[13] A. Barraud, O. Kahn and J-P. Launay, *Science et Technologie*, no. 15, May (1989) .

[14] A. Ruaudel-Texier, M. Vandevyver and A. Barraud, *Mol. Cryst. Liq. Cryst*., 120 (1985) 319.

[15] T. Nakamura, M. Matsumoto, F. Takei, M. Tanaka, T. Sekiguchi, E. Manda and Y. Kawabata, *Chem. Lett*., 1986, 709.

[16] S. A. Dhindsa, R. J. Ward, M. R. Bryce, Y. M. Lvov, H. S. Murno and M. C. Petty, *Synth. Met*., Vol. 35, 1990, 307 ; A. S. Dhindsa, M. R. Bryce, H. Ancelin, M. C. Petty and J. Yarwood, *LANGMUIR*, 6, 1990 1680; M. R. Bryce and M. C. Petty, *NATURE*, Vol. 374, No. 27, 1995, 771.

[17] F. Rustichelli, S. Dante, P. Mariani, I. V. Myagkov and V. I. Troitsky, *Thin Solid Films*, Vol. 242, 1994, 267; V. I. Troitsky, T. S. Berzina, A. Petrigliano and C. Nicolini, *Thin Solid Films*, in press.

[18] T. Suzuki, H. Yamochi, G. Srdanov, K. Hinkelmann and F. Wudl, *J. Am. Chem. Soc*., Vol. 111, 1989, 3108; T. Nakamura, G. Yunome, R. Azumi, M. Tanaka, H. Tachibana, M. Matsumoto, S. Horiuchi, H. Yamochi and G. Saito, J. Phys. Chem., Vol. 98, 1994, 1882.

[19] V. M. Yartsev and C. S. Jacobsen, *Phys. Rev*. B, Vol. 24, No.10 1981 6167.

[20] J-P. Bourgoin, A. Ruaudel-Texier, M. Vandevyver, M. Roulliay, A. Barraud, M. Lequan and R-M. Lequan, *Makromol. Chem., Macromol. Symp.*, 46 1991 163.

[21] C. Dourthe, M. Izumi, Ch. Garrigou-Lagrange, Th. Buffeteau, B. Desbat and P. Delhaes, *J. Phys. Chem.*, Vol. 96, 1992, 2812.

[22] T. Nakamura, G. Yunome, R. Azumi, M. Yumura, M. Matsumoto, S. Horiuchi, H. Yamochi and G. Saito, *Synth. Met.*, 56, 1993, 3853.

[23] M. Izumi, H. Ohnuki, H. Yamaguchi, H. Oyanagi and P. Delhaes, *Synth. Met.*,Vol. 56 1993 2560; H. Ohnuki, M.I zumi, K. Kitamura, H. Yamaguchi, H. Oyanagi and P. Delhaes, *Thin Solid Films*, Vol. 243 1994, 415.

[24] H. Ohnuki, K. Kitamura, M. Izumi, H. Yamauchi and R. Kato, *Synth. Met.*, Vol. 71 1995 pp.2077-2078; H. Ohnuki, K. Kojima, M. Izumi, H. Yamaguchi, H. Oyanagi and R. Kato, *Thin Solid Films*, inpress.

[25] L. Aguilhon, J-P. Bourgoin, A. Barraud and P. Hesto, *Synth. Met.*, Vol. 71 1995, 1971; P. Hesto, L. Aguilhon, G. Trembly, J-P. Bourgoin, M. Vandevyver and A. Barraud, *Thin Solid Films*, Vol. 242, 1994, 7.

[26] J. Richard, M. Vandevyver, A. Barraud, J-P. Morand and P. Delhaes, *J. Colloid Interface Sci.*, Vol. 129, 1989, 254.

[27] K. Yamaji, *Solid State Commun.*, Vol. 61, 1987, 413.

[28] J. B. Torrance, *Acc. Chem. Res.*, Vol. 12, 1979, 79.

[29] A. I. Kitaigorodskii, *Organic Chemical Crystallography*, (Consultants Bureau, New York, 1961); A. I. Kitaigorodskii, Molecular Crsytals and Molecules, (Academic Press, London, 1973).

[30] M. Dressel, G. Grüner, J-P. Pouget, A. Breining and D. Schweitzer, J. Phys. I France Vol. 4, 1994 579 ; M. Meneghetti, R. Bozio and C. Pecile, *J. Physique*, Vol. 47, 1986, 1377.

[31] H. Ohnuki, K. Kitamura, K. Uchinokura and M. Izumi, unpublished.

[32] M. J. Rice, V. M. Yartsev and C. S. Jacobsen, *Phys. Rev.* B, Vol. 21, 1980, 3437.

[33] J. E. Eldridge, C. C. Homes, J. M. Williams, A. M. Kini and H. H. Wang, *Spectrochimica Acta*, Vol. 51A, 1995, pp. 947-960; J. C. R. Faulhaber, D. Y. K. Ko and P. R. Briddon, *Synthetic Metals*, Vol. 60, 1993, pp. 227-232; M. E. Kozlov, K. I. Pokhodnia and A. A. Yurchenko, Spectrochimica Acta, Vol. 43 A, 1987, pp. 323-329.

[34] P. Delhaes and Ch. Garrigou-Lagrange, *Phase Transitions*, Vol. 13, 1988, pp. 87-99 ; Ch. Garrigou-Lagrange, E. Dupart, J-P. Morand and P. Delhaes, *Synth. Met.*, Vol. 27, 1988, pp. B537-B542.

Part IV

Cluster Materials

Synthesis and Structural Characterization of a Quaternary Zintl Phase Material from the Rb:Ge:In:As System: $Rb(Ge_{1.5}In_{0.5})As_2(As_{1.5}Ge_{0.5})$

Julie L. Shreeve-Keyer and Robert C. Haushalter

NEC Research Institute, 4 Independence Way, Princeton, NJ 08540, USA

Abstract

In our investigations of new materials with potentially interesting electronic properties, we have synthesized the quaternary Zintl phase material $Rb(Ge_{1.5}In_{0.5})As_2(As_{1.5}Ge_{0.5})$, (**1**), from Rb_3As_7, Ge and In at 700°C. The single crystal structure of **1** has been determined, and has been found to be a layered compound comprised of sheets of indium, germanium, and arsenic with the layers separated by interlamellar rubidium cations. The compounds $Rb_xGe_{2-x}In_xAs_4$, (**2**), and $K_xGe_{2-x}Tl_xAs_4$, (**3**) have also been synthesized and were also found to have layered structures. The structure of **1** was determined by single crystal *X*-ray diffraction: orthorhombic, $Cmc2_1$ (#36) with a = 3.799 (3) Å, b = 16.50 (2) Å, c = 12.24 (1) Å, V = 767 (1) Å^3, and D_{calcd} =4.589 g/cm^3 for Z = 4. Least squares refinement of the model based on 322 observed reflections ($I > 3.0\sigma(I)$) and 45 variable parameters converged to $R(R_w)$ = 2.2% (2.6%). The tendency of the layered Zintl phases to undergo partial elemental substitution to achieve a closed shell electronic configuration is discussed.

1. Introduction

Since the size of physical features in current electronic devices are approximately midway between macroscopic length measurements (millimeters) and atomic dimensions, there is still a great deal of diminution in size possible. As the device dimensions shrink, chemically prepared self-assembled systems and thin films will likely play am increasingly important role in fabrication schemes. In order to understand the behavior of electrons on atomic length scales, it would be desirable to prepare classes of synthetic materials with one-dimensional (1-D) and two-dimensional (2-D) structural elements present. Several Zintl phases and Zintl anion materials could potentially offer an attractive platform to investigate the confinement of electrons in very small domains. To this end, we have begun a synthetic investigation of 1-D metal polymers and lamellar materials with atomically thin layers. In this paper we discuss some new materials in the Group 1-Ge-As system and compare their structures to other solids with layers comprised of elements that typically form semiconducting compositions.

The Zintl phase materials are an interesting class of solids that can, in a certain sense, be considered as possessing properties intermediate between those of ionic salts and metals or intermetallic phases. In the synthesis of these compounds the reaction between electropositive elements, such as an alkali or alkaline earth metal, and a heavier post-transition element results in electron transfer from the more electropositive elements to the electronegative elements. The resulting isolated anions and polyanions often possess extensive element-element bonding to complete the octet of electrons in

Springer Proceedings in Physics, Vol. 81
Materials and Measurements in Molecular Electronics
Editors: K. Kajimura · S. Kuroda

their valence orbitals. Zintl alloys are frequently prepared by heating a direct combination of the elements, or the elements with simple binary phases. For example, KSn may be prepared by mixing molten potassium and molten tin to give KSn. [1a,b] Similar synthetic techniques can be used to produce KPb, [1c] KSb, [1d] KTl, [1b] K_5In_8, [1e] K_5Ga_8, [1f] and others, which may be isolated and identified by powder diffraction.

Recently there has been increasing interest in Zintl phase materials with a number of structurally fascinating ternary materials being prepared. Investigations by von Schnering, [2] Schäfer, [3] Corbett, [4] and others [5] have shown that Zintl phases possess a huge number of novel structure types, bonding schemes, and unusual physical properties. While most of these studies have focused on the preparative and structural aspects of the materials, there are a few examples of the low temperature preparation of metastable materials, such as thin films [6] and amorphous metallic spin glasses. [7] Synthesis of ternary I-III-V Zintl phase materials as a preliminary step in the investigation of new routes to the preparation of III-V type semiconductors has also been of interest. [5g]

While there are several examples of ternary Zintl phase materials, [2-5] there are fewer examples of materials containing four elements. We have undertaken an exploratory synthetic investigation of Group 1-13-14-15 quaternary Zintl phases to determine if it is possible to incorporate such a large number of elements into a crystalline material and how the structure and composition would influence the physical properties. Recently we have reported the syntheses and single crystal structure of three compounds in this new class of materials, $K_8In_8Ge_5As_{14}(As_3)$, [8] $K_5In_5Ge_5As_{10}(As_4)$, [8] and $K_9In_9GeSb_{22}$. [9] We report here the synthesis and single crystal *X*-ray structure of another material in this newly emerging class of compounds, $RbGe_2In_{0.5}As_{3.5}$, (**1**). The structure of **1** contains In-Ge-As layers interleaved with Rb^+ ions. The compounds $Rb_xGe_{2-x}In_xAs_4$, (**2**), and $K_xGe_{2-x}Tl_xAs_4$, (**3**), have also been prepared.

In recent investigations of the layered compounds $K_9In_9GeSb_{22}$, [9] and $K_5In_5Ge_5As_{10}(As_4)$, [8], it was shown by a combination of structural studies and extended Hückel calculations, that relatively small amounts of the fourth element would specifically substitute into certain crystallographic sites in order for the polyanionic layer to achieve a closed shell electronic configuration. In fact, an examination of the Zintl phase literature shows that in essentially all cases the polyanions are electron precise. This tendency appears to be so strong, that one needs to seriously question stoichiometries that would give rise to electron imprecise formulations.

2. Experimental

General. The compounds described below are air- and moisture sensitive. All manipulations were performed under an inert atmosphere using standard glovebox, Schlenk, and vacuum line techniques.

Preparation of $Rb(Ge_{1.5}In_{0.5})As_2(As_{1.5}Ge_{0.5})$, (1). The solids Rb_3As_7 (0.500g, 0.640 mmol), In (0.294g, 2.55 mmol), and Ge (0.186, 2.56 mmol), were combined in a mole ratio of 1:4:4 and then sealed in an evacuated quartz ampoule. This mixture was heated to 700°C over 12 h (55.8°/hour), then slowly cooled at a linear rate to 500°C over 48 hours (4.2°/hour), followed by cooling to ambient temperature over 12 h. This method gave a heterogeneous mixture, which contained *X*-ray quality crystals of **1**, which appear as silver needles throughout the reaction mixture. EDS analysis of crystals of **1** showed Rb, In, Ge, and As in an approximate 1:1:8:11 ratio. No attempt has been made to optimize the yield or phase purity at this time.

***X*-ray Diffraction Analysis of 1.** A silver needle-shaped crystal of approximate dimensions 0.05 x 0.10 x 0.30 mm^3 was sealed into a glass capillary under an inert atmosphere of helium. The *X*-ray intensity data were collected on a Rigaku AFC7R automatic four-circle diffractometer equipped with MoK_α radiation and an RU300 18kW rotating anode generator. Cell constants and an orientation matrix for data collection were obtained from the setting angles of 25 carefully centered reflections in the range $21.52° \leq 2\theta \leq 25.88°$, and corresponded to a *C*-centered orthorhombic cell with the unit cell dimensions given in Table 1. Based on the systematic absences of hkl: $h + k \neq 2n$, h0l: $l \neq 2n$, packing considerations, a statistical analysis of the intensity distribution, and the solution and refinement of the structure, the space group was determined to be $Cmc2_1$ (#36). An empirical absorption correction using the program DIFABS [10] was also applied in the final stages of refinement, and resulted in transmission factors ranging from 0.78 to 1.00. The data were also corrected for Lorentz and polarization effects. The structure was solved by direct methods [11] and refined on *F* by full-matrix least-squares using the *teXsan* crystallographic software package of Molecular Structure Corporation. [12] Further details of the *X*-ray structural analysis of compound **1** are given in Table 1. The final positional and equivalent isotropic displacement parameters are given in Table 2.

TABLE 1. EXPERIMENTAL CRYSTALLOGRAPHIC DETAILS FOR $Rb(Ge_{1.5}In_{0.5})As_2(As_{1.5}Ge_{0.5})$, (**1**).

Temperature (°C)	20	$\mu(MoK_\alpha)$ (cm^{-1})	311.21
Crystal color, Habit	silver, needle	Diffractometer	Rigaku AFC7R
Crystal Dimensions (mm^3)	*ca.*0.05x0.10x0.30	Scan Type	ω-2θ
Crystal System	orthorhombic	Scan Rate (deg/min)	16.0
Space Group	$Cmc2_1$ (#36)	$2\theta_{max}$	60.2°
a (Å)	3.799(3)	Reflections Measured	Total: 375
b (Å)	16.50(2)	Observations ($I>3.00\sigma(I)$)	322
c (Å)	12.24(1)	No. of Variables	45
Volume (Å^3)	767(1)	R, R_w	2.2%, 2.6%
Z	4	Goodness of Fit	1.18
D_{calc} (g/cm^3)	4.589	Max. Peak in Final Diff. Map	0.56 eÅ^{-3}
F_{000}	932.00	Min. Peak in Final Diff. Map	-0.59 eÅ^{-3}

Preparation of $Rb_xGe_{2-x}In_xAs_4$, (2). The solids Rb_3As_7 (0.500g, 0.640 mmol), In (0.294g, 2.56 mmol), Ge (0.186, 2.56 mmol), and As (0.239g, 3.19 mmol) were combined in a mole ratio of 1:4:4:5 and then sealed in an evacuated quartz ampoule. This mixture was heated to 700°C over 12 h (55.8°/hour), then slowly cooled at a linear rate to 500°C over 48 hours (4.2°/hour), followed by cooling to ambient temperature over 12 h. This method gave a heterogeneous mixture, which contained *X*-ray quality crystals of **2**, which appear as silver plates throughout the reaction mixture. EDS analysis of crystals of **2** showed Rb, In, Ge, and As in an approximate 0.05:0.05:1:2 ratio. We were unable to prepare **2** as a single phase product.

X-ray Diffraction Analysis of 2. A silver plate-shaped crystal of approximate dimensions 0.10 x 0.05 x 0.20 mm^3 was sealed into a glass capillary under an inert atmosphere of helium. The X-ray intensity data were collected on a Rigaku AFC7R automatic four-circle diffractometer equipped with MoK_α radiation and an RU300 18kW rotating anode generator. Cell constants and an orientation matrix for data collection were obtained from the setting angles of 24 carefully centered reflections in the range $20.39° \leq 2\theta \leq 29.40°$, and corresponded to a primitive orthorhombic cell with the unit cell dimensions a = 10.4058 (9) Å, b = 16.579 (4) Å, c = 3.750 (2) Å, V = 646.9 (3) Å^3.

TABLE 2. ATOMIC COORDINATES AND B_{eq} (Å^2) FOR $Rb(Ge_{1.5}In_{0.5})As_2(As_{1.5}Ge_{0.5})$, (**1**).

atom	x	y	z	occupancy	B(eq)
Rb(1)	1.5000	0.7648(1)	0.1533(1)	0.500	2.64(3)
As(1)	0.0000	0.59558(9)	0.2808(1)	0.500	1.72(2)
As(2)	0.5000	0.56173(9)	0.5586(1)	0.500	1.56(2)
As(3)[a]	0.5000	0.3765(1)	0.34442(10)	0.500	1.74(2)
As(4)[a]	0.0000	0.34968(9)	0.4704(1)	0.500	1.63(2)
In(1)	0.0000	0.4693(2)	0.5978(2)	0.250[b]	1.80(2)
Ge(1)	0.0000	0.4693(2)	0.5978(2)	0.250[c]	1.80(2)
Ge(2)	0.5000	0.52456(9)	0.36431(9)	0.500	1.39(2)

[a] 1/2 Ge occupancy spread over the As(3) and As(4) sites. This was not refined, as Ge and As are essentially indistinguishable by X-ray diffraction.
[b] 1/2 In occupancy at the Wyckoff a site
[c] 1/2 Ge occupancy at the Wyckoff a site

Preparation of $K_xGe_{2-x}Tl_xAs_4$, (3). The solids K_3As_7 (0.500g, 0.779 mmol), Tl (0.636g, 3.11 mmol), Ge (0.226g, 3.11 mmol), and As (0.409g, 5.46 mmol) were combined in a mole ratio of 1:4:4:7 and then sealed in an evacuated quartz ampoule. This mixture was heated to 750°C over 12 h (60°/hour), then slowly cooled at a linear rate to 558°C over 48 hours (4.0°/hour), followed by cooling to ambient temperature over 12 h. This method gave a heterogeneous mixture, which contained X-ray quality crystals of **3**, which appear as silver plates throughout the reaction mixture. We were unable to prepare **3** as a single phase product.

X-ray Diffraction Analysis of 3. A silver plate-shaped crystal of approximate dimensions 0.10 x 0.01 x 0.30 mm^3 was sealed into a glass capillary under an inert atmosphere of helium. The X-ray intensity data were collected on a Rigaku AFC7R automatic four-circle diffractometer equipped with MoK_α radiation and an RU300 18kW rotating anode generator. Cell constants and an orientation matrix for data collection were obtained from the setting angles of 24 carefully centered reflections in the range $20.49° \leq 2\theta \leq 29.65°$, and corresponded to a primitive monoclinic cell with the unit cell dimensions a = 3.71 (2) Å, b = 10.25 (2) Å, c = 16.49 (5) Å, β = 90.1 (3)°, V = 626 (3) Å^3.

3. Discussion

Thermal treatment of a mixture of Rb_3As_7, In and Ge in evacuated, sealed quartz ampoules yielded *X*-ray quality single crystals of the unusual compound $RbGe_2In_{0.5}As_{3.5}$ (**1**). Compound **1** was isolated as crystals from a heterogeneous mixture of products. Thermal treatment of Rb_3As_7, In, Ge and As or K_3As_7, Tl, Ge and As gave $Rb_xGe_{2-x}In_xAs_4$, (**2**) and $K_xGe_{2-x}Tl_xAs_4$, (**3**), respectively. Compounds **2** and **3** were also isolated as single crystals from a heterogeneous mixture of products. No attempt has been made to optimize the yields or phase purities of these materials at this time.

The compound $RbGe_2In_{0.5}As_{3.5}$, (**1**), contains certain atomic sites which are occupied by two atom types. The original formula $RbGe_2As_4$ assigned in the early stages of the structural analysis of **1** is electron imprecise. Thus, the $RbGe_2As_4$ formulation is unlikely to be correct, and ways of reducing one electron per formula unit (e.g., a partial deficiency in the rubidium sites, a partial occupancy of In^{3+} in the Ge^{4+} sites or other substitutions) should be considered. The *X*-ray data suggests there is a partial occupancy of indium in one of the germanium sites, Ge(1), in **1**. This was modeled crystallographically by assigning that site to be occupied by both indium and germanium in a 1:1 ratio. It was clear by refinement of the crystallographic data and the observed thermal parameters, that the In substitution was limited to the Ge(1) site and not the Ge(2) site. The substitution of germanium and indium atoms at a single site has been definitively observed previously in the compound $K_9In_9GeSb_{22}$, [9] where a partial occupancy of germanium in one of the indium sites resulted in an electron precise formulation. This crystallographic model of **1** led to a formula of $RbGe_{1.5}In_{0.5}As_4$, which is still not electronically precise, but is unbalanced by one-half of an electron per formula unit. This one-half electron could be accounted for by either partial occupancy of the Rb cation, which could most likely be detected crystallographically, or partial occupancy of Ge in the As sites, which could not be apparent from the *X*-ray data. Refinement of the Rb sites showed them to be fully occupied. Therefore, this final one-half of an electron must be accounted for by periodic substitution of Ge along the As chain, such that 0.5 Ge atoms partially occupies the two As sites, As(3) and As(4). This gives an overall formula of $Rb(Ge_{1.5}In_{0.5})As_2(As_{1.5}Ge_{0.5})$, or $RbGe_2In_{0.5}As_{3.5}$, which is electron precise.

Compound **1** is a Zintl phase material. As previously mentioned, Zintl phase compounds frequently form bonds between the different elements to give complex polyanions that are electron precise, closed-shell systems with a complete octet of electrons in the outer valence shells. For a Zintl phase compound consisting of several different elements, an *X*-ray structure determination can frequently lead to several different structural formulae with acceptable thermal parameters and *R*-factors. In these cases, it is important to examine several aspects of the structure to find a suitable model. Firstly, whether or not the formulae are electron precise must be considered. Extended Hückel tight binding calculations [13] and DFT-MO calculations [14] have been performed on $K_9In_9GeSb_{22}$ [9] and $K_5In_5Ge_5As_{10}(As_4)$, [8], respectively. These studies showed that in these systems formulae which were structurally reasonable, but not electronically precise, led to systems in which one electron per formula unit was placed into an antibonding level, and therefore was unlikely to be correct. These calculations further stressed the importance of having an electron precise formula in these systems. Secondly, the bond distances in the compound must be examined carefully to see if there are any unusual lengthening or shortening effects. Examination of the M-As (M = Ge, Ge(In)) bond distances in **1** revealed that the average Ge(1)/In(1)-As distance is 2.49 (1) Å, which is slightly longer than the 2.456 (8) Å Ge(2)-As distance in **1**, and longer than the 2.43 (2) Å average Ge-As distance found in $K_8In_8Ge_5As_{14}(As_3)$, [8] which has no partial occupancy of indium. This lengthening in **1** is consistent with indium occupancy in the germanium site, as the atomic radius of indium is larger than germanium. Thirdly, the refinement of the structure was consistent with the partial occupancy of In at the Ge(1) site. By refining this model, the *R*-factor dropped slightly, and the thermal parameters became more uniform for the atoms in the layer. Finally, microprobe analysis (EDS) of the data crystal of **1** showed there to be indium in the crystal.

The structure of **1** consists of layers of Ge, In, and As which run parallel to the plane containing the **a** and **c** axes and are insulated from one another by interleaving regions of Rb^+ cations. The layers are formed by $Ge(In)As_4$ tetrahedra and infinite arsenic chains which are linked together to form the two-dimensional, covalently bonded sheets. When **1** is viewed down the **a** axis (**Figure 1**), a view parallel to the layers, one can see the five membered rings that are formed by the two Ge(In) and three As atoms. The rings are linked through the Ge(In) atoms by one arsenic atom. These rings are further condensed to form the layers.

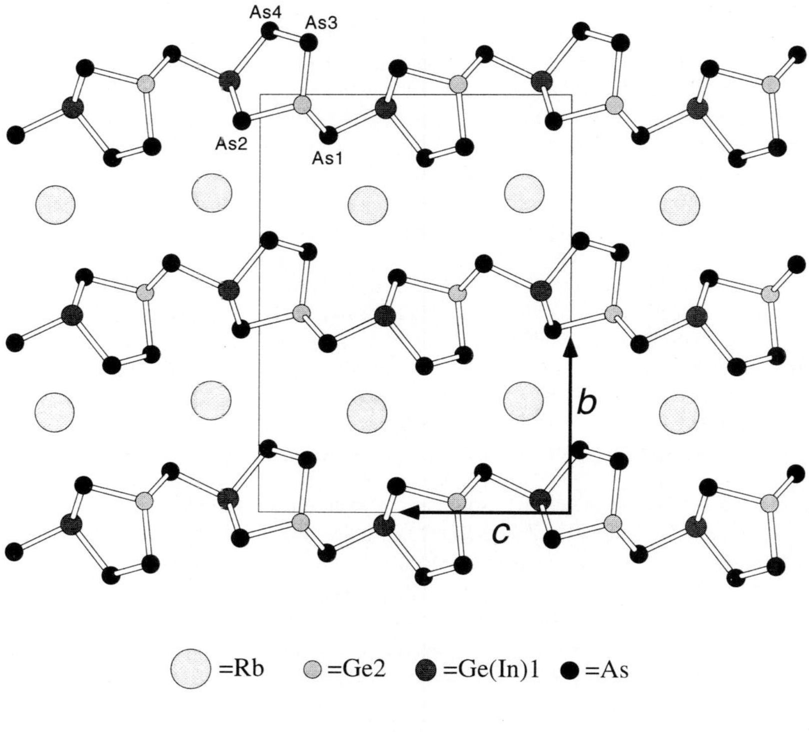

Figure 1. View of the $RbGe_2In_{0.5}As_{3.5}$, (**1**), structure showing the layers comprised of germanium, indium and arsenic atoms.

A polyhedral representation of a single layer in **1** is shown in **Figure 2**. Each layer consists of rows of Ge(In)As_4 tetrahedra, which are linked to neighboring Ge(In)As_4 tetrahedra through corner-shared interactions. Each tetrahedra has three of its four arsenic atom corners linked to other tetrahedra, forming As-Ge(In)-As interactions. The fourth arsenic atom is bonded to another arsenic atom, from the adjacent row of tetrahedra, forming an infinite chain of arsenic atoms running along either side of the layer. This infinite chain is an interesting feature, which is also seen in $K_5In_5Ge_5As_{10}(As_4)$, [8] as shown in **Figure 3**. The single crystallographically unique As-As distance along the chain is 2.486(2) Å, which is slightly shorter than the 2.529 (4) Å As-As distance in $K_5In_5Ge_5As_{10}(As_4)$, [8] and is well within the distance for an As-As bond. Periodic substitution of germanium in these sites would be expected to have little measurable effect on the bond lengths. Other bond lengths for **1** are given in Table 3. There is a repeat within the layer of **1**, such that there are two rows of Ge(In)As_4 tetrahedra which contain an infinite arsenic chain on one side of the layer linked to another set of rows which the arsenic chain is n the other side of the layer. This causes an alternation of the chains on either surface of the layer.

TABLE 3. SELECTED BOND DISTANCES (Å) AND ANGLES (°) FOR $Rb(Ge_{1.5}In_{0.5})As_2(As_{1.5}Ge_{0.5})$, (**1**).

atom	atom	distance	atom	atom	distance
Ge(1)/In(1)	As(1)	2.483(2)	As(2)	Ge(2)	2.456(2)
Ge(1)/In(1)	As(2)	2.483(1)	As(3)	As(4)	2.486(2)
Ge(1)/In(1)	As(4)	2.516(2)	As(3)	Ge(2)	2.457(2)
As(1)	Ge(2)	2.455(2)			

atom	atom	atom	angle
As(1)	Ge(1)/In(1)	As(2)	116.05(4)
As(1)	Ge(1)/In(1)	As(4)	102.75(5)
As(2)	Ge(1)/In(1)	As(2)	99.83(7)
As(2)	Ge(1)/In(1)	As(4)	111.25(4)
Ge(1)/In(1)	As(1)	Ge(2)	99.78(6)
Ge(2)	As(1)	Ge(2)	101.37(9)
Ge(1)/In(1)	As(2)	Ge(1)/In(1)	99.83(7)
Ge(1)/In(1)	As(2)	Ge(2)	91.92(5)
As(4)	As(3)	As(4)	99.65(8)
As(4)	As(3)	Ge(2)	96.63(6)
Ge(1)/In(1)	As(4)	As(3)	104.18(6)
As(3)	As(4)	As(3)	99.65(8)
As(1)	Ge(2)	As(1)	101.37(9)
As(1)	Ge(2)	As(2)	106.51(6)
As(1)	Ge(2)	As(3)	115.71(6)
As(1)	Ge(2)	As(2)	106.51(6)
As(1)	Ge(2)	As(3)	115.71(6)
As(2)	Ge(2)	As(3)	110.15(7)

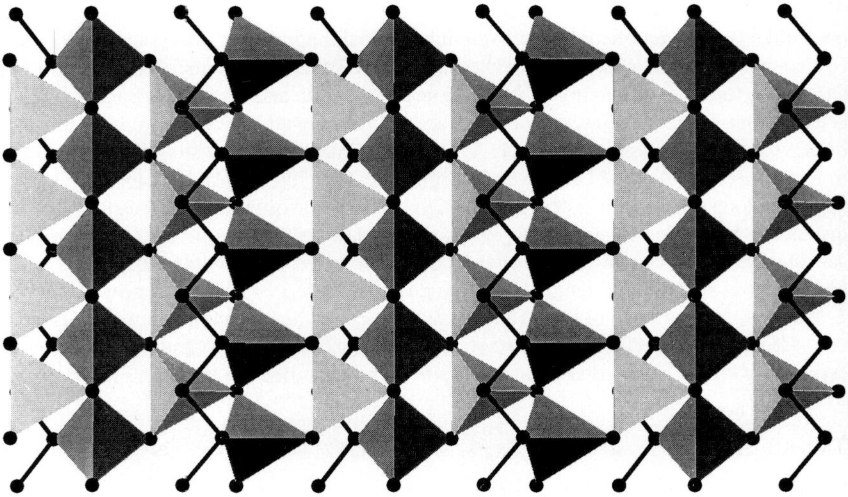

Figure 2. Polyhedral representation of a single layer in $RbGe_2In_{0.5}As_{3.5}$, (**1**).

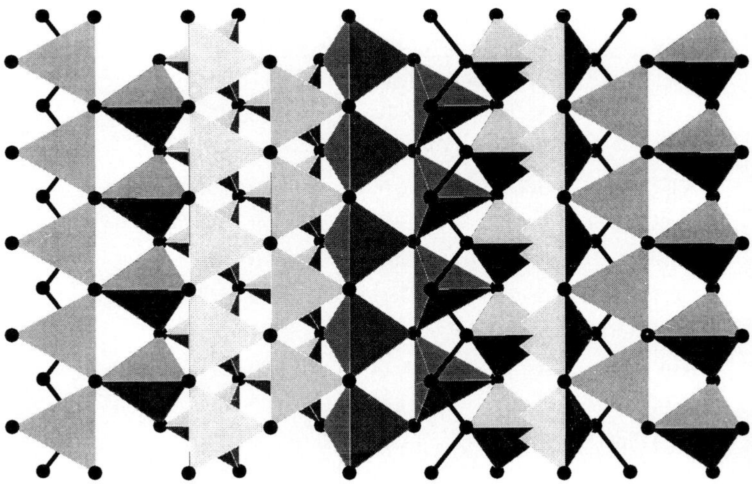

Figure 3. Polyhedral representation of a single layer in $K_5In_5Ge_5As_{10}(As_4)$.

Compounds $Rb_xGe_{2-x}In_xAs_4$, (**2**), and $K_xGe_{2-x}Tl_xAs_4$, (**3**), were found to contain layers which are similar to those in compound **1**. The main difference appears to be in the registry of the layers and the locations of the alkali cations within the structure. Unfortunately we were unable to develop a totally adequate crystallographic model for **2** and **3**, due to the inferior crystal quality and some space group ambiguities. However, we were able to determine the connectivity in these systems to allow for some comparison to **1**.

Crystals of **1-3**, as well as $K_8In_8Ge_5As_{14}(As_3)$, [8] $K_5In_5Ge_5As_{10}(As_4)$, [8] and $K_9In_9GeSb_{22}$, [9] are all extremely air- and moisture-sensitive. The compound $K_8In_8Ge_5As_{14}(As_3)$ was also found to be exceedingly reactive towards and type of oil or grease, causing an immediate exfoliation of the layers and swelling of the crystals. This extreme sensitivity, coupled with the small very size of the crystals has made it difficult to perform conductivity experiments thus far.

The formation of In-Ge-As(Sb) layers as in **1** has also been seen in other Group 1-13-14-15 systems, including $K_8In_8Ge_5As_{14}(As_3)$, [8] $K_5In_5Ge_5As_{10}(As_4)$, [8] and $K_9In_9GeSb_{22}$. [9] In all of these cases, materials are formed which contain layers composed of the main group elements which are separated from each other by regions of alkali metal cations. All of these structures are further similar in that they contain puckered rings of indium, germanium, and arsenic or antimony which are condensed into the two-dimensional sheets. The structure of **1** shares similarities with $K_5In_5Ge_5As_{10}(As_4)$ in both the formation of an infinite arsenic or arsenic/germanium chain, but also in that $K_5In_5Ge_5As_{10}(As_4)$ contains five-membered rings of Ge, In, and As as shown in **Figure 4.** The structure of $K_5In_5Ge_5As_{10}(As_4)$ is more complicated, in that it also contains six-membered rings which form other portions of the layer. In $K_5In_5Ge_5As_{10}(As_4)$ the infinite arsenic chains are also formed from two adjoining As atoms in the five-membered rings.

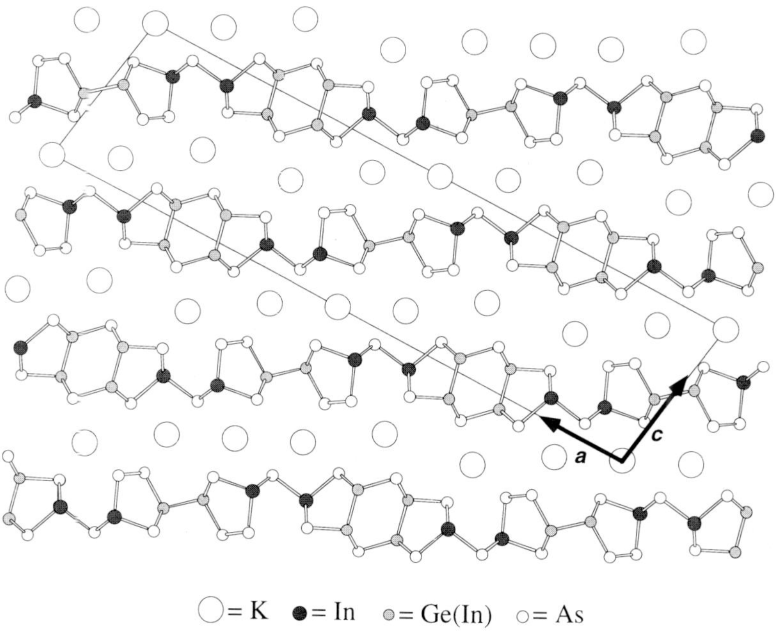

Figure 4. View of the layered structure of $K_5In_5Ge_5As_{10}(As_4)$, down the **b** axis.

The five membered-ring structure in **1** is similar to the ring structure in KSi_3As_3. [15] The structure of KSi_3As_3 consists of Si-As sheets which are separated by potassium cations. These sheets are formed by puckered rings of the metalloid elements which are linked together to form chains, which are further condensed to form the layers, as shown in **Figure 5**. However, in KSi_3As_3 the two five-membered rings are linked to one another through a Si-Si bond, rather than through an arsenic atom as is seen in **1**.

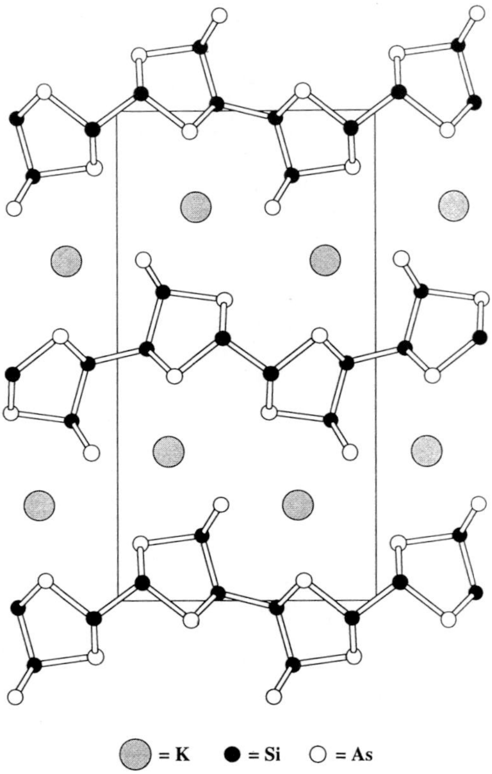

Figure 5. Structure of KSi_3As_3.

4. Conclusions

The arsenide $Rb(Ge_{1.5}In_{0.5})As_2(As_{1.5}Ge_{0.5})$, (**1**), is a novel quaternary Zintl phase material in this newly discovered class of quaternary Zintl phase materials. Compound **1** was synthesized by thermal methods, and is only the fourth example, to our knowledge, of a crystallographically characterized Group 1-13-14-15 compound. The compounds $Rb_xGe_{2-x}In_xAs_4$, (**2**), and $K_xGe_{2-x}Tl_xAs_4$, (**3**), are further examples of new quaternary materials. To date, all of these compounds have formed layered structures with layers comprised of the main group elements separated by regions of alkali metal cations.

In conclusion, we have successfully prepared the novel, quaternary Zintl phase materials $Rb(Ge_{1.5}In_{0.5})As_2(As_{1.5}Ge_{0.5})$, (**1**), $Rb_xGe_{2-x}In_xAs_4$, (**2**), and $K_xGe_{2-x}Tl_xAs_4$, (**3**), as well as $K_8In_8Ge_5As_{14}(As_3)$, [8] $K_5In_5Ge_5As_{10}(As_4)$, [8] and $K_9In_9GeSb_{22}$, [9] which were previously prepared in our laboratories. Their structures have been determined by single crystal *X*-ray diffraction. These compounds are representative of a new, potentially abundant, class of layered materials from the Group 1-13-14-15 system. In all of these compounds, layered structures which contain atomically thin, polyanionic layers of the main group atoms that are separated by a region of alkali cations. In **1**, the electron precise nature is obtained by introducing partial occupancies of indium and germanium, and germanium and arsenic in specific sites, consistent with the tendency of Zintl phase materials to completely fill their valence shells. Further studies to determine the electronic properties of these materials are currently underway in our laboratories.

Acknowledgment: Travel support via the Invitation Program for Overseas Visiting Researchers from Industrial Science and Technology, Ministry of International Trade and Industry, Japan, is gratefully acknowledged.

References:

[1.] a) I. F. Hewaidy, E. Busmann, W. Klemm, *Z. Anorg. Allg. Chem.*, 1964, *328*, 283. b) M. Hansen, *Constitution of Binary Alloys*, McGraw-Hill, New York, 2nd edn., 1958. c) B. F. Alblas, C. van der Marel, W. Geertsma, J. A. Meijer, A. B. van Oosten, J. Dijkstra, P. C. Stein, W. van der Lugt, *J. Non-Cryst. Solids*, 1984, *61*, 201. d) E. Busmann, S. Lohmeyer, *Z. Anorg. Allg. Chem.*, 1961, *312*, 53. e) R. Thummel, W. Klemm, *Z. Anorg. Allg. Chem.* 1970, *376*, 44. G. Bruzzone *Acta Crystallogr.*, 1969, *25B*, 1206. f) E. Rinch, P. Feschote *C. R. Acad. Sci.*, 1961, *252*, 3592. S. Yatsenho, *Kristallografiya*, 1983, *28*, 809.

[2] a) **$K_2NaInSb_2$:** W. Carrillo-Cabrera, N. Caroca-Canales, H. G. von Schnering, *Z Anorg. Allg. Chem.* 1993, *619*, 1717. b) **$K_2NaGaAs_2$, $K_2NaInAs_2$:** M. Somer, K. Peters, H. G. von Schnering, *Z. Anorg. Allg. Chem.* 1992, *613*, 19. c) **$K_2NaInAs_2$:** M. Somer, K. Peters, E. M. Peters, H. G. von Schnering, *Z. Kristallogr.* 1991, *195*, 97. d) **Cs_6InAs_3:** W. Blase, G. Cordier, K. Peters, M. Somer, H. G. von Schnering, *Angew. Chem. Int. Ed. Engl.* 1991, *30*, 326.

[3] a) **$Ca_3In_2As_4$:** G. Cordier, H. Schäfer, M. Stelter, *Z. Naturforsch. B.* 1986, *41B*, 1416. b) **$Ca_5Sn_2As_6$:** B. Eisenmann, H. Jordan, H. Schäfer, *Z. Anorg. Allg. Chem.* 1985, *530*, 74.

[4] a) **$A_3Na_{26}In_{48}$ (A = K, Rb, Cs):** S. C. Sevov, J. D. Corbett, *Inorg. Chem.* 1993, *32*, 1612. b) **$Na_{15}In_{27.4}$:** S. C. Sevov, J. D. Corbett, *J. Solid State Chem.* 1993, *103*, 114. c) **Rb_2In_3:** S. C. Sevov, J. D. Corbett, *Z. Anorg. Allg. Chem.* 1993, *619*, 128. d) **$Na_7In_{11.8}$:** S. C. Sevov, J. D. Corbett, *Inorg. Chem.* 1992, *31*, 1895. e) **K_8In_{11}:** S. C. Sevov, J. D. Corbett, *Inorg. Chem.*. 1991, *30*, 4875.

[5] Some examples include: a) **K_6InAs_3:** W. Blase, G. Cordier, M. Somer, *Z. Kristallogr.* 1993, *206*, 141. b) **$K_2In_2As_3$, $K_3In_2As_3$:** G. Cordier, H. Ochmann, *Z. Kristallogr.* 1991, *197*, , 293, 295. c) **Na_3InAs_2:** G. Cordier, H. Ochmann, *Z. Kristallogr.* 1991,

195, 105. d) **Cs_6InAs_3:** W. Blase, G. Cordier, M. Somer, Z. *Kristallogr.* 1991, *195*, 117. e) **$Na_{10}Ge_2As_6$:** B. Eisenmann, J. Klein, M. Somer, Z. *Kristallogr.* 1991, *197*, 265. f) **Na_4KGeAs_3:** B. Eisenmann, J. Klein, Z. *Kristallogr.* 1991, *197*, 279. g) **$K_4In_4As_6$:** T. L. T. Birdwhistell, C. L. Klein, T. Jefferies, E. D. Stevens, C. J. O'Connor, *J. Mat. Chem.* 1991, *1*, 555. h) **$K_8In_6Ge_{40}$:** S. Sportcuch, M. Tillard-Charbonnel, C. Belin, Z. *Kristallogr.* 1994, *209*, 541.

[6.] R. C. Haushalter, L. J. Krause, *Polyimides*, 1984, *2*, 735. M. M. J. Treacy, R. C. Haushalter, S. B. Rice, *Ultramicroscopy*, 1987, *23*, 135. R. C. Haushalter, M. M. J. Treacy, S. B. Rice, *Angew. Chem.* 1987, *99*, 1172. R. C. Haushalter, D. P. Goshorn, M. G. Sewchok, C. B. Roxlo, *Mat. Res. Bull.* 1987, *22*, 761.

[7.] R. C. Haushalter, C. J. O'Connor, A. M. Umarji, G. K. Shenoy, C. K. Saw, *Solid State Commun.* 1984, *49*, 929. J. W. Foise, C. J. O'Connor, R. C. Haushalter, *Solid State Commun.* 1987, *63*, 349.

[8.] J. L. Shreeve-Keyer, R. C. Haushalter, Y. -S. Lee, S. Li, S.; C. J. O'Connor, D. -K. Seo, M. -H. Whangbo, *Chem. Mater.*, submitted.

[9.] J. L. Shreeve-Keyer, R. C. Haushalter, D. -K. Seo, M. -H. Whangbo, *J. Solid-State Chemistry*, in press.

[10.] DIFABS: N. Walker, N. Stuart, *Acta Crystallogr.*, *39A* (1983) 158.

[11.] DIRDIF-94: P. T. Beurskins, G. Admiraal, G. Beurskins, W. P. Bosman, R. de Gelder, R. Isreal, J. M. M. Smits, (1994). The DIRDIF -94 program system, Technical Report of the Crystallography Laboratory, University of Nijmegen, The Netherlands.

[12.] TEXSAN Single Crystal Analysis Software Package. Version 1.7-1, March 31, 1995, Molecular Structure Corporation, The Woodlands, TX, 77381, USA.

[13.] M.-H. Whangbo, R. Hoffmann, *J. Am. Chem. Soc.* 1978, *100*, 6093.

[14.] ADF Release 1.1.3, Department of Chemistry, Vrije Universiteit, Amsterdam.

E. J. Baerends, D. E. Ellis, P. Ros, *Chem. Phys.* **1973**, *2*, 4. c)G. te Velde,

E. J. Baerends, *J. Comp. Phys.* **1992**, *99*, 84.

[15.] W. -M. Hurng, J. D. Corbett, S. -L. Wang, A. Jacobson, *Inorg. Chem.* 1987, *26*, 2392.

Electronic Properties of Cluster Crystal Incorporated in Zeolite Crystals

Yasuo Nozue[1], Tetsuya Kodaira[2], Osamu Terasaki[1,3] and Harutoshi Takeo[2]

[1]Department of Physics, Graduate School of Science, Tohoku University, Sendai 980-77, Japan
[2]National Institute for Advanced Interdisciplinary Research, Tsukuba 305, Japan
[3]Center for Interdisciplinary Research, Tohoku University, Sendai 980-77, Japan

Abstract

When alkali metals are loaded into zeolite, their clusters are stabilized in the cage. The average loading density of guest alkali metal can be widely controlled without structural change of zeolite framework. When potassium metal is loaded into zeolite LTA, K clusters with the effective size of 11 Å are generated in α cage. They are arrayed in a simple cubic structure. They show an itinerant electron ferromagnetism at low temperature depending on the guest potassium density. On the other hand, the diamagnetic Na clusters are generated in the cage of LTA. At a dilute Na loading density, clusters are generated in β cage of LTA. The paramagnetic Na cluster with the C_{3v} symmetry is generated by the photoexcitation of diamagnetic cluster.

1 Introduction

Cluster can have wide freedom in the symmetry, size and chemical composition. Resultant electronic state can have the variety in the quantum levels including the spin state. When we make a cluster crystal in which clusters are arrayed regularly with the mutual interaction, resultant electronic properties as well as the structure may be quite different from those expected from the bulk materials of composite atom.

Electrons confined in the cluster have a mutual interaction due to the Coulomb repulsion as well as the Pauli's exclusion principle. The former interaction leads to the energy splitting of the various spin states, such as the Hund rule in atoms. In some clusters, an electronic state cause the displacement of composite atoms followed by the change in the electronic state. This phenomenon plays an important role in the structure change of cluster as the electron-phonon interaction. When the synthesized clusters are stable, such as C_{60}, we can crystallize them as the regular array of cluster. Furthermore, a complex crystal, such as K_xC_{60}, have a wide variety in the structure and electronic properties.

In the present paper, we show a regular array of clusters stabilized in the porous nano-space of zeolite. Generally, the most easiest method for the preparation of small particles is given by the deposition in a glass matrix. This method has the advantage of the wide range control of particle size. However, as for the clusters, this method is not effective, because the chemical reaction with matrix makes a serious problem at the surface boundary of the particle, and we can not generate particles with a unique size. In order to solve these problems, the zeolite method has the advantage of the size control and the reduction of the undesired chemical reaction with the matrix. Furthermore, micropores with the nanometer size are regularly arrayed in zeolite. Zeolite provides the space for the mutually interacting clusters, which leads to the macroscopic phenomenon. Hence, zeolite crystals have remarkable features as the container for the fabrication of cluster crystal.

The framework material of zeolite is usually made of alminosilicate which is transparent between infrared and ultraviolet regions. This point is very important in the optical study of guest materials. Zeolite is crystalline. For example, in zeolite LTA, cages are arrayed in a simple cubic structure with the lattice constant of 12.3 Å, and the number density of cage amounts to 5.4×10^{20} cm^{-3}. Zeolite is stable at high temperatures. It is possible to introduce different materials, such as metals and semiconductors, into zeolite cages by means of different methods. In Fig. 1, clusters arrayed in the space of LTA framework is schematically illustrated. Structurally unstable clusters can be supported in nanometer size pores. Besides the cage type zeolite, there are a lot of channel type zeolite. One dimensional chain structure can be stabilized these.

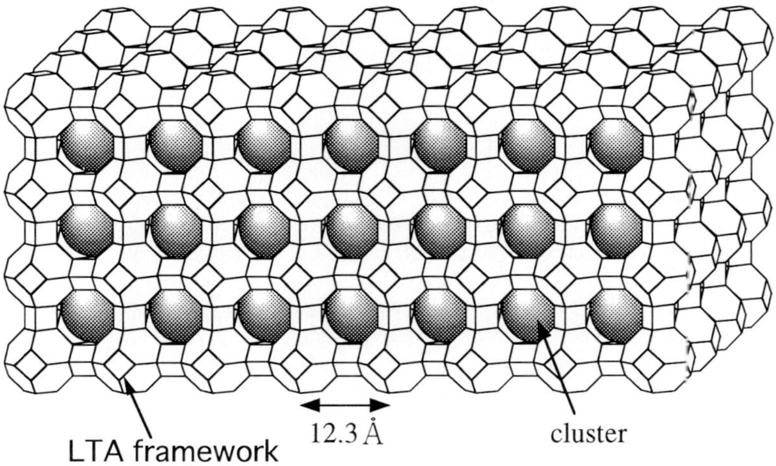

Fig. 1 Schematic representation of arrayed clusters in zeolite LTA. Clusters are generated in each cage (α cage), and arrayed in a simple cubic structure. There are different structure types in zeolite.

2 Zeolite Crystals as the Container of Cluster

The quality of zeolite crystal is recently further improved, and zeolite increasingly plays an important role as the container of guest materials as well as the molecular sieve, catalysis, adsorbent, builder, etc. Zeolite research field is still developing in the synthesis of new structure type of zeolite, the large single crystal and different chemical compositions. Very recently, new type of mesoporous materials, such as FSM and MCM-41, are synthesized. They have a size-tunable space up to ~10 nm.

Many types of zeolite crystals are synthesized under various growth conditions. The most typical zeolite is alminosilicate. A general chemical formula of alminosilicate zeolite is given by

$$M_{m/r} \cdot Al_m Si_n O_{2(m+n)} \cdot xH_2O, \quad (m \leqq n),$$

where $Al_m Si_n O_{2(m+n)}$ constructs the framework of zeolite, and M is an r-valent cation which is sitting at the site near the framework. A lot of water molecules are captured in the space of framework, when zeolite is kept in the air. The framework network is constructed by the sharing of O atom of SiO_4 or AlO_4 tetrahedra. The most typical cation is Na^+. We can easily change Na^+ to many other cations, such as K^+, Ca^{2+}, Cd^{2+}, Ag^+, etc. One hundred kinds of zeolite framework structure types are already known. Various types of zeolite can be synthesized in the different chemical composition with the common framework structure, because zeolites are synthesized in different Si to Al ratios with different kinds of cation. Zeolite with different chemical compositions of framework are synthesized, such as the $AlPO_4$ etc.

LTA is one of the most typical zeolites. In Fig. 2, LTA framework structure is shown schematically, where cations are neglected in the figure. Open and closed circles indicate O and (Si, Al) atoms, respectively. Cations are not shown here. A basic unit cage is called the β cage or the sodalite cage. The β cage has the effective inner size of 6 Å. The β cages are connected at 4-membered rings, and arrayed in a simple cubic structure. Among β cages, a new cage, called the α cage (or the supercage of LTA), exists. In other words, the α cages are connected by the sharing of 8-membered rings in a simple cubic structure. The α cage has the effective inner size of 11 Å. The 8-membered ring is called the window of α cage. In LTA, the Si to Al ratio is usually unity, and Si and Al atoms are alternately ordered by the sharing of the intermediate O atom. We denote Na type zeolite LTA with Si to Al ratio of unity as Na-LTA(1), hereafter. Zeolite LTA is sometimes called a different name, A, and the Na type is called NaA.

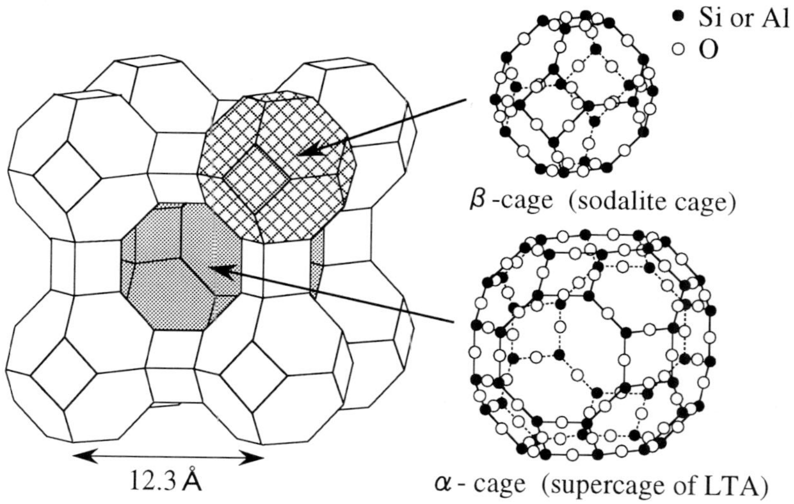

Fig. 2 Framework structure of zeolite LTA. Closed and open circles indicate (Si of Al) and O, respectively. The β cages are connected at the 4-membered rings in a simple cubic structure with the lattice constant of 12.3 Å. The α cage (or the supercage of LTA) is formed among β cages. In the figure, cations are not shown.

Besides LTA, there are FAU (faujasite) and SOD (sodalite) zeolites which have the β cage as the basic unit cage. In FAU, β cages are connected at 6-membered rings, and arrayed in a diamond structure. In SOD, β cages are connected by the sharing 4-membered rings, and arrayed in a body centered cubic structure. Zeolite MOR (mordenite) has a different type of framework, and has parallel channels with an inner diameter of about 7 Å.

3 Ferromagnetism of Free Electrons

We usually assume the free electron gas as the simplest model for alkali metals. In this model, electrons with parallel spin do not have the same quantum state due to the Pauli's exclusion principle. This lead to the exchange hole in the Hartree-Fock approximation. However, electrons with anti-parallel spin have no correlation in this model. According to the Bloch's calculation[1] based on the first order perturbation, free electrons show ferromagnetism at lower electron concentrations:

$$r_s > 5.4531, \qquad (1)$$

where r_s is the radius of the Wigner-Size cell. This is called the Bloch's ferromagnetism. The dilute electron condition is satisfied in Cs, but no ferromagnetism but Pauli s spin paramagnetism is observed.

In real free-electrons gas, electrons with anti-parallel spin have a finite Coulomb interaction. This interaction leads to the change in the motion and spin state of electrons. The effect caused by this interaction is called the electron correlation. The electron correlation effect can be included in the higher-order perturbation calculation. According to Pines[2] and Brueckner and Sawada [3], no ferromagnetism occurs at any electron density in the free electron metal. In conclusion, the electron correlation unstabilizes the ferromagnetism of free electrons. On the contrary to the above expectation, ferromagnetism has been observed in potassium clusters arrayed in zeolite LTA[4]. Magnetic element, such as transition metal, is not included in this material. This is the first observation of *s*-electron ferromagnetism. In potassium cluster in LTA, *s*-electrons are partly confined in the regular nanospace of clusters.

Some metals, such as Fe, Co and Ni, show itinerant electron ferromagnetism. The simplest model of the itinerant ferromagnetism is given by the Stoner model. The Stoner condition for ferromagnetism is given by

$$\frac{U\rho(E_F)}{N} > 1, \qquad (2)$$

Where U is the on-site Coulomb energy, $\rho(E_F)$ is the density of state at the Fermi energy, and N is the number density of site. In this model, the long range Coulomb interaction is neglected. The Stoner condition seems to be satisfied easily in many metals, but the effective value of U is known to be much smaller than the simply estimated value. According to the Kanamori's theory[5], itinerant electron ferromagnetism occurs at the special condition that both the wide band width and the high density of state at the Fermi surface are satisfied at the same time. According to this theory, the ferromagnetism of metal can be satisfied in the exceptional case. Usually, antiferromagnetism or Pauli's spin paramagnetism occurs in metals.

4 Potassium Clusters in Zeolite LTA

4.1 Optical Properties

When the guest alkali metal is adsorbed into completely dehydrated zeolite, the specific coloration occurs depending on the combination of zeolite and alkali metal. The color varies also depending on the loading density of guest metal. When Na is dilutely adsorbed into Na-Y, pink or red coloration occurs[6,7], where Y means the FAU type zeolite with higher Si to Al ratio. The electronic state in materials can be elucidated by the analysis of optical spectrum. In Fig. 3, the absorption spectrum is shown for the dilutely K-loaded into K-LTA(1)[8]. A significant absorption band appears at infra-red region 1.2 eV. There appears no absorption at visible and UV region. This band is assigned to the K cluster in α cage.

The effective potential for *s*-electron of K cluster in α cage is simplified as a spherical well. In Fig. 4, one-electron quantum states are illustrated for the spherical well potential with the infinite depth. The states named 1*s*, 1*p*, 1*d*, ··· have the degeneracy of 2, 6, 10, ···, respectively, as indicated in the each parenthesis. The absorption peak energy of alkali metal cluster has been well explained by assuming the size of the effective potential well for *s*-electron. The peak energy 1.2 eV in Fig. 3 can be explained by the 1*s* to 1*p* transition of electron with the effective radius of 11 Å. This size just coincides with the inner size of the α cage. Hence, we can conclude that K clusters are generated in the α cage. The reason why such a simple potential is available for the alkali metal cluster is ascribed to the Coulomb potential due to the many cations and negatively charged framework. In order to calculate a detailed potential of cluster, we need to know the cation position. The cation position of dehydrated K-LTA(1) has been determined by the x-ray diffraction analysis[9]. According to this analysis, K^+ cation occupies each center of 6-membered ring and 8-membered ring. The numerical calculation shows that the deeper potential for *s*-electron is realized in the α cage with the size of about 11 Å. However, the potential barrier between adjacent cage is finite. Especially, the barrier height is lowered at the center of 8-membered ring. In Fig. 3, the absorption spectrum has the fine structure. The structure can be explained by the electron transfer model between adjacent cages. The finite barrier height and open electronic shell leads to the metallic state of arrayed cluster.

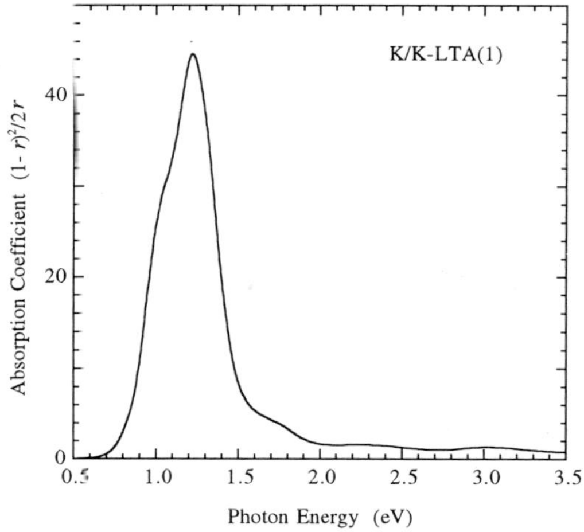

Fig. 3 Absorption spectrum of dilutely K-loaded K-LTA(1). A remarkable absorption band is observed at the peak energy of 1.2 eV. This is assigned to the K-cluster generated in the α cage of LTA with the effective size of 11 Å.

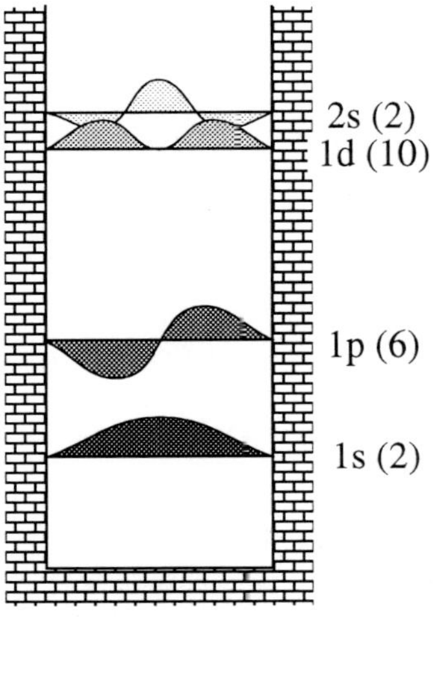

Fig. 4 One-electron quantum states calculated in the spherical well potential with the infinite depth. The states of 1*s*, 1*p*, 1*d* etc. appear in the increasing order of energy. The number in the parenthesis indicates the degeneracy including the spin. When more than one electrons occupy the state, the symmetry and energy should be calculated according to the total spin quantum number.

When the loading density of guest K metal is increased, the optical spectrum changes systematically[8]. When the number density of s-electrons increases, the optical response due to the collective motion, such as the surface plasmon oscillation, is observed in the reflection spectrum[8].

4.2 Ferromagnetism

In order to elucidate the intercluster interaction as well as the spin state of cluster, we have measured magnetic properties. When K metal is adsorbed into K-form LTA, a ferromagnetism has been observed[4]. The ferromagnetism strongly depends on the average electron concentration per cluster. As shown in Fig. 5, an *ac* magnetic susceptibility at 2 K increases and decreases in several orders as a function of the electron concentration per cluster. Because of the powder form of sample, the measured *ac* susceptibility is saturated at about 0.3 due to the demagnetization field. The compensated value is much larger than unity at the maximum. When we assume a paramagnetic cluster with the spin 1/2 in each α cage, *ac* magnetic susceptibility χ is in the order of 10^{-4} emu/cm^3 even at liquid helium temperature. The observed huge value is a manifestation of ferromagnetism. The observed *M-H* curve shows a very soft ferromagnetism.

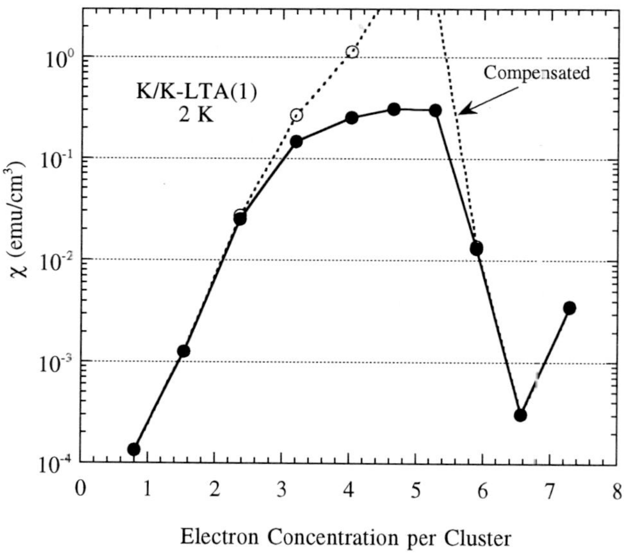

Fig. 5 *Ac* magnetic susceptibility χ of K loaded K-LTA(1) as a function of the average electron concentration per cluster. The value of χ increases and decreases in the several orders.

The magnetization at 1.7 K under the external magnetic field of 100 Oe is plotted as a function of the electron concentration per cluster in Fig. 6[4]. The magnetization increases above the electron concentration of 2, and has the peak around 5 and decreases at higher concentrations. The electron concentration at which a significant magnetization is observed is in the region between 2 and 8 electrons per cluster. This region is corresponding to the configuration that the 1p state is partly occupied with electrons. The electron concentration showing the peak magnetization is around 5 per cluster. This value just coincides with the two electron for 1s state and three electrons for 1p state. In this electron configuration, the maximum spin state is expected, because of the Hund rule of $1p$-state electrons. The maximum magnetic moment per cluster is estimated to be $0.24\mu_B$ from the maximum magnetization 1.17 G. The increase and the decrease in the magnetization in Fig. 6 is quite similar to the Slater-Pauling curve for $3d$-transition metal alloys.

The Curie temperature also strongly depends on the average electron concentration. The maximum Curie temperature is about 8 K.

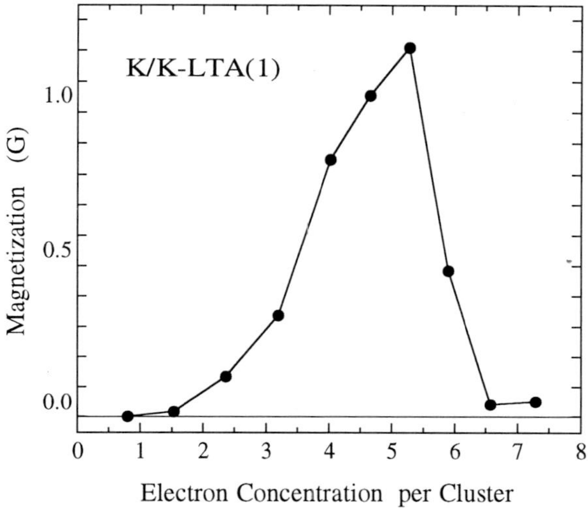

Fig. 6 Magnetization of K loaded K-LTA(1) at 1.7 K at the external magnetic field of 100 Oe. Magnetization increases and decreases depending on the average electron concentration per cluster. The peak is seen around 5 electrons per cluster which value just coincides with the half-filling of $1p$ quantum state of cluster: two electrons for $1s$ state and three electrons for $1p$ state.

The temperature dependence of *ac* susceptibility partly shows a spin-glass phenomenon, but basically shows Curie-Weiss law like behavior. From the Curie constant, the local magnetic moment is estimated to be about $1\mu_B$ per cluster. This value is much larger than the value estimated from the magnetization. This is common behavior of itinerant electron ferromagnetism. The relation between average magnetic moment and the local moment estimated from the Curie constant is one of the important problem of the itinerant ferromagnetism.

Chowdhury and Nasu[10] have pointed out that, antiferromagnetic phase is usually more stable than ferromagnetic phase at the just half-filled condition, i.e. 5 electrons per cluster, according to the tight-binding model of clusters. When we assume only the nearest neighbor transfer term of electrons, the nesting of the Fermi surface makes antiferromagnetic phase stable. In Fig. 6, there appears no indication of antiferromagnetic phase, but ferromagnetic phase. They explained the stability of ferromagnetism at the just half-filled condition within the mean-field approximation by assuming the second nearest neighbor transfer terms, especially its off-diagonal terms.

When Na clusters are generated in Na-form LTA, diamagnetism has been observed at any electron concentration[11,12]. This is explained by the electron-phonon interaction. The cation displacement in Na cluster may deepen the electronic potential which overcomes the repulsive interaction between electrons. The essential difference resides in the stronger ionization potential of Na atom, which makes the electron-phonon interaction intense. As the result of the strong electron-phonon interaction, electrons in zeolite cages make a pair of up- and down-spin electrons in the $1s$ state. Even at the electron concentration more than 2, observed magnetism is diamagnetic. This is ascribed to the Jahn-Teller effect on the $1p$ state[11,12]. This behavior is essentially similar to those expected in free cluster with 4 electrons[13,14]. In Rb clusters, ferromagnetism has been observed, but the Curie temperature is slightly lower than that observed in K clusters[11,15].

5 Na Clusters in Zeolite LTA

5.1 Photochromism

When Na is loaded into Na-form LTA, only diamagnetic Na clusters are generated. In the diamagnetic cluster, even number of electrons are paired to have the singlet state. The strong repulsive interaction, however, exists between two electrons. In this case, some cages are empty for cluster. This can be called bipolaron glass. This pairing is stabilized by the electron-phonon interaction. When the Na loading density is low, optical absorption bands named D_1 and D_2 have been observed at 3.1 and 2.8 eV, respectively[16]. These bands are assigned to the Na cluster generated in β cage. When the Na loading density increases, Na clusters are generated in α cage[16]. Na clusters in α cage also show diamagnetism[12].

Absorption spectra are shown for dilutely Na loaded Na-LTA(1) in Fig. 7[17]. Solid and dotted curves are the spectra before and under the irradiation at 3.2 eV, respectively. Under the irradiation, absorption bands D_1 and D_2 are weakened, and photochromic bands C_1 and C_2 appear at 2.0 and 1.65 eV, respectively. The integrated absorption intensity of these spectra is conserved. The origin of bands C_1 and C_2 is assigned to the common cluster, because the relative intensity of C_1 to C_2 is independent of the irradiation energy and its integrated intensity.

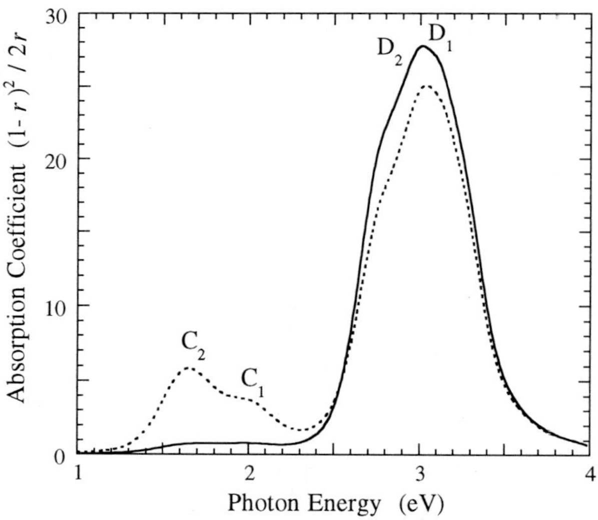

Fig. 7 Absorption spectra of dilutely Na-loaded Na-LTA(1). The solid curve is for original sample, and dotted curve under the light irradiation at 3.2 eV.

A lot of Na cations are distributed in the space of framework before the adsorption of Na atoms. Generally, a potential for s-electron of cluster is given by the attractive force of Na cations and the repulsive force of the negatively charged framework. The excitation energy of alkali metal clusters is well explained by the effective size of potential for s-electron[8,16]. If we assign the origin of D_1 and D_2 bands to the transition from the $1s$ orbital to the $1p$ orbital, the effective diameter of the spherical potential is estimated to be 6.5 Å. This value coincides well with the inner diameter of β cage.

The energies of bands C_1 and C_2 are lower than those of bands D_1 and D_2. Usually, the energy difference between quantum electronic states is reduced by the increase in the effective size of potential. Hence, the lower excitation energy of bands C_1 and C_2 is ascribed to the larger size of cluster potential. This cluster is expected in α cage which is larger than the β cage. According to the numerical calculation, however, the excitation energy is estimated to be 1.2 eV in the spherical potential with the diameter of 11 Å. This is true in case of K cluster. The excitation energy of band C_1 or C_2, however, is much higher than those of K cluster. Hence, the effective confinement size of the cluster of bands C_1 and C_2 is smaller than the inner size of α cage but larger than the inner size of β cage. The splitting of the bands C_1 and C_2 is due to the lower symmetry of the cluster, as discussed later.

5.2 Photoinduced ESR spectrum

Curves (a) and (b) in Fig. 8 indicate the first and the second derivatives of photoinduced ESR spectra in Na loaded Na-LTA(1), respectively, under the white light irradiation of Xe-lamp. The light irradiation is effective only at D_1 and D_2 bands region. In curve (a), we can see hyper-fine-structures (HFS) with super-hyper-fine-structures (SHFS). The similar paramagnetic clusters are observed before the irradiation. This result suggests that dilute paramagnetic clusters are already distributed thermally in the sample before the irradiation. As marked by closed circle in curve (b), four lines of SHFS are separated by the period of 5.8 ± 0.2 G. The HFS is consisted of ten lines separated by the period of 37.2 ± 0.1 G. The curve (c) indicates the simulation spectrum according to the following model.

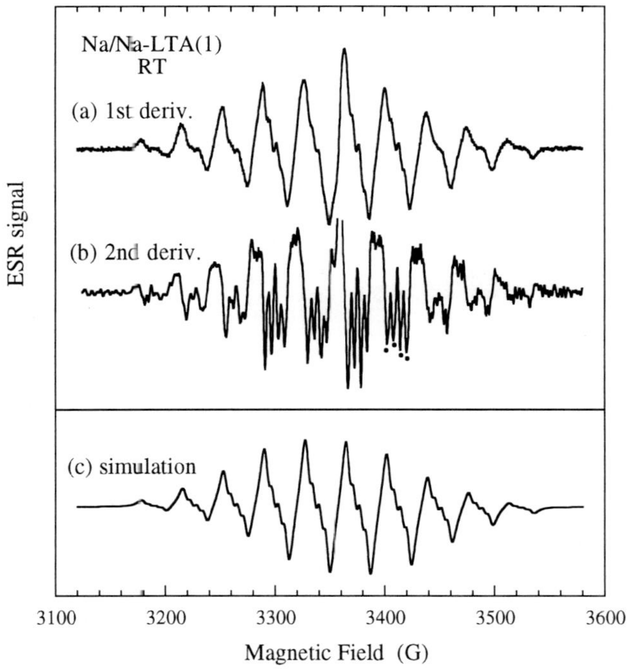

Fig. 8 Photoinduced ESR spectrum of Na loaded Na-LTA(1). Curves (a) and (b) are 1st and 2nd derivative, and curve (c) is the result of simulation.

The nuclear spin of Na is 3/2. When we assume three nuclear spins with the equivalent contribution, the HFS consists of ten lines whose relative intensity is 1:3:6:10:12:12:10:6:3:1. When we assume an extra nuclear spin 3/2 with the small contribution, each HFS line can have the four SHFS with the equivalent intensity. The simulation spectrum shown by curve (c) in Fig. 8 well coincides with the observed photoinduced spectrum of curve (a). According to this simulation, the photogenerated paramagnetic cluster consists of three equivalent Na cations with an extra one. Therefore, the most probable cluster is Na_4^{3+} with the C_{3v} symmetry. By the way, an ESR signal with the HFS of thirteen lines has been observed in NaY. This cluster is assigned to Na_4^{3+} cluster with the T_d symmetry, and four nuclear spins have the equivalent contribution[18, 19]. The present photogenerated cluster with the C_{3v} symmetry is different from above one.

5.3 Structure of photogenerated cluster

According to the structural analysis of dehydrated Na-LTA(1) by the X-ray diffraction, an Na cation is located near the each center of 8-membered ring and at the each center of 6-membered ring[9]. As mentioned above, the ESR active cluster is expected to have the C_{3v} symmetry. The three fold rotational symmetry axis can exist along the [111] direction of LTA framework structure. The photogenerated cluster has the same rotational axis, where an extra Na cation exists on the [111] axis in order to conserve the three-fold rotational symmetry. The most plausible position of the cations and the *s*-electron wave function are illustrated in Fig. 9. In the figure, oxygen atoms of the framework are omitted. The Na cations which do not participate in the cluster are also omitted. The three equivalent Na cations are located near the center of 8-membered ring. The main part of *s*-electron wave function is captured by these three cations, and the small part is extended to the cation located near the center of 6-membered ring. Namely, the cluster is generated partly across the 6-membered ring. The cation at the 8-membered ring may be shifted to the center of cluster due to the captured *s*-electron.

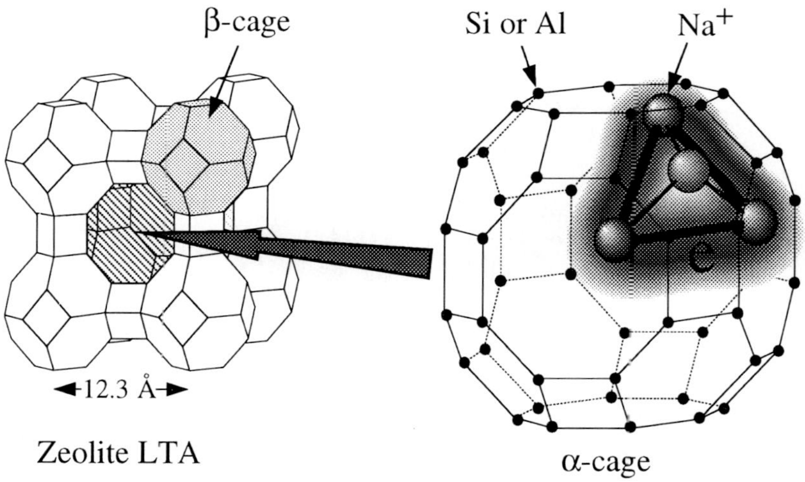

Fig. 9 Expected structure of photogenerated paramagnetic cluster with the C_{3v} symmetry. The rotational axis is along the [111] direction of LTA framework.

The Na loading density of the present sample is estimated to be about 0.2 per unit cell. The loading density is equivalent to that of *s*-electron. The observed spin density, however, is estimated to be of the order of 10^{-3} per unit cell. This value is much smaller than the *s*-electron density. Hence, the most of s-electrons are in the spin singlet state due to the pairing of two electrons, and become ESR silent. The appearance of HFS of ESR spectrum means that the observed electron spin is well localized within a paramagnetic cluster. It is elucidated also from the magnetic susceptibility measurement that the major part of Na clusters are diamagnetic at any loading density of Na[12].

A small part of paramagnetic clusters are already distributed before the irradiation. These paramagnetic clusters may have the energy slightly higher than that of diamagnetic cluster, and are excited thermally. The occurrence of the single component of photoinduced ESR signal suggests that the excitation of the β cage clusters finally leads to the generation of the unique paramagnetic clusters. If one electron is transferred from β cage to α cage, the other electron is left behind in β cage, which means the generation of new paramagnetic cluster in β cage. However, no extra-ESR signal is observed for such paramagnetic cluster. Hence, the one-photon excitation of the β cage cluster leads to the generation of two equivalent paramagnetic clusters in α cages.

In order to understand the above situation, we consider the stability of the diamagnetic cluster in β cage. It is concluded that the electron-phonon interaction overcomes the electron-electron repulsion interaction in the diamagnetic Na cluster in β cage. This mechanism is simply interpreted as follows. If the first electron occupies the β cage, the electron is shared with some Na cations, because the distance between Na cations are short enough to be overlapped in their 3s atomic orbital. The electron causes the displacement of Na cations due to the attractive interaction between them. This displacement makes the potential deep for 3s-electron, and the total energy can be minimized. However, the paramagnetic cluster is unstable when the second electron occupies the β cage. If the two electron occupy the β cage, the potential becomes much deeper than the single electron potential, because of the further displacement of Na cations. If the gain by such electron-phonon interaction overcomes the electron-electron repulsive interaction, the diamagnetic cluster can be stabilized. Finally, a diamagnetic cluster can be more stable than two independent paramagnetic clusters.

According to above discussion, the reason why unique paramagnetic clusters are generated by the light irradiation is ascribed to the instability of paramagnetic cluster in β cage. On the other hand, in NaY, Na_4^{3+} paramagnetic cluster has been observed as the most stable state. This may be due to the lower cation density. In K clusters in K-LTA(1), diamagnetic cluster can not be stabilized, because the electron-phonon interaction is weaker that that in Na cluster[11].

If the cluster has the C_{3v} symmetry, the triply degenerated 1p-like orbital splits into two orbitals, one of which is doubly degenerated. The ratio of the optical transition probability from 1s orbital to two 1p-like orbitals is 1 to 2. The intensity ratio of C_1 to C_2 well coincides with this model.

6 Summary

Alkali metal clusters show different optical and magnetic properties depending on the alkali metals in zeolite LTA. In K clusters, a metallic state is realized. They shows itinerant ferromagnetism of s-electron at liquid He temperature. In Na clusters, diamagnetism is observed. The paramagnetic clusters are generated by the photoexcitation of diamagnetic cluster generated in β cage. This is a manifestation of the unstability of diamagnetic Na cluster, where two electrons are paired to have the singlet state, although the Coulomb repulsion exists between them.

References

[1] F. Bloch, Z. Phys. 57 (1929) 545.

[2] D. Pines, Phys. Rev. 95 (1954) 1090.

[3] K. A. Brueckner and K. Sawada, Phys. Rev. 112 (1958) 328.

[4] Y. Nozue, T. Kodaira, S. Ohwashi, T. Goto and O. Terasaki, Phys. Rev. B**48** (1993) 12253.

[5] J. Kanamori, Prog. Theor. Phys. 30 (1963) 275.

[6] P. H. Kasai, J. Chem. Phys. 43 (1965) 3322.

[7] J. A. Rabo, C. L. Angell, P. H. Kasai and V. Schomaker, Disc. Faraday Soc. **41** (1966) 328.

[8] T. Kodaira, Y. Nozue, S. Ohwashi, T. Goto and O. Terasaki, Phys. Rev. B**48** (1993) 12245.

[9] R. Y. Yanagida, A. A. Amaro and K. Seff, J. Phys. Chem. **77** (1973) 805.

[10] A. Z. Chowdhury and K. Nasu, J. Phys. Chem. Solids, **56** (1995) 1193.

[11] Y. Nozue, T. Kodaira, S. Ohwashi, N. Togashi, T. Monji and O. Terasaki, Stud. Surf. Sci. Cat. **84** (1994) 837.

[12] T. Kodaira, Y. Nozue, S. Ohwashi, N. Togashi and O. Terasaki, Proc. 7th Int. Symp. Small Particles and Inorganic Clusters (ISSPIC7), 1994, Kobe, to be published in Surf. Rev. Lett.

[13] B. K. Rao, S. N. Khanna and P. Jena, Phys. Rev. B**36** (1987) 953.

[14] S. N. Khanna, B. K. Rao, P. Jena and J. L. Martins, *Physics and Chemistry of Small Clusters*, ed. P. Jena, B. K. Rao and S. N. Khanna, Plenum Publishing Corp. 1987, p. 435.

[15] Y. Nozue, T. Kodaira, S. Ohwashi, N. Togashi and O. Terasaki, Proc. 7th Int. Symp. Small Particles and Inorganic Clusters (ISSPIC7), 1994, Kobe, to be published in Surf. Rev. Lett.

[16] T. Kodaira, Y. Nozue, and T. Goto, Mol. Cryst. Liq. Cryst. **218** (1992) 55.

[17] T. Kodaira, Y. Nozue, M. Kaise, O. Terasaki, and H. Takeo, to be submitted.

[18] P. P. Edwards, M. R. Harrison, J. Klinowski, S. Ramdas, J. M. Thomas, D. C. Johnson and C. J. Page, J. Chem. Soc., Chem. Commun. (1984) 982.

[19] M. R. Harrison, P. P. Edwards, J. Klinowski and J. M. Thomas, J. Solid State Chem. **54** (1984) 330.

Conducting fulleride polymers

András Jánossy[1], Sándor Pekker[2], Gábor Oszlányi[2], László Korecz[3] and László Forró[4]

1. Institute of Physics, Technical University Budapest, H-1521 Budapest, Hungary
2. Research Institute for Solid State Physics, H-1525 Budapest, POB. 49, Hungary
3. Central Research Institute for Chemistry, H-1525 POB. 17 Hungary
4. Laboratoire de Physique des Solides Semicristallins, IGA, Departement de Physique, Ecole Polytechnique Federale de Lausanne 1015 Lausanne, Switzerland

Abstract

The alkali fullerides, $(AC_{60})_n$ with A= K, Rb, Cs are conducting crystalline polymers in which covalently bonded ions form parallel linear chains. The polymerization depolymerization transition at about 400 K is reversible. Unlike other alkali fulleride compounds the polymers are stable in air. Structural aspects are discussed together with electron and nuclear magnetic resonance and frequency dependent conductivity. At ambient temperatures all polymers are strongly correlated metals. $(KC_{60})_n$ remains metallic to low temperatures. $(RbC_{60})_n$ and $(CsC_{60})_n$ have a metal insulator transition below 50 K. Magnetic data indicate that these systems may have quasi one dimensional electronic structures and that the ground state is an ordered spin density wave state. The opposing view believes $(AC_{60})_n$ are 3D conductors like doped $(CH)_x$ and $(SN)_x$ which are anisotropic but are not quasi one dimensional conductors in spite of the linear chain polymeric structure.

Introduction

The fullerene molecule C_{60} [1] has a remarkable stability and it came as a surprise that the molecules can be rather easily bonded together by photopolymerization. It may be of even greater surprise that charged C_{60}^- ions polymerize spontaneously at relatively high temperatures. Maybe because it was so unlikely, it took some time after the discovery of the A_3C_{60} superconductors in 1990 and a number of other fulleride salts until the existence of the orthorhombic $(AC_{60})_n$ polymers (A=K, Rb, Cs) were identified in 1994. These polymers are stable below 350 K and do not react with air - a unique exception among alkali fulleride compounds. The synthesis of the $(AC_{60})_n$ polymers does not require special techniques. On the contrary, when e.g. the superconducting A_3C_{60} compounds are prepared care has to be taken to complete the reaction to avoid the formation of some polymer phase. A strong and narrow conduction electron spin resonance line easily detected in usual ESR spectrometers at ambient temperatures is the characteristic fingerprint of the polymer. The reversible polymerization - depolymerization transition at 400 K is also easily followed by the change in linewidth. The polymer is crystalline but the powder X ray diffraction lines of samples synthesized until present are broad due to the poor quality and/or small size of crystallites. A further difficulty in identifying the polymer phase is the complicated phase diagram and the rich variety of metastable phases.

Winter and Kuzmany [2] observed the monomer fcc KC_{60} compound above 350 K in the Raman spectrum and its structure was determined by Zhu et al.[3]. Janossy et al. suggested [4] that the low temperature phase of RbC_{60} is a metal at ambient temperatures. Chauvet et al.[5] showed that this phase is derived from the high temperature fcc phase by a cubic distortion in which the separation of C_{60} ions along the (110) direction is decreased by 10 % while separations in other directions are

Springer Proceedings in Physics, Vol. 81
Materials and Measurements in Molecular Electronics
Editors: K. Kajimura · S. Kuroda

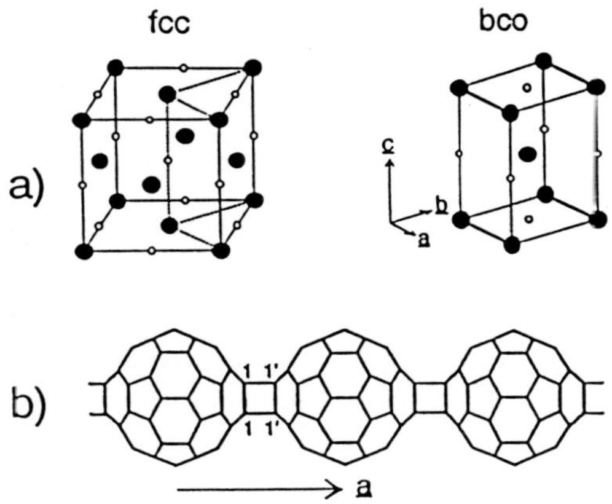

Figure 1. a.) Polymerization shortens one of the face diagonals of the high temperature fcc cell by 10 % and it becomes orthorhombic (bco). Solid circles: C_{60} ions; open circles: alkali metal ions. b) idealized structure of the polymer chain of $(AC_{60})_r$ compounds.

hardly changed (Fig.1a). The metallic nature of RbC_{60} was confirmed and the chainlike structure and ESR measurements of a phase transition below 50 K suggested a quasi one dimensional electronic structure. Pekker et al. [6] raised the possibility that the C_{60}^- ions are covalently bonded and that the orthorhombic phase consists of conducting linear polymers (Fig.1b). A refinement of the X ray diffraction patterns of RbC_{60} by Stephens et al. [7] and the fibrous nature and insolubility of KC_{60} prepared using a new coevaporation technique shown by Pekker et al. [8] proved finally the existence of the polymeric phase. An interesting recent development is the finding by Zhu [9] that under pressure the ternary compounds, Na_2RbC_{60} and Na_2CsC_{60} are also polymers.

As mentioned above, even the restricted phase diagram of A_xC_{60} compounds about $x \approx 1$ is complicated and we shall outline here some major features only. The fcc AC_{60} compounds are structurally similar to the high temperature fcc pure C_{60} phase. The C_{60}^- ions are rapidly reorienting and the alkali ions occupy the octahedral sites. At 400 K all of these sites are occupied. In a narrow temperature range above 350 K KC_{60} separates [2] into K_3C_{60} and almost pure C_{60}. No phase separation is observed in RbC_{60} and CsC_{60}. Polymerization takes place between 270 and 350 K. A moderately fast cooling prevents phase separation in KC_{60} and almost pure polymers are made with all three alkalis. A number of metastable phases are formed by various heat treatments e.g. a monomeric phase may be established at low temperatures by rapid cooling.

Role of electron correlations

The difficulty in describing the electronic properties of AC_{60} compounds arises from the large onsite Coulomb interaction compared to the relatively small overlap between C_{60} ions. Lof et al. [10] suggested that the ground state of an ideal, stoichiometric A_3C_{60} compound would be an insulator due to the splitting of the half filled conduction band by a Hubbard gap. This argument could be valid for $(AC_{60})_n$ polymers also since they have a half filled conduction band.

The monomeric fcc AC_{60} phase, the precursor of the polymer, is on the borderline of the metal insulator transition driven by electron localization. The high temperature stable form is an insulator

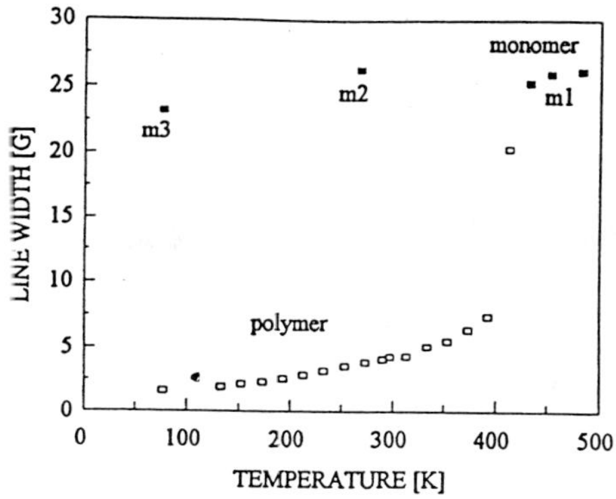

Figure 2. ESR linewidth vs. temperature in the polymeric $(RbC_{60})_n$ (open symbols) and monomeric RbC_{60} (full symbols) phases. The polymer is obtained by slowly cooling from 500K. The monomeric phases at high and low temperatures have slightly different cubic crystal structures which hardly affects the ESR linewidth. m1:, freely rotating C_{60}; m2: quenched from 500K to 77K and rapidly heated to 268 K, C_{60} reorientation hindered; m3: quenched from 500K to 77K; no C_{60} reorientation. The monomeric phase is metallic up to 268 K and has a small gap above 350 K.

since a Curie like spin susceptibility is observed above 400 K by NMR [11] and ESR [12]. The magnitude of the susceptibility corresponds to approximately an S=1/2 spin per molecular unit. The gap is small and electrical conductivities near 400 K of the semiconducting fcc AC_{60} and metallic $(AC_{60})_n$ polymers are not very different. A metastable monomeric sc AC_{60} (simple cubic) phase may be prepared by quenching from 500 K to low temperatures. Kosaka et al [13] found a temperature independent susceptibility showing that this phase is a metal. Fig.2 shows that the ESR linewidths of the monomer phase of RbC_{60} are about the same in the stable high temperature and in the metastable low temperature phases. However, in the metastable phase the spin susceptibility is nearly the same at 77 K and 268 K confirming that it remains a metal up to 268 K. The susceptibility is however quite large, more than that of the polymer. The structures of the fcc and sc phases are similar [14]. In the fcc phase C_{60} molecules are freely rotating while the sc phase is orientationally ordered. Thus a possible explanation for the metallic behaviour at low temperatures and a small gap insulator at high temperatures is a localisation by Coulomb interactions. In the high temperature phase the overlap, t between neighbouring ions may be smaller than in the sc phase since the separation of C_{60} ions is larger. Another factor is that the rotation averages the strongly orientation dependent overlap integrals. The unscreened onsite Coulomb interaction, U is the same for the fcc and sc phase and the qualitative difference in the paramagnetic susceptibility may show that U/t is somewhat too small to localise electrons in the sc phase.

Dimensionality of the electronic structure

One of the most interesting questions is whether the $(AC_{60})_n$ polymers are anisotropic conductors or have a quasi one dimensional (1D) electronic structure. The choice is between a 3D or quasi 1D metal, the structure does not allow 2D. The parallel running chains with an unfilled conduction electron band suggests a quasi 1D Fermi surface. In this case the ground state should be a three dimensionally ordered state with broken symmetry. Chauvet et al. [5] interpreted the transition in

$(RbC_{60})_n$ below 50 K as an instability against a spin density (SDW) or charge density wave (CDW) ground state. This would distinguish it from other conducting polymers like doped polyacetylene, $(CH)_x$ or polysulfur nitride, $(SN)_x$ in which the interchain overlap between parallel running conducting polymeric chains is strong enough to rend them quasi 2D metals. There is no low dimensional instability in these polymers, doped $(CH)_x$ remains a metal to very low temperatures while $(SN)_x$ is superconducting below Tc=0.3 K [15]. In the $(AC_{60})_n$ polymers there is no unambiguous evidence for a quasi 1D electronic structure. Band calculations of Erwin, Krishna and Mele [16] suggest that in $(AC_{60})_n$ polymers the interchain overlap is comparable to the overlap along chains in spite of the covalent bonds and thus they have a 3D electronic structure. No single crystals are available at present which would be suitable for a study of the anisotropy of physical properties, e.g. the electrical conductivity. No anisotropy was observed by Pekker et al. [8] in the transmission of light through fibers of a $(KC_{60})_n$ polymer grown by coevaporation. These fibers are not single crystals as originally assumed but have a rather complicated morphology [17] for which no optical anisotropy is expected.

It is too early to give a definite answer yet and in this paper we shall only try to give arguments pro and contra a quasi 1D electronic structure. We discuss the implications of the crystal structure, the magnetic and electrical properties.

sp^3 character of interball bonding carbon atoms

During polymerization the fcc monomer lattice contracts 10% as the polymer chains are formed while there are only small changes of the lattice parameters in other directions [18]. Bonds along the chains which link the balls are covalent and are only slightly larger than 1.5 Angstroms but they may have a dominantly σ character as it is the case for usual (2+2) cycloadducts [6]. The charge of the balls is redistributed and the electronic configuration of carbon atoms of the interball bonds changes from sp^2 to sp^3. Electronic structure calculations of dimers with the appropriate (2+2) cycloaddition bonding [23] show little difference between neutral and charged systems in the electronic structure at the bonding sites. It has been suggested by Pekker et al.[6] that the largest intrachain overlap is not between the bonding sites since the conduction band is formed from delocalised π electrons and the σ bonds do not allow for conduction. Next neighbour carbon carbon distances along the chain are similar to interchain nearest neighbour distances thus from the structure one cannot rule out that the polymers are anisotropic 3D metals.

Recent Magic Angle Spinning NMR studies [19,20] identified unambiguously the sp^3 carbon sites linking the balls in $(RbC_{60})_n$, $(CsC_{60})_n$ and $(KC_{60})_n$. The analysis shows negligible spin density for the sp^3 sites as expected for carbon atoms linked to the four neighbours by σ bonds. On the other hand, compared to usual planar molecules bonds are strained in the polymer. The refinement of the crystal structure [7] indicates longer interball bonds than expected for a pure σ bond and this favours the picture of a larger anisotropy in the electronic structure. The heat of polymerization is relatively small [21], and this also indicates a mixed character for the interball bonds.

Implications of band structure calculations

Band structure calculations of single chains [22-24] show very narrow bands. Calculations of Surjan et. al. [22] and Stafstrom et al. [23] find a doubling of the unit cell along the chain with an alternating larger and smaller charge near the interball bonds. Crystal fields in the polymer lift the three fold orbital degeneracy of the lowest unoccupied t_{1u} level of the free C_{60} molecule and the band filling is not 1/6 but 1/2. This clearly favours a CDW in the crystal with a doubling of the unit cell along the chains. The CDW gap is small, only 0.13 eV. Nearly dispersionless highest occupied bands is a common feature of the single chain calculations. Bandwidths of the order of 0.01 eV are found and -as pointed out by Tanaka et al. [24] - the onsite Coulomb repulsion would certainly rend such systems insulating even if the Peierls mechanism is too weak to open a gap.

Contrary to expectations from isolated single chains, the band structure of the 3D crystal calculated by Erwin et al. [16] within the local density approximation shows about the same band widths of 0.5 eV along and perpendicular to the chains. The band filling is -more or less accidentally- about 1/2 but of course there is no nesting wave vector and a transition to a CDW or SDW cannot occur. They suggest that the paramagnetic 3D metals at high temperature undergo a transition to a semimetallic ground state with a small but finite density of states at the Fermi level. The ground state in this model has a rather particular magnetic order with ferromagnetically aligned spins along the chains and an antiferromagnetic alignment for spins on neighbouring chains [25]. A rather large magnetic moment of $0.5\mu_B$ per ball is predicted. According to this calculation 1D behaviour would occur only in a diluted system where some kind of intercalant holds the chains apart.

Magnetic and electric properties

The physical properties of $(KC_{60})_n$ are rather different from those of $(RbC_{60})_n$ and $(CsC_{60})_n$ in spite of the rather similar crystal structures. The interball distances along the chain [18] are the same for all three, only the second neighbour interchain distances are slightly different, that of $(KC_{60})_n$ is the smallest. The microwave and optical conductivity in the far infrared of $(KC_{60})_n$ remains metallic to low temperatures [26] (Fig. 3). The resistivity decreases with temperature in a broad range and is probably determined by electron phonon scattering like in normal metals. The spin susceptibility measured by ESR and ^{39}K Knight shift is temperature independent. Thus most experimental evidence suggests that $(KC_{60})_n$ is a 3D metal. A small increase of resistivity observed below 50 K in the dc conductivity is the only exception. Hone et al. [27] interpreted this as a partial opening of the Fermi surface due to a charge density wave transition. In e.g. the quasi 1D conductor Nb_3Se_3 the CDW instability leads to a partial opening of the gap. 3D systems have no CDW instabilities and a partial

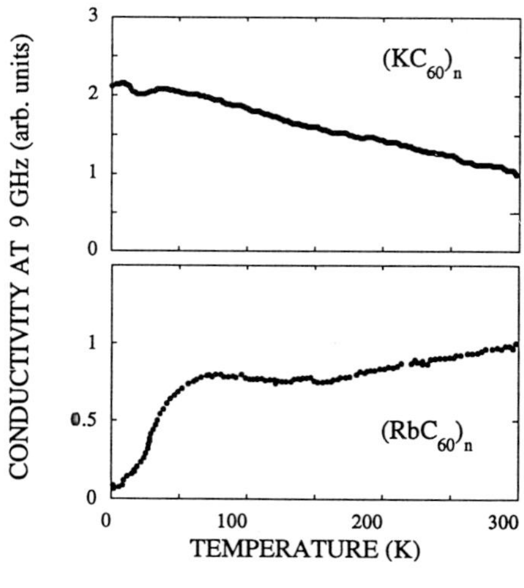

Figure 3. Microwave conductivity at 9 GHz of $(KC_{60})_n$ and $(RbC_{60})_n$ versus temperature normalized at 300K. $(KC_{60})_n$ has a metallic, $(RbC_{60})_n$ has an insulating ground state.

opening of a gap in $(KC_{60})_n$ would support the idea of a quasi 1D character. The corresponding decrease in the spin susceptibility has not yet been, however, observed.

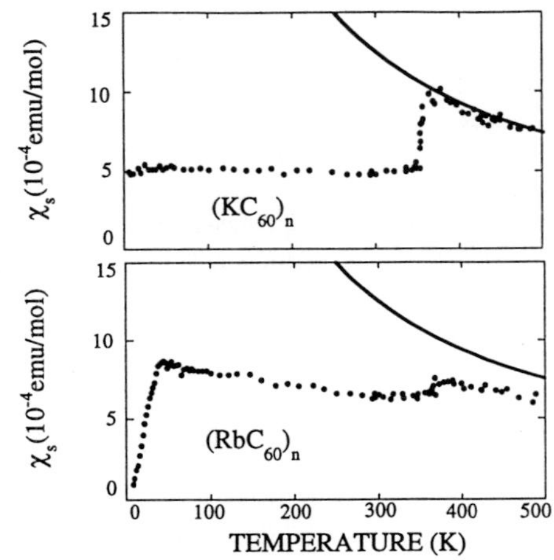

Figure 4. Spin susceptibility of KC_{60} and RbC_{60} measured by ESR. The high temperature fcc phase- polymerizes into $(KC_{60})_n$ and $(RbC_{60})_n$ below T_0 = 350K. The transition is reversible but has a large hysteresis. Data shown are in cooling. The solid line is the Curie dependence of a free S=1/2 system with 1 spin/mole. At low temperatures $(RbC_{60})_n$ magnetically orderes and the ESR data do not follow the static susceptibility.

At ambient temperatures $(RbC_{60})_n$ and $(CsC_{60})_n$ are conducting but unlike $(KC_{60})_n$ the conductivity slowly decreases with decreasing temperature. Above 50 K the spin susceptibility measured by ESR increases with decreasing temperature (Fig. 4). The MAS spectrum shows [19,20] the corresponding temperature dependent ^{13}C shift. An increase of the ^{13}C and alkali NMR spin lattice relaxation rates above 50 K has been assigned to antiferromagnetic fluctuations in a broad temperature range [28]. As shown in Fig. 3 for $(RbC_{60})_n$ below about 50 K the microwave conductivity decreases rapidly and the ground state is insulating. The metal insulator transition decreases the optical conductivity in the far infrared for energies below 10 meV [26]. Below 50 K the metal insulator transition is accompanied by a magnetic ordering. The gradual disappearance of the ESR, a shift and an increase of the linewidth were the first indications of the magnetically ordered ground state. These features resemble the spin density wave transition of the organic charge transfer salt $TMTSF_2PF_6$ [29]. The static spin susceptibility measured [30] by SQUID remains large at low temperature as expected for an antiferromagnetic transition. Small internal magnetic fields have been observed [31] by μSR at low temperature. A gradual transition is observed [32] by NMR also, in this case the internal fields are large and indicate large magnetic moments localized on C_{60} balls.

Effect of dimensionality on the CESR linewidth

The narrow CESR linewidth of $(RbC_{60})_n$ has been interpreted as the signature of a quasi 1D electronic structure. High quality $(RbC_{60})_n$ polymer has a 9 GHz CESR width of 4.9 G which decreases to 1.7 G at 77 K. This may be contrasted with the ambient temperature linewidth ΔH=550 G of fcc Rb_3C_{60}. In normal pure metals the Elliott mechanism of spin scattering on the random potential of defects or phonons is the dominant source of spin relaxation [33]. In this case the spin relaxation rate measured by the ESR linewidth and the momentum relaxation rate measured by the electrical resistivity are proportional. In the organic quasi 1D salts the conduction electron spin relaxation rate is long (and thus the ESR linewidths are narrow) because the Elliott mechanism is not effective in 1D [34]. Normal metals have much larger linewidths even at low temperatures due to spin orbit scattering on small concentrations of impurities. On the other hand, quasi 1D organic conductors like $Qn(TCNQ)_2$ or TTF-TCNQ have linewidths of 1 to 10 Gauss.

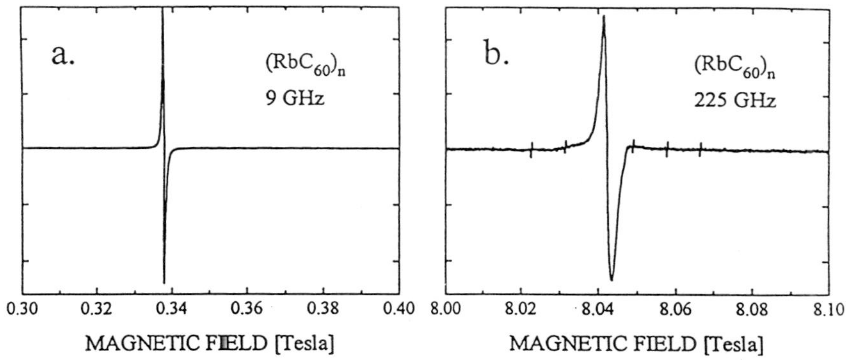

Figure 5. Conduction electron spin resonance spectrum of $(RbC_{60})_n$ at low and high magnetic fields a) resonance frequency f = 9.6 GHz; T=300K. b) 225 GHz; T=230K (The narrow six lines are Mn/MgO g-factor markers). Note the broader linewidth at 225 GHz.

The actual homogeneous linewidth of $(RbC_{60})_n$ may be even somewhat less than 1.7 G. At 225 GHz (resonance field 8.0 Tesla) the CESR width is ΔH=20G (Fig.5) at 230 K and decreases only slightly with decreasing temperature [35]. A plausible explanation for the field dependent linewidth is that the intrinsically large g-factor anisotropy of $(RbC_{60})_n$ is motionally narrowed by the fast momentum scattering but this narrowing is not complete. A homogeneous width of ΔH=1G is obtained from an extrapolation to zero magnetic field.

We emphasize that the 225 GHz lineshape is very different from that of the powder spectrum of insulators. In an orthorhombic system the g-factor has extrema in 3 different directions which give rise to characteristic peaks in the ESR powder spectrum. Although the linewidth of $(RbC_{60})_n$ at 225 GHz is dominated by the field dependent term the shape has no singularities.

Pressure dependence of the CESR linewidth of $(KC_{60})_n$ and $(RbC_{60})_n$

The idea that the 1D character of the electronic structure reduces the CESR spin relaxation rate is supported by the pressure dependence of the linewidth measured at ambient temperatures. As shown in Fig.6 the linewidth is increased with increasing pressure [36] and the increase is much stronger in $(RbC_{60})_n$ than in $(KC_{60})_n$. Under pressure the interchain separation is reduced and consequently the

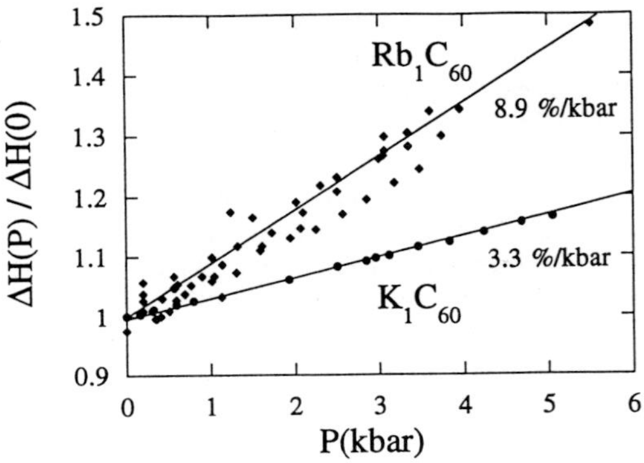

Figure 6. Pressure dependence of the CESR linewidth of $(KC_{60})_n$ and $(RbC_{60})_n$ at room temperature. The broadening of the linewidth with pressure is characteristic of the decrease of anisotropy in quasi-one dimensional conductors.

overlap perpendicular to the chains is enhanced. The reduction of the interball separation along the chains is most likely much less. The anisotropy of the thermal contraction is an indirect proof for this: as expected, the thermal contraction is much less along the chains than in other directions [18]. The increased perpendicular overlap reduces the anisotropy and the Elliott mechanism becomes more and more effective. A similar increase of the CESR spin relaxation rate with pressure is observed in some quasi 1D organic charge transfer salts [37]. The stronger pressure induced broadening in $(RbC_{60})_n$ supports the view that it has a more anisotropic electronic structure than $(KC_{60})_n$.

Conclusion

In spite of a large number of experiments in several laboratories the fundamental question of whether the $(AC_{60})_n$ polymers are quasi 1D or 3D conductors can not be yet answered. Band structure calculations and the sp^3 character of bonding carbon atoms are in favour of a 3D electronic structure. Some physical properties, in particular the enhanced NMR and conduction electron spin lattice relaxation rates as the metal insulator transition is approached and the small and pressure dependent linewidths support the idea that $(RbC_{60})_n$ and $(CsC_{60})_n$ are quasi 1D conductors.

Acknowledgements

Work in Budapest has been supported by grant OTKA T 015984. Work in Lausanne is supported by the Swiss National Science Foundation for Scientific Research.

References

[1]. H.W. Kroto, J.R. Heath, S.C. O`Brien, R.F. Curl and R.E. Smalley, Nature 318,162 (1985).
[2] J. Winter and H. Kuzmany, Solid State Commun., 84, 935 (1992).
[3] Q. Zhu, O. Zhou, J.E. Fischer, A.R. McGhie, W.J. Romanow, R.M. Strongin, M.A. Cichy and A.B. Smith III, Phys. Rev. B47, 13948 (1993);
[4] A. Janossy, O. Chauvet, S. Pekker, J.R. Cooper and L. Forro, Phys. Rev. Lett. 71 , 1091 (1993);

[5] O. Chauvet, G. Oszlanyi, L. Forro, P.W. Stephens, M. Tegze, G. Faigel, and A. Janossy, Phys. Rev. Lett. 72 , 2721 (1994);
[6] S. Pekker, L. Forro, L. Mihaly and A. Janossy, Solid State Commun., 90, 349 (1994);
[7]. P.W. Stephens, G. Bportel, G. Faigel, M. Tegze, A. Janossy, S. Pekker, G. Oszlanyi and L. Forro, Nature 370, 636 (1994);
[8]. S. Pekker, A. Janossy, L. Mihaly, O. Chauvet, M. Carrard and L. Forro, Science 265 ,1077 (1994);
[9] Q. Zhu, Phys. Rev. B52, R723 (1995);
[10] R.W. Lof., M.A. van Veenendalal, B. Koopmans, H.T. Jonkman, and G.A. Sawatzky, Phys. Rev. Lett. 68 , 3924 (1992);
[11] R. Tycko, G. Dabbagh, D.W. Murphy, Q. Zhu and J.E. Fischer, Phys. Rev. B52, R723 (1995);
[12] G. Oszlanyi, G. Bortel, G. Faigel, M. Tegze, L. Granasy, S. Pekker, P.W. Stephens, G. Bendele, R. Dinnebier, G. Mihaly, A. Janossy, O. Chauvet and L. Forro, Phys. Rev. B51, 12228 (1995);
[13] M. Kosaka, K. Tanigaki, T. Tanaka, T. Atake, A. Lappas and K. Prassides, Phys. Rev. B51 , 12018 (1995);
[14] A. Lappas, M. Kosaka, K. Tanigaki and K. Prassides, J. Am. Chem. Soc.117, 7560 (1995)
[15] R.L. Greene, G.B. Street and L.J. Suter, Phys. Rev. Lett. 34 , 577 (1975);
[16] S.C. Erwin, G.V. Krishna and E.J. Mele, Phys. Rev. B51, 7345 (1995);
[17] M. Carrard, L. Forro, L. Mihaly and S. Pekker, Mol. Cryst. Liq. Cryst., submitted
[18] G. Oszlanyi, G. Bortel, G. Faigel, M. Tegze, P.W. Stephens, and L. Forro, in Physics and Chemistry of Fullerenes and its Derivatives, H. Kuzmany, J. Fink M. Mehring and S. Roth, editors, World Scientific, p.323 (1995);
[19] T. Kalber, G. Zimmer and M. Mehring, Phys. Rev. B51, 16471 (1995); K.F. Thier, G. Zimmer, M. Mehring and F. Rachdi, preprint;
[20] H. Alloul, V. Brouet, E. Lafontaine, L. Malier and L. Forro, preprint;
[21] L. Granasy, T. Kemeny, G. Oszlanyi, G. Bortel, G. Faigel, M. Tegze, S. Pekker, L. Forro, and A. Janossy, Solid State Commun., 97, 573 (1996);
[22] P.R. Surjan and K. Nemeth, Solid State Commun., 92, 407 (1994);
[23] S. Stafström, M. Boman and J. Fagerstrom, Europhys. Lett. 30, 295 (1995);
[24] K. Tanaka, Y. Matsuura, Y. Oshima and T. Yamabe, Solid State Commun., 93, 163 (1995);
[25] E.J. Mele, G.V. Krishna and S.C. Erwin, Phys. RevB52, 12493 (1995);
[26] F. Bommeli, L. Degiorgi, P. Wachter, O. Legeza, A. Janossy G. Oszlanyi O. Chauvet and L. Forro, Phys. Rev. B51, 14791 (1995);
[27] J. Hone, M.S. Fuhrer, K. Khazeni, and A. Zettl, Phys. Rev. B52, R8700 (1995);
[28] V. Brouet, H. Alloul, Y. Yoshinari and L. Forro, in Physics and Chemistry of Fullerenes and its Derivatives, H. Kuzmany, J. Fink M. Mehring and S. Roth, editors, World Scientific, p.366 (1995);
[29] H.J. Pedersen, J.C. Scott, and K. Bechgaard, Solid State Commun., 35, 207 (1980);
[30] G. Baumgartner, H. Alloul, unpublished;
[31] Y.J. Uemura, K. Kojima, G.M. Luke, W.D. Wu, G. Oszlanyi, O. Chauvet and L. Forro, Phys. Rev. B52, R6991 (1995); W.A. MacFarlane, R.F. Kiefel, S. Dunsiger, J.E. Sonier and J.E. Fischer, Phys. Rev. B52, R6995 (1995);
[32] V. Brouet, H. Alloul, Y. Yoshinari, and L. Forro, preprint;
[33] R.J. Elliott, Phys. Rev. 96, 266 (1954);
[34] Y. Tomkiewicz, E.M. Engler and T.D.Schultz, Phys. Rev. Lett. 35, 456 (19975
[35] G. Oszlanyi, A. Janossy, S. Pekker, L. Forro (to be published)
[36] G. Baumgartner, O. Chauvet, A. Sienkiewicz, S. Pekker and L. Forro, (to be published);
[37] L. Forro, J.R. Cooper, G. Sekretarczyk, M. Krupski, K. Kamaras, J. Physique (1984);

Development of Superconductors based on C_{60}

Katsumi Tanigaki
Fundamental Research Laboratories, NEC Corporation
34-Miyukigaoka, Tsukuba 305

Abstract

Development in an alkali-metal C60 fulleride superconductor family is described. An overview of conductivity and superconductivity in these materials is given, focusing on the crystal structure and the behavior of conduction electrons in the system. Usual and unusual properties of this new class of materials are discussed when they are compared to the conventional metals and superconductors.

1. Introduction

When we look back the history of conductors and superconductors [1] from the material science point of view, the second generation superconductor family would be Nb_3M_{IV} (M_{IV} denotes Ge, Sn and Pb etc.) and V_3M_{III} (M_{III} denotes the Ga and Al etc.). Although these were new materials, the observed metallic properties and superconductivity with a critical temperature (Tc) of about 20 K were conventional. In 1970's $(Ba,K)BiO_3$ oxides were found and Tc was improved to 30 K. The situation was rapidly changed after the appearance of two copper oxides superconducting families, $La_{2-x}M_xCuO_{4-y}$ (M=Ca, Sr and Ba) with Tc's of about 40 K and $LnBa_2Cu_{7-\delta}$ (Ln=Y and/or other lanthanide's) with Tc's of about 90 K [2]. As a different approach to find out new conductors and superconductors, boron (B) system compounds have also been surveyed and Tc has been so far raised up to 23 K [3]. A unique $(SN)_x$ inorganic polymer superconductor was also found in 1975 [4]. Much differently from the trend of the research, a new approach on molecular conductors and superconductors started by being triggered by the idea of the exiton mechanism proposed by Little [5]. Charge-transfer type molecular compounds have been studied for realizing this idea and such efforts have led to a finding of the first $(TMTSF)PF_6$ (TMTSF: tetramethyl-tetraselena-fulvalene) organic superconductor followed by BEDT-TTF (BEDT-TTF: bisethylenedithiotetrathiafulvalene) system superconductors [6].

One of the most important issues in the course of development of molecular conductors and superconductors was concerned with dimensionality. Because of the low dimensionality, many organic conductors exhibited a metal-insulator transition at low temperatures because of the occurrence of CDW or SDW etc. before showing superconductivity. When C_{60} was found [7,8], it might have been natural that a surge of interests arose since the C_{60} crystal shows three-dimensionality as well as high symmetry [8]. Metallic properties in C_{60} were found when alkali-metal elements were doped into C_{60} solids in order to introduce electron carriers [9] and subsequently superconductivity with a Tc of 18 K was found in potassium doped C_{60} [10]. Since then doping by introducing various other elements, such as other alkali-metals, alkaline-earth metals, rare-earth metals, into C_{60} solids has been extensively studied for carrier injection and Tc has reached at 30 to 40 K so far [11,12]. Hole doping using electron accepting elements has also been studied, but it has not yet succeeded.

It should now be very important to think what the usual or unusual properties observed in C_{60}-based materials are, when they are compared to the conventional metals and the normal superconducting materials. This will give many important ideas into the future development of molecular (cluster-based) superconductors with higher Tc. Variety of experiments have been performed for this new family of materials in these five years after the discovery of high conductivity [1] and superconductivity [2] with relatively high-Tc in alkali-metal C_{60} fullerides. Some of them have been agreed by many scientists and some other ones have been denied. In this

Springer Proceedings in Physics, Vol. 81
Materials and Measurements in Molecular Electronics
Editors: K. Kajimura · S. Kuroda

article I would like to describe the specific features in alkali-metal C_{60} fullerides from the viewpoint of structural as well as metallic and superconducting properties, using reliable experimental data.

2. Basic Concept of Metals and Superconductors in C_{60} Fullerides

C_{60} is a cluster-type molecule consisting of sixty carbon atoms and has a spherical shape with I_h symmetry and radius of about 0.8 nm as shown in Fig.1. The HOMO level of the C_{60} molecule has five degenerate levels with h_u symmetry and the LUMO level with t_{1u} symmetry is triply degenerate. Since the five HOMO levels are completely occupied by ten electrons, C_{60} has a closed-shell electronic structure. The orbitals forming these levels are p-type and the electrons delocalize over the molecule. Therefore, in principle the properties of the C_{60} molecule are determined by the p-electron characteristics.

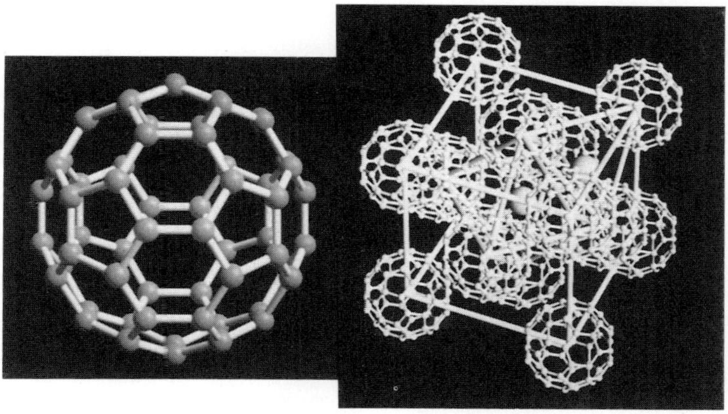

Fig.1 Structure of C_{60} and structure of fcc C_{60} solid.

Reflecting the fact that C_{60} is a closed shell molecule with a relatively large gap between HOMO and LUMO, the C_{60} solid is a van der Waals type crystal with face-centered cubic structure. In solids the h_u HOMO and t_{1u} LUMO levels form bands. The higher edge of the valence band consists of the h_u-derived levels and the lower edge of the conduction band is made of the t_{1u}-derived levels. The band gap of the C_{60} solid is about 1.8 eV and is categorized as a typical semiconductor.

Electron or hole carriers must be introduced into this solid for achieving electronically intriguing properties. Two methods are generally considered for carrier injection. One is replacement of some of C_{60} molecules in the lattice by electron-rich or electron-poor molecules. This is the same technique as that used in silicon technology. For example, B or P doping replaces some of Si thereby generating P-type and N-type semiconductors. As a candidate molecule for replacing C_{60}, BC_{59} and NC_{59} could be considered for hole and electron injection and endohedral materials such as $La^{3+}C_{82}{}^{3-}$ [13-15] could also be used. However, there are no such experimental reports at the present stage. The other promising method is intercalation like that used for graphite. Many different elements can be intercalated into the spaces of graphite layers, and one of the graphite intercalations KC_8 is known to be superconducting [16]. In contrast to graphite having a two-dimensional (2D) character, C_{60} solids have a three-dimensional (3D) character. Instead of the interlayer spacings of graphite, the 3D interstitial site spacings can be used for C_{60}. Such examples are actually reported in combination with alkali metals [9,10], alkaline earth metals [17,18] and some of other rare-earth elements [19,20].

Among various crystal phases reported for electron-carrier doping so far, the structure of the prototypical A_3C_{60} (A=K and Rb) fullerides is face-centered cubic (fcc) as shown in Fig.1. In this structure one electron is transferred from alkali metal element to C_{60}, resulting to a three minus-charged C_{60}^{3-} molecule. This is caused by the small ionization energy of A and the relatively large electron affinity of C_{60} [21,22]. In the lattice the C_{60}^{3-} ions adopt two different molecular orientations related by a 90° rotation about the <001> crystal axis (merohedral disorder) [23]. The A_3C_{60} fullerides can be considered to result from a three-dimensional (3D) intercalation process of fcc C_{60} solid, being compared to the two-dimensional (2D) intercalation process of graphite for which the first and the second stage intercalation compounds can generally be synthesized [24]. Considering C_{60} as an element, the A_3C_{60} compounds can be thought of as ionic salts comprising of C_{60}^{3-} and A^+ ions in the ratio of one to three. As it has been demonstrated by many studies, C_{60} ionic salts with a variety of crystal structures and different stoichiometries can be made using C_{60}^{n-} as a quasi element as shown in Fig.2. This results from the ease of reduction of C_{60} to many oxidation states, ranging from -1 to -6, principally because of the degeneracy of the molecular LUMO t_{1u} level.

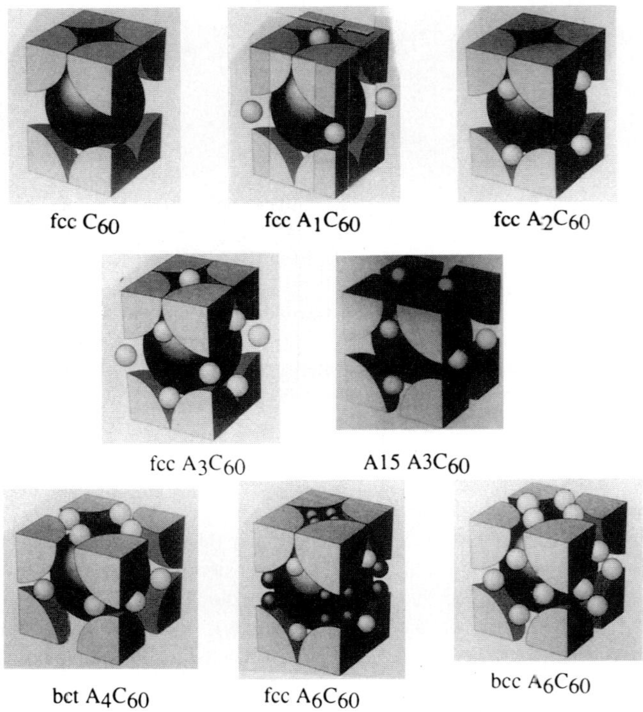

Fig.2 Various crystal phases found in C_{60} fullerides. In fcc A_6C_{60} the Na_4 cluster is accommodated in the larger octahedral site. A15 phase is found when alkaline-earth elements are intercalated. Cs_3C_{60} is classified into either A15 or cation-vacant type bct. Reproduced from reference 11 with modifications. Copyright 1992, Pergamon Press.

In the fcc pristine C_{60} solid, there are two types of unoccupied interstitial holes: the smaller tetrahedral (T-) (two per C_{60} unit) and the larger octahedral (O-) (one per C_{60}) sites. Depending on the relative occupancy of the T- and O-sites and ignoring orientational order of the fullerene units, four types of crystalline structures can result: (i) the rock salt type (A_1C_{60}) in which one alkali atom is accommodated in every O-site; (ii) the antifluorite type (A_2C_{60}) in which all T-sites are occupied; this structure is encountered in ionic compounds like CaF_2; (iii) when all T- and O-sites are fully occupied, fcc A_3C_{60} forms (cryolite structure); and (iv) when alkali atoms reside in half the available T-sites, the well known zinc-blend (ZnS) structure results; this has only been observed as a metastable Na_1C_{60} phase.

The fcc structure of pristine C_{60} is modified into additional structural types when more than three ions per C_{60} are incorporated. At present, there is a number of A_xC_{60} families synthesized with x=1-6, where A denotes alkali-metal [11,12], alkaline-earth metal [17,18] and some of the rare-earth metal (Yb and Er) elements [19,20]. The adopted structural type in A_xC_{60} is sensitively dependent on the number x as well as on the type of A used. For instance, when A= K, Rb or Cs are used in the reactions, body-centered tetragonal (bct) A_4C_{60} [25] and body-centered cubic (bcc) A_6C_{60} [26] phases form in addition to fcc A_3C_{60} [23]. On the other hand, when Na is used, the fcc structure persists to higher values of x. In this case, the most stable stoichiometries are those of x=2 and 6, the latter having the novel feature of a Na_4 cluster accommodated in the larger octahedral site [27]. Na intercalation has been shown to be possible up to x=11 [28]. For Cs and divalent Ba, an A15 phase forms for stoichiometries Cs_3C_{60} [29] and Ba_3C_{60} [30], in addition to the stable bct A_4C_{60} [25] and bcc Cs_6C_{60} [26] and Ba_6C_{60} [18] phases.

It should be noted that for alkaline-earth metal elements, superconductivity is encountered at higher concentrations than those in alkali-doped fullerides, namely for primitive cubic Ca_5C_{60} [17], bcc $(Sr,Ba)_6C_{60}$ [18] or orthorhombic Ba_4C_{60} [31]. As for Yb and Er, conducting phases form around the stoichiometry of three metal ions per C_{60} [19,20] and superconductivity is reported for Yb intercalation [19]. In the case of hole-carrier doping, iodine was first tried and found to form either I_2C_{60} or/and I_4C_{60} [76,77]. However, any intriguing metallic/superconducting or even more magnetic properties are not observed from these fullerides.

3. Usual and Unusual Features in A_3C_{60} Fullerides

3.1 Stability and lattice parameters of A_3C_{60}

It has been clarified, from NMR studies [32] focusing on nuclear spins of alkali metals and detailed X-ray diffraction analyses [33], that the site selectivity of alkali metals is generally observed. That is the larger cations are accommodated in the O-sites and the smaller ones in the T-sites. Considering the above simple picture of fcc A_3C_{60} and the experimental results so far obtained, the following conclusions for the phase stability of A_3C_{60} fullerides can be deduced:

[i] The stability of the fcc A_3C_{60} phase is increased by the O-site occupation with larger alkali-metals as follows: $A(T)_2Cs(O)C_{60} > A(T)_2Rb(O)C_{60} > A(T)_2K(O)C_{60}$.

[ii] The T-site occupation with Cs^+ (r+=1.70 Å) makes the fcc A_3C_{60} phases unstable. Actually no fcc stable phase cannot be obtained for $Cs(T)_2Cs(O)C_{60}$ and $Na(T)Cs(T)Cs(O)C_{60}$ and the $K(T)Cs(T)Cs(O)C_{60}$ is unstable in the case of a standard synthesis method. The only exception can be found for $Rb(T)Cs(T)Cs(O)C_{60}$. In this fulleride, the Rb^+ accommodated in one of the T-sites expands the other T-site so that it can be filled by the Cs^+ cations without too much loss of stability. This composition gives the highest Tc of 33 K found so far in the C_{60} superconductor family.

[iii] In the case of Na- and Li-containing fullerides, $Li(T)_2A(O)C_{60}$ and $Na(T)_2A(O)C_{60}$ are the predominant stable phases, since the occupation of O-site with either Li^+ or Na^+

with small ionic radii leads to instability in the fcc A_3C_{60} phase.

From this simple argument as well as the geometrical handling of A_3C_{60}, we also expect that the lattice parameters will be controlled by the ionic radius of alkali cations occupied in the T-sites [34]. In order to check this the lattice constants in A_3C_{60} were plotted as a function of ionic radius of alkali metals accommodated in the T-site as shown in Fig.3. A good correlation can be seen as expected. It should also be noted that the reported ammonium doped fulleride $Na(NH_3)_4(O)Na(T)Cs(T)C_{60}$ [35] also lies on the line depicted in this figure, supporting that the consideration above is generally valid.

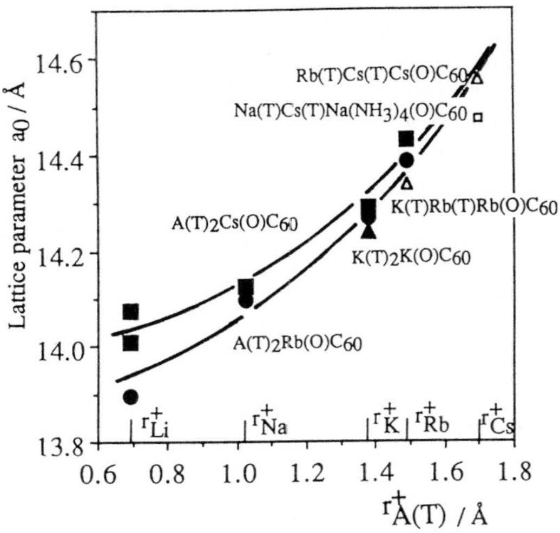

Fig.3 The lattice parameters of a_0 as a function of the alkali-metal ionic radius in T-site. The lattice size is well parametrized by the alkali-metal ionic radius in T-site with little influence of it in O-site.

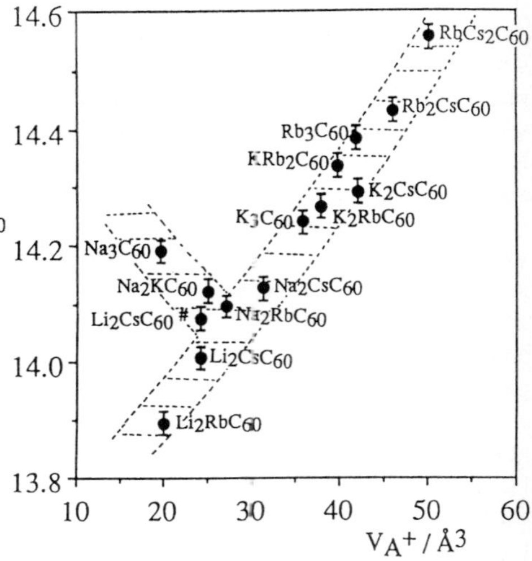

Fig.4 The relationship between the cubic lattice constant a_0 and the total volume (V_{A^+}) of the intercalated alkali cations in A_3C_{60}. The fcc cell is taken as an equivalent bct cell with $a_{bct}=a_{fcc}/\sqrt{2}$.

A large variation in lattice parameter for A_3C_{60} fullerides can be achieved by changing the type or the combination of alkali metals. Such a lattice parameter change is depicted in Fig.4 as a function of total volume of the alkali cations used [27,34]. Generally we can see a good correlation between these two parameters. However, if we glance at this picture a little bit more carefully, the following things should be noted. First, stable fcc phases of both Cs_3C_{60} and Li_3C_{60} are not formed by mixing stoichiometric amount of these alkali-metal elements with C_{60} using the conventional thermal diffusion methods. A similar situation can be seen for fcc KCs_2C_{60}, which can be made but the crystal quality is extremely poor and the fraction of the obtained fcc phase is small. It should be noted that a success has been recently made on Cs_3C_{60} by suppressing the synthesis temperature below 100 K using ammmonium solution [29], although its structure is not fully determined. Second, although the lattice parameters seem to decrease monotonically as the total cationic volume of the used alkali metals decreases, a significant deviation can be observed for Na_3C_{60} [27]. This deviation starts from Na_2KC_{60} as can be seen also in this figure.

3.2 Theoretical Consideration

Many band structure calculations have been carried out in order to understand the electronic states of A_3C_{60} [36-38]. Some important results when three alkali-metal elements per C_{60} are introduced are summarized here. First, almost a full electron per alkali metal atom is transferred from the s-orbital to the t_{1u}-derived bands of C_{60} solids, i.e., three electrons per C_{60} molecule. Consequently, no considerable distribution of conduction electrons is observed around the A elements. The lower part of the conduction bands is roughly half-filled with these electrons, resulting in a metallic ground state for A_3C_{60}. Second, the band originating from the s-orbitals of the alkali metals is located far above (about 2 eV) the Fermi level of the t_{1u}-conduction bands. Hence the metallic properties of A_3C_{60} can be best described as being primarily governed by the solid C_{60} states, while alkali cations have little influence and their role is only to supply three electron carriers per C_{60}. Third, when the C_{60} lattice expands upon the introduction of alkali atoms with large ionic radii, density of states at the Fermi level increases through band narrowing. This has most confidently been confirmed by ^{13}C-NMR measurements [39,40] and is now believed to be the most important parameter for controlling T_c in A_3C_{60}.

In the framework of the phonon-mediated BCS formalism, Tc can simply be expressed by the BCS relation:

$$Tc = 1.13h/2\pi <\omega>/k_B \, Exp[-1/(N_{Ef}V)] \; .$$

Here, $<\omega>$ is the cut-off frequency of related phonons and V denotes the coupling constant between phonons and electrons. The features of C_{60} based superconductors arise in principle from the very weak overlap of the wave functions of the C_{60} molecules. The width of the conduction band derived from the molecular t_{1u} levels is rather narrow and, as a consequence, N_{Ef} is large and the intramolecular phonons have little dispersion leaving high frequency modes. These have provided researchers with a very successful guideline in the attempts to raise the Tc of A_3C_{60} metallic fullerides [11,12] by lattice expansion through the increase of the ionic radii of the ions in the two types interstitial sites (O- and T-sites) of C_{60} solid. Hereby the higher Tc in C_{60} based superconductors has simply been sought for alkali-metal C_{60} fulleride metals by changing the alkali-metal dopants that can be accommodated in the interstitial T- and O-sites.

3.3 Experimental Evidences of Conventional/Unconventional Features

Almost all experimental studies employed so far seem to support such a typical metallic behavior for A_3C_{60} (especially K_3C_{60} and Rb_3C_{60})--- Photoemission spectroscopy shows the existence of density of states (N_{Ef}) at the Fermi level for the stoichiometry A_3C_{60} [41,42]; far-infrared reflection spectroscopy shows a Drude behavior [43,44]; thermoelectric power measurements displays a linear dependence on temperature essential for metals [45]; ^{13}C-NMR spectroscopy shows the Korringa relationship, $1/T_1T$=constant, consistent with metallic behavior [39,40], although the three band configurations arising from the three symmetry-inequivalent carbon atoms of the C_{60} molecule might have to be taken into account for detailed interpretation of the data [39,46]; SQUID measurements depict a Pauli-like behavior in the magnetic susceptibilities [47]. These results are generally consistent with and support the band structure calculations [36-38].

The relationship between T_c and lattice parameter (a_0) is depicted in Fig.5. As it has been already exemplified, T_c varies from 2.5- 3.5 K (Na_2KC_{60}, Na_2RbC_{60}) [48,49] to 33 K ($RbCs_2C_{60}$) [50], the latter being only surpassed by the high-T_c copper oxide superconductors. In the first glance we can see a simple picture in superconductivity, in this figure, that Tc is

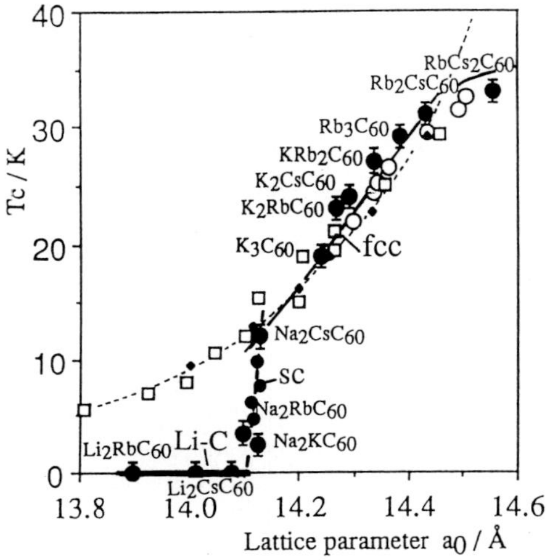

Fig.5 The relationship between the superconducting transition temperature T_c and the cubic lattice constants a_0 of A_3C_{60} (A=Li, Na, K, Rb, Cs and their mixtures) salts over a wide range of a_0. Data indicated by open and closed circles are experimental data. The solid line is the fitted curve to the both sets of data. The open triangles and squares are the relationships for K_3C_{60} and Rb_3C_{60} obtained high pressure experiments and the dotted line is the T_c-a_0 relationship expected from the simple BCS theory using N_{Ef} values by LDA calculations. Three classifications are seen in this figure: fcc A_3C_{60} in the large lattice size region, sc A_3C_{60} in the small lattice size region where some of them show extremely low Tc's and non-superconducting Li_2AC_{60} fullerides.

mainly controlled by lattice expansion since the density of states at the Fermi level increases through band narrowing [51,52] when the C_{60} lattice expands upon the introduction of alkali atoms with larger ionic radii. This also implies that the appearance of superconductivity in C_{60} fullerides has been most reasonably interpreted in terms of electron-phonon coupling to the high frequency intramolecular vibrations which is nearly invariable as a function of temperature.

In order to confirm whether the scenario of the superconducting mechanism described earlier is valid, physical parameters such as change in density of states (DOS) at the Fermi level and influence of phonons must be studied in detail.

In order to obtain the relative DOS values among the A_3C_{60} superconductor family, the most reliable data could be estimated from ^{13}C-NMR since in this method is needed only one parameter of the average electron density at a carbon nucleus for an orbital at the Fermi level that is the proportional constant connecting experimental T_1T values to N_{Ef} [39,40,53]. On the other hand, in both SQUID and ESR the fraction of metallic phases is essentially needed, which is difficult to estimate and varies depending on the preparation conditions as well as the type of alkali-metals used. ^{13}C-NMR studies show that N_{Ef} values increase as the lattice expands, supporting the scenario that N_{Ef} is the most important parameter for controlling Tc in A_3C_{60}. A similar conclusion can be obtained from thermoelectric power measurements [45] and optical reflection spectroscopy [43].

The information of phonons can be studied directly by either raman [54,55] or neutron [56,57] spectroscopy. Significant changes in the phonon peak positions and intensities are observed principally for the intramolecular Hg vibrational modes, both in the high-energy tangential (130-200 meV) and the low-energy radial (~50 meV) regions. Furthermore, the coupling strength (V) between phonons and electrons can be reduced from the phonon linewidth changes, being reported to be 45.6 meV for K_3C_{60}. These observations, when they are combined with the observed Tc, the renomarized Coulomb interaction parameter ($\mu^*=0.15$) and the density of states (N_{Ef}=14 states eV^{-1} mole C_{60}^{-1}), lead to the large average mean-phonon-frequency as high as 1000 K [56,57].

Information related to the associated phonons can also be available from the isotope effects in Tc of both ^{13}C carbon and isotoped dopants like $^{87}Rb/^{85}Rb$. The phonon frequencies are modulated well by introducing ^{13}C into the $^{12}C_{60}$ cage. The importance of the C_{60} intramolecular high frequency phonons for making superconducting electron pairs can also be strongly supported by the fact that substitution of ^{13}C to ^{12}C in C_{60} leads to the reduction in Tc as shown in Fig.6, so called positive ^{13}C isotope effect [58,59]. It should be noted that the estimated value of important isotope parameter α scatters strongly depending on the experiments carried out so far, and accordingly that further measurements would be anticipated for the future study again. In contrast to the ^{13}C isotope effect, no isotope effect on Tc was observed in the case of $^{87}Rb/^{85}Rb$ [60] , indicating little contribution of alkali-metal elements.

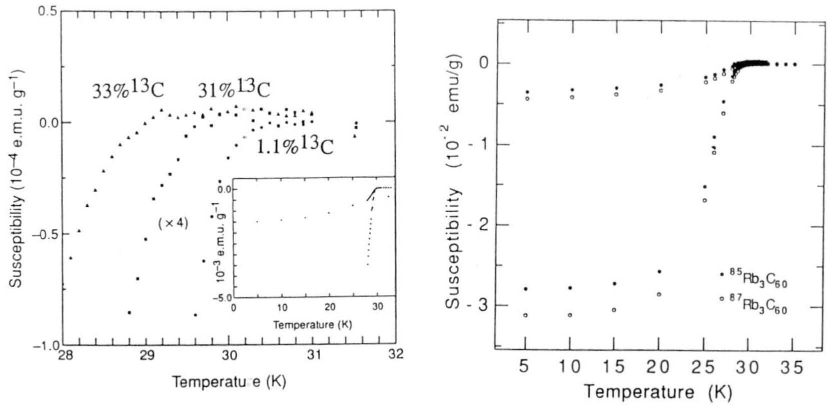

Fig.6 Isotope effect of ^{13}C and ^{87}Rb on superconducting transition temperature in Rb_3C_{60}.

Both vibrational spectroscopy and isotope effects of ^{13}C and ^{87}Rb seem to support an important role of high frequency intra-molecular phonons on the paring mechanism in these superconductors. As has been described in this article, the experimental results so far reported seem to support that the superconducting mechanism is most likely phonon-mediated BCS, where the change in N_{Ef} most efficiently controls Tc with less influence of phonon frequency. This interpretation can be supported by the experimental fact that Tc decreases monotonically with increasing pressure [61-63], since lattice parameters can mainly be contracted with the other parameters being less influenced and therefore N_{Ef} can be reduced leading to the monotonic reduction in Tc.

While the simple band model seems to give a fairly good picture for understanding the metallic properties observed in A_3C_{60} in the first order of approximation, unusual behaviors unexpected from conventional metals have also been observed. First, much differently from the simple picture described earlier, fcc Na_2C_{60} and bct phases with the stoichiometry of A_4C_{60} are not metallic [41,42,64]. Second, the experimental facts both of a broad absorption peak at the

Fermi level different from the expected conduction band peak [65] and the unexpectedly large temperature dependence of the t_{1u} conduction bands observed in photoemission experiments [66] seem to imply that the band structure is more complicated and delicately modulated upon geometrical parameters than expected from a free electron dispersed band model. As described earlier, it was argued that the high density of states at the Fermi level, N_{Ef} arising from the weak overlap of the wavefunctions of neighboring C_{60} molecules, leads to K_3C_{60} and Rb_3C_{60} behaving as molecular conductors with high conductivity. However, we have to keep in mind that this stems from the resulting rather narrow conduction band and as a consequence strong electron-electron correlations should also be generally taken into account. Such situations can be expected if one is reminded that the Hubbard U in A_3C_{60} is approximately 0.5-0.8 eV and of the same order of magnitude as the t_{1u} bandwidth (ΔW) of approximately 0.5 eV. Thus taking into account the almost identical values between U and ΔW, we expect that the A_3C_{60} salts will be near the borderline to Mott insulator behavior and that the band structure will be very sensitive to the C_{60} intermolecular separation. The third important and different behavior from that in normal metals is the fact that transport measurements show that the evolution of conductivity is not linear as a function of temperature and instead follows a T^2- functional form [67,68,69] although conductivity of K_3C_{60} and Rb_3C_{60} increases as temperature decreases as expected for standard metals. Discussion is still continuing for the interpretation of the temperature dependence of conductivity [70,71], but one of the conceivable possibilities is again that the A_3C_{60} compounds are strongly electron-correlated systems [72].

When we look at the Tc-a_0 relationship more carefully again, three features different from the simple description are apparently evident. One is the unexpected steep decrease in T_c observed for A_3C_{60} in the small lattice parameter region. The T_c's of 2.5-3.5 K observed for Na_2KC_{60} and Na_2RbC_{60} are considerably low far from the expectation simply extrapolated using the data of pressure experiments on K_3C_{60} and Rb_3C_{60}. Their Tc's should be around 8-10 K from extrapolation. The other is the saturation in T_c occurring in A_3C_{60} with large lattice parameters. The slightly higher T_c could be expected than the T_c values of Rb_2CsC_{60} (T_c=31 K) and $RbCs_2C_{60}$ (T_c=33 K) considering their lattice parameters. Finally, both Li_2CsC_{60} and Li_2RbC_{60} are not superconducting, despite the apparent band filling by electron transfer into C_{60} solids. These phenomena cannot be explained by the simple description itself and seem to be related to the unusual metallic properties described earlier.

If we see the relaxation times of conduction electrons by ESR [73-76], unusual properties are apparent. The observed ESR spectra for K_3C_{60}, K_2RbC_{60}, KRb_2C_{60} and Rb_3C_{60} are shown in Fig.7. The most striking feature of this figure is the extent to which the ESR linewidth is so much dependent on the alkali metals present. For instance, the linewidth of K_3C_{60} is very narrow ($DH_{1/2}$=25.5 G: halfwidth of the Lorentzian lineshape), while that of Rb_3C_{60} is 744 G. K_2RbC_{60} and KRb_2C_{60} show linewidths of 98 and 363 G, straddling those of K_3C_{60} and Rb_3C_{60}. The ESR studies clearly show that both narrow and wide linewidths are the intrinsic CESR signals of the metallic A_3C_{60} phases. The relationship between T_1 and λ (where T_1 is the spin-lattice relaxation time of conduction electrons and λ is the spin-orbit coupling constant of alkali-metal elements) has now been established for a variety of A_3C_{60} salts [77].

The influence of doped alkali-metal elements presented above has also been pointed out in the case of graphite intercalations (GIC's) and conducting polymers, but with its less emphasis on organic molecular conductors. When electron spins remain on the s-orbitals of the alkali metals, the Korringa law for metals (${}^{N}T_1T$ = constant, where ${}^{N}T_1$ is the spin-relaxation time for nuclear spins) can also be expected for ^{87}Rb NMR studies and the fraction of the remaining s-electrons can be estimated from the Knight shift observed. Actually ^{87}Rb NMR shows that such a constant term is observed in a certain temperature region. However, the fact that the observed Knight shift (ca. 0.03 %) is much smaller than that of Rb metal (0.652 %) indicates that the fraction of the remaining s-electrons on Rb is only at most of the order of a few % [78].

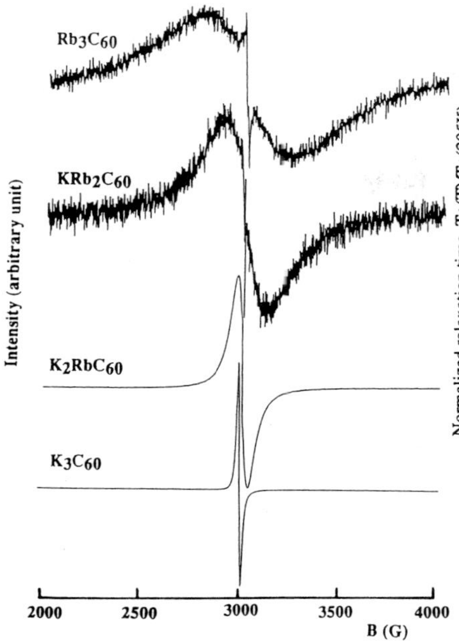

Fig.7 ESR spectra observed for $K_xRb_{3-x}C_{60}$ fullerides.

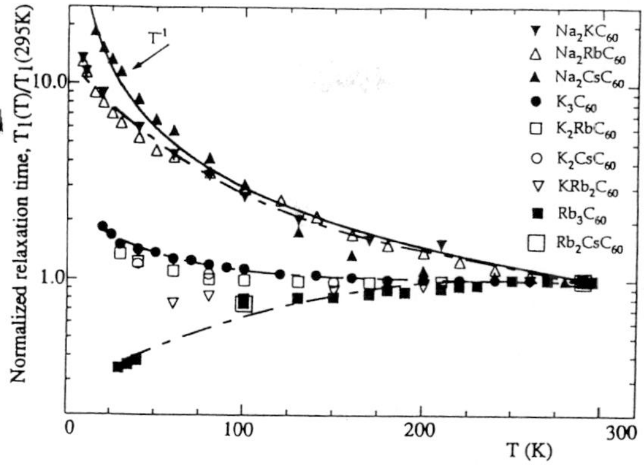

Fig.8 The change in the spin-lattice relaxation times, T_1 as a function of temperature for nine A_3C_{60} superconductors. T_1 is normalized by its value at 295 K. The solid line shows the T^{-1} function. The dotted lines are guides to the eye, emphasizing the three different temperature dependencies found.

Therefore, it is indeed curious that this small fraction of the remaining s-electrons on the alkali ions has such a large influence on the electron spin relaxation times and leads to the observed large variation in ESR linewidths. This contradicts the simple but quite generally accepted view since the early days of fullerene research, that the conduction properties in the A_3C_{60} fullerides can be satisfactorily described considering only the partially filled t_{1u} derived bands of C_{60}. Unexpectedly much larger influence observed in A_3C_{60} than in GIC's seems to show importance in the specialty of the conduction band stemming from the degenerate t_{1u} molecular orbitals of highly symmetrical C_{60}.

When the temperature evolution of the T_1 values is also examined, another important trend is immediately apparent. The temperature evolution of T_1 has been studied for nine metallic $A_2A'C_{60}$ phases with T_c's ranging between 2.5 K and 31 K. The results, normalized to the room temperature values, are displayed in Fig.8 with the fulleride salts listed in order of decreasing lattice size [77]. The temperature dependence changes gradually away from the simple inverse temperature law for small a_0 towards a behavior, where T_1 increases with increasing temperature in the fullerides as lattice expands. It is remarkable that such deviation is so strongly evident already in K_3C_{60}. The relative magnitude of these changes is in good accord with the size of a_0, that is, Na_2KC_{60}, $Na_2RbC_{60} < Na_2CsC_{60} < K_3C_{60} < K_2RbC_{60} < K_2CsC_{60} < KRb_2C_{60} < Rb_3C_{60} < Rb_2CsC_{60}$.

In conventional metallic systems and within the framework of the free electron model, the change in T_1 as a function of T can be derived by recalling Elliot's expression [79]. The T_1 can be related to the resistivity $\rho = m/[ne^2T_r]$ if the g-value shift is nearly temperature-independent as usually observed in standard metals, where m is the mass of the electron, e the electron charge, n the electron carrier density and T_r the relaxation time of electrical conductivity. Since ρ is determined by the phonon scattering, the T_1 versus temperature relationship takes the form $T_1 \sim T^{-1}$ above the Debye temperature for conventional metals. The implication of the observed unusual opposite temperature dependence is that a description of metallic fullerides in terms of solid dispersion metal bands [37,38,80] can only be safely applied in the small a_0 regime. The free electron band model starts to be increasingly inapplicable upon lattice expansion. The reason has not yet been clear, but either band reconstruction as a function temperature (in this case g-value will change as a function of temperature) or increase in electron correlations sensitively upon lattice expansion would be most likely [81].

4. Concluding remarks

New finding of high conductivity and superconductivity with high Tc only surpassed by copper oxide superconductivity for C_{60} based materials has resulted in a surge of interests in fundamental aspects in physics and chemistry as well as in a search for technological applications. In alkali-metal C_{60} fulleride superconductors, we had a good guideline for elevating superconducting transition temperature (Tc). Since the phonon frequency and the coupling constant making superconducting electron-pairs are nearly invariable and the density of states at the Fermi level can be monotonically varied upon lattice parameter change, higher Tc can be achieved for A_3C_{60} fullerides with larger lattice parameter by introducing alkali-metals with large ionic radii. Remarkably, the situation in C_{60} superconductors is very similar to the one encountered in the development of organic superconductors even though the dimensionality is different. It seems that the lasting increase in Tc is saturating to date.

Because of the simplicity of the C_{60} geometry and variety of electronic properties found in different crystal structures, a number of studies will continuously be carried out in the future even though difficulties in handling intercalated compounds arise by their air sensitivity. Details of these new solids will be expected to give ideas for further development of a new superconducting family.

5. Acknowledgments

I would like to acknowledge I. Hirosawa, S. Kuroshima, T. W. Ebbesen, J. Mizuki, M. Kosaka, O. Z. Zhou, H. Yoshikawa, T. Manako, Y. Shimakawa, Y. Kubo, J. S. Tsai for collaboration, all of them are in NEC. I am grateful to K. Prassides and his group members for collaboration in structural analyses. I also would like to express my thanks to Y. Maniwa, K. Mizoguchi, K. Kume and T. Saito for collaboration in NMR study and to Y. Iwasa for raman study. E. Özdas is acknowledged for discussion while he is staying in our laboratory as a visiting scientist.

6. References

[1] For example, M. Tinkham, "Introduction to Superconductivity", McGrraw-Hill, Inc., 1975.

[2] J. G. Bednorz and K. A. Müller, Z. Phys. B., Condensed Matter Vol.64, 1986, pp.189-193; for a review , N. Tsuda, K. Nasu, A. Fujimori and K. Siratori, "Electron Conduction i Oxides", Springer-Verlag, Heidelberg, 1991.

[3] R. J. Cava, H. Takagi, B. Batlogg, H. W. Zandbergen, J. J. Krajewski, W. F. Peck, Jr, R. B. van Dover, R. J. Felder, T. Siegrist, K. Mizuhashi, J. O. Lee, H. Eisaki, S. A. Carter and S. Uchida, Nature Vol.367, 1994, p.146.

[4] R. L. Greene, G. B. Street and L. J. Suter, Phys. Rev. Lett. Vol.34, 1975, p.577.

[5] W. Little, Phys. Rev. Vol.A134, 1964, p.1416.

[6] D. Jerome, A. Mazaud, M. Ribault, J. Bechgaad, J. Phys. Lett. Vol.L-95, 1980, p.41.

[7] H. W. Kroto, J. R.Heath , C. O'Brien, R. F. Curl and R. E. Smalley, Nature, 318, 162, 1985.
[8] W. Krätschmer, L. D. Lamb, K. Fostiropoulos and D. R. Huffman, Nature Vol. 347, (1990), p.354.
[9] R. C. Haddon, A. F. Hebard, M. J. Rosseinsky, D. W. Murphy, S. J. Duclos, K. B. Lyons, B. Miller, J. M. Rosamilia, R. M. Fleming, A. R. Kortan, S. H. Glarum, A. V. Makhija, A. J. Muller, R. H. Elick, S. M. Zahurak, R. Tycko, G. Dabbagh and F. A. Thiel, Nature, Vol.350, 1991, p.320.
[10] A. F. Hebard, M. J. Rosseinsky, R. C. Haddon, D. W. Murphy, S. H. Glarum, T. T. M. Palstra, A. P. Ramirez and A. R. Kortan, Nature, Vol.350, 1991, p.600; A. F. Hebard, Physics Today, November 1992, p.26.
[11] D. W. Murphy, M. J. Rosseinsky, R. M. Fleming, R. Tycko, A. P. Ramirez, R. C. Haddon, T. Siegrist, G. Dabbagh, J. C. Tully and R. E. Walstedt, J. Phys. Chem., Vo.53, 1992, p.1321.
[12] K. Tanigaki, I. Hirosawa, T. W. Ebbesen, J. Mizuki and J. S. Tsai, J. Phys. Chem. Solids, Vol.54, 1993, p.1645.
[13] Y. Chai, T. Guo, C. Jin, R. E. Haufler, L. P. F. Chibante, J. Fure, L. Wang, J. M. Alford and R. E. Smalley, J. Phys. Chem. Vol.95, 1991, p.7564.
[14] H. Shinohara, H. Sato, M. Ohkohchi, Y. Ando, T. Kodama, T. Shida, T. Kato and Y. Saito, Nature Vol.357, 1992, p.52.
[15] R. D. Jhonson, M. S. Varies, J. Salem, D. S. Bethune and C. S. Yannoni, Nature Vol.355, 1992, p.239.
[16] P. Lauginie, H. Estrade, J. Conard, D. Duérard, P. Lagrange and M. E. Makrini, Physica Vol.B99, 1980, p.514.
[17] A. R. Kortan, N. Kophylov, S. Glarum, E. M. Gyorgy, A. P. Ramirez, R. M. Fleming, F. A. Thiel and R. C. Haddon, Nature Vol.355, 1992, p.529.
[18] A. R. Kortan, N. Kophylov, S. Glarum, E. M. Gyorgy, A. P. Ramirez, R. M. Fleming, O. Zhou, F. A. Thiel, P. L. Trevor and R. C. Haddon, Nature Vol.360, 1992, p.566.
[19] E. Özdas, A. R. Kortan, N. Kopylov, A. P. Ramirez, T. Siegrist, K. M. Rabe, H. E. Bair, S. Schuppler and P. H. Citrin, Nature Vol.375, 1995, p.126.; E. Ozdas, A. R. Kortan, N. Kopylov, S. M. Glarum, R. C. Haddon, R. M. Fleming, A. P. Ramirez, abstract of MRS Fall meeting, 1994, G.4.7.
[20] H. Yoshikawa, S. Kuroshima, I. Hirosawa, K. Tanigaki and J. Mizuki, "Eu Fullerides Formation Studied by Photoemission Spectroscopy", Chem. Phys. Lett. Vol.239, 1995, p.103.
[21] Y. Wang, D. Tománek, G. F. Bertsch, R. S. Ruoff, Phys. Rev. B., Vol.47, 1993, p.6711.
[22] A. Rösen and B. Wästberg, J. Chem. Phys. Vol.90, 1989, p.2525.
[23] P. W. Stephens, L. Mihaly, P. L. Lee, R. L. Whetten, S.-M. Huang, R. Kaner, F. Deiderich and K. Holczer, Nature Vol.351, 1991, p.632.
[24] A. R. Ubbelohde, L. A. Lewis, "Graphite and its crystal compounds," Oxford (1960).
[25] R. M. Fleming, M. J. Rosseinsky, A. P. Ramirez, D. W. Murphy, J. C. Tully, R. C. Haddon, T. Siegrist, R. Tycko, S. H. Glarum, P. Marsh, G. Dabbagh, S. M. Zahurak, A. V. Makhija and C. Hampton, Nature, Vol.352, 1991, 701.
[26] Q. Zhu, D. E. Cox, J. E. Fischer, K. Kniaz, A. R. McGhie and O. Zhou, Nature Vol.355, 1992, p.712.
[27] M. J. Rosseinsky, D. W. Murphy, R. M. Fleming, R. Tycko, A. P. Ramirez, T. Siegrist, G. Dabbagh and S. E. Barrett, Nature, Vol.356, 1992, p.416; For a review, J. Materials Chemistry, Vol.5, 1995, p.1497.
[28] T. Yildirim, O. Zhou, J. E. Fischer, N. Bykovetz, R. A. Strongin, M. A. Cichy, A. B. Smith III, C. L. Lin and R. Jelinek, Nature Vol.360, 1992, p.568.
[29] T. T. M. Palstra, O. Zhou, Y. Iwasa, P. E. Sulewski, R. M. Fleming and B. R. Zegarski, Solid State Commun. Vol.92, 1994, 71.
[30] T. Yildirim, O. Zhou, J. E. Fischer, N. Bykovetz, R. A. Strongin, M. A. Cichy, A. B. Smith III,

C. L. Lin and R. Jelinek, Nature Vol.360, 1992, 568.
[31] M. Baeniz, M. Heinze, K. Lüders, H. Werner, R. Schlägl, M. Weiden, G. Sparn and F. Steglich, Solid State Comun. Vol.96, 1995, p.539.
[32] I. Hirosawa, K. Tanigaki, J. Mizuki, T. W. Ebbesen, Y. Shimakawa, Y. Kubo, and S. Kuroshima, Solid State Commun., Vol.82, 1992, p.979.
[33] Y. Maniwa, K. Mizoguchi, K. Kume, K. Tanigaki, T. W. Ebbesen, S. Saito, J. Mizuki, J. S. Tsai, Y. Kubo, Solid State Commun., Vol.82, 1992, p.783.
[34] K. Tanigaki, I. Hirosawa, J. Mizuki and T. W. Ebbesen, Chem Phys. Letters Vol.213, 1993, 395.
[35] O. Zhou, R. M. Fleming, D. W. Murphy, M. J. Rosseinsky, A. P. Ramirez, R. B. van Dover and R. C. Haddon, Nature Vol.362, 1993, p.433.
[36] S. C. Erwin and W. E. Pickett, Science Vol.254, 1991, p.842; For a review, M. P. Gelfand, Superconductivity Review, Vol.1, 1994, p.103.
[37] S. Saito and A. Oshiyama, Phys. Rev. B Vol.44, 1991, p.11536.
[38] W. Andreoni, P. Giannozzi and M. Parrinello, Phys. Rev. B. Vol.51, 1994, p.2087.
[39] Y. Maniwa, T. Saito, A. Ohi, K. Mizoguchi, K. Kume, K. Kikuchi, I. Ikemoto, S.Suzuki, Y. Achiba, M. Kosaka, K. Tanigaki and T. W. Ebbesen, J. Phys. Soc. Jpn. Vol.63, 1994, p.1139.
[40] R. Tycko, G. Dabbagh, M. J. Rosseinsky, D. W. Murphy, R. M. Fleming, A. P. Ramirez and J. C. Tulley, Science Vol.253, 1991, p.884.
[41] P. J. Benning, J. L. Martins, J. H. Weaver, L. P. E. Chibante and R. E. Smalley, Science Vol.252, 1991, p.1417.
[42] T. Takahashi, S. Suzuki, T. Morikawa, H. Katayama-Yoshida, S. Hasegawa, H. inokuchi, K. Seki, K. Kikuchi, S. Suzuki, K. Ikemoto and Y. Achiba, Phys. Rev. Lett. Vol.68, 1992, p.1232.
[43] Y. Iwasa, K. Tanaka, T. Yasubda, T. Koda and S. Koda, Phys. Rev. Lett. Vol.69, 1992, 2284.
[44] L. Degiorgi, G, Gruner, P. Wachter, S.-M. Huang, J. Wiley, R. L. Whetten, R. B. Kaner, K. Holczer and F. Diedrich, Phys. Rev. B Vol.46, 1992, p.11250.
[45] T. Inabe, H. Ogata, Y. Maruyama, Y. Achiba, S. Suzuki, K. Kikuchi and I. Ikemoto, Phys. Rev. Lett. Vol.69, 1992, p.3797.
[46] K. Holczer, S. Klein, H. Alloul, Y. Yoshinari, F. Hippert, S.-M. Huang, R. B. Kaner and R. L.Whetten, Europhys. Lett. Vol.23, 1993, 63.
[47] A. P. Ramirez, M. J. Rosseinsky, D. W. Murphy and R. C. Haddon, Phys. Rev. Lett. Vol.69, 1992, p.1687; For a review, Superconductivity Review, Vol.1, 1994, 1.
[48] K. Tanigaki, I. Hirosawa, T. W. Ebbesen, J. Mizuki, Y. Shimakawa, Y. Kubo, J. S. Tsai and S. Kuroshima, Nature Vol.356, 1992, p.419.
[49] M. J. Rosseinsky, D. W. Murphy, R. M. Fleming, R. Tycko, A. P. Ramirez, T. Siegrist, G. Dabbagh and S. E. Barrett, Nature Vol.356, 1992, 416.
[50] K. Tanigaki, T. W. Ebbesen, S. Saito, J. Mizuki, J. S. Tsai, Y. Shimakawa, Y. Kubo and S. Kuroshima, Nature Vol.352, 1991, p.222.
[51] M. J. Rosseinsky, A. P. Ramirez, S. H. Glarum, D. W. Murphy, R. C. Haddon, A. F. Hebard, T. T. M. Palstra, A. R. Kortan, S. M. Zahurak, and A. V. Makhija. Phys. Rev. Lett. Vol.66, 1991, p.2830.
[52] R. M. Fleming, A. P. Ramirez, M. J. Rosseinsky, D. W. Murphy, R. C. Haddon, S. M. Zahurak and A. V. Makhijia, Nature Vol.352, 1991, p.787.
[53] M. Kosaka, K. Tanigaki, T. W. Ebbesen, Y. Nakahara and K. Tateishi, Appl. Phys. Lett. Vol.63, 1993, p.2561.
[54] J. Winter and H. Kuzmany, Solid State Commun. Vol.84, 1992, p.935.
[55] P. C. Ekuland, P. Zhou, K. A. Wang, G. Dresselhaus, M. S. Dresselhaus, J. Phys. Chem. Solids Vol.53, 1992, p.1391.
[56] K. Prassides, J. Tomkinson, C. Christides, M. J. Rosseinsky, D. W. Murphy and R. C. Haddon, Nature Vol.354, 1991, p.462.

[57] K. Prassides, C. Christides, M. J. Rosseinsky, J. Tomkinson, D. W. Murphy and R. C. Haddon, Europhys. Letters, Vol.19, 1992, p.629.
[58] A. P. Ramirez, A. R. Kortan, M. J. Rosseinsky, S. J. Duclos, A. M. Mujsce, R. C. Haddon, D. W. Murphy, A. V. Makhija, S. M. Zahurak and K. B. Lyons, Phys. Rev. Lett. Vol.68, 1992, p.1058.
[59] T. W. Ebbesen, J. S. Tsai, K. Tanigaki, J. Tabuchi, Y. Shimakawa, Y. Kubo, I. Hirosawa and J. Mizuki, Nature Vol.355, 1992, p.620.
[60] T. W. Ebbesen T. W., J. S. Tsai, K. Tanigaki, H. Hiura, Y. Shimakawa, Y. Kubo, I. Hirosawa and J. Mizuki, Phys. C., Vol.203, 1992, p.163.
[61] O. Zhou, G. B. M. Vaughan, Q. Zhu, J. E. Fischer, P. A. Heiney, N. Coustel, J. P. MacCauley Jr. and A. B. Smith III, Science 255, 833 (1992).
[62] Sparn, Thompson J.D., Whetten R.L., Huang S.-M., Kaner R. B., Diederich F., Gruner G. and Holzer K., Phys. Rev. Lett. Vol.68, 1992, p.1228.
[63] G. Sparn, J. D. Thompson, S.-M. Huang, R. B. Kaner, F. Diederich, R. L. Whetten, G. Grune and K. Holzer, Science Vol.252, 1991, p.1829.
[64] G. Zimmer, M. Helmle, F. Rachdi and M. Mehring, Europhys. Lett. Vol.27, 1994, p.543..
[65] P. J. Benning, F. Stepniak and J. H. Weaver, Phys. Rev. B Vol.48, 1993, p.9086.
[66] M. Knupfer, M. Merkel, M. S. Golden and J. Fink, Phys. Rev. B Vol.47, 1993, p.13944.
[67] T. T. M. Palstra, A. F. Hebard and R. C. Haddon, Phys. Rev. Lett. Vol.68, 1992, p.1054.
[68] X.-D. Xiang, J. H. Hou, G. Briceno, W. A. Vareka, R. Mostovoy, A. Zettle, V. H. Crespi and M. L. Cohen, Science Vol.256, 1992, p.1190: X.-D. Xiang, J. H. Hou, V. H. Crespi, A.Zettl and M. L. Cohen, Nature Vol.361, 1993, p.54.
[69] Y. Maruyama, T. Inabe, H. Ogata, Y Achiba, S. Suzuki, K. Kikuchi and I. Ikemoto, Chem. Lett., 1991, p.1849: Y. Maruyama, T. Inabe, M. Ogata, Y. Achiba, K. Kikuchi, S. Suzuki and I. Ikemoto, Mater. Sci. and Eng. B Vol.19, 1993, p.162.
[70] M. L. Cohen, Mater. Sci. and Eng. B Vol.19, 1993, p.111.
[71] W.A. Vareka and A. Zettl, Phys. Rev. Lett. Vol.72, 1994, p.4121.
[72] T. T. M. Palstra and R. C. Haddon, Solid State Commun. Vol.92, 1994, 71.
[73] W. H. Wong, M. E. Hanson, W. G. Clark, G. Grüner, J. D. Thompson, R. L. Whetten, S. M. Humng, R. B. Kaner, F. Diedrich, P. Petit, J. J. Andre and K. Holczer, Europhys. Lett. Vol.18, 1992, p.79.
[74] A. Jànossy, O. Chauvet, S. Pekker, J. R. Cooper and L. Forró, Phys. Rev. Lett. Vol.71, 1993, p.1091.
[75] M. Kosaka, K. Tanigaki and I. Hirosawa, Phys. Rev. Lett. Vol.72, 1994, p.3130.
[76] K. Tanigaki, M. Kosaka, T. Manako, Y. Kubo, I. Hirosawa, K. Uchida and K. Prassides, Chem. Phys. Letters, Vol.240, 1995, p.627.
[77] K. Tanigaki and K. Prassides, J. Materials Chemistry, Vol.5, 1995, p.1515.
[78] T. Saito, Y. Maniwa, K. Kume, I. Hirosawa, M. Kosaka and K. Tanigaki, J. Phys. Soc. Jpn., Vol.64, 1995, p.4513.
[79] R. J. Elliott, Phys. Rev. Vol.96, 1954, p.266.
[80] A. Oshiyama and S. Saito, Solid State Commun., Vol.82, 1992, p.41.
[81] S. Suzuki, A. Nakao, Phys. Rev. B, Vol. 52, 1995, p.14206.

Part V

Organic Conductors

Optical Properties of Low-Dimensional Molecular Crystals and LB Films

V. M. Yartsev

Centro de Física, Instituto Venezolano de Investigaciones Científicas (IVIC),
Apartado 21 287, Caracas 1020-A, VENEZUELA

Abstract

A general theory for calculating spectroscopic properties of low-dimensional molecular crystals is developed on the basis of the cluster approach. Electronic correlations and the coupling of electrons to intramolecular vibrations are explicitly taken into account. The relative role of different model parameters is investigated and both effects are found to be equally important. This result predicts a possibility to modify physical properties via the choice of constituents, stoichiometry, or external conditions (temperature, pressure, photoexcitation, *etc.*). The theory explains well the experimental visible and infrared spectra and proves that the optical response in molecular charge-transfer crystals has a local nature. Charge-transfer salts prepared in the form of Langmuir-Blodgett (LB) films are also discussed. The cluster approach analysis of the experimental absorption spectra of LB films is shown to provide information about molecular organization and can offer a way to modify the optical response as required by specific practical applications.

1. Introduction

Molecular electronics is generally regarded as a possible way to achieve the further miniaturization of electronic devices down to the characteristic scale of the order of nanometer for controlled modifications of physical properties of interest. Neutral-ionic phase transition, switching and changes in the optical properties under external influences (temperature, pressure, illumination, etc.) have been observed in charge-transfer (CT) ion-radical salts. Hence, these salts, which can be prepared in the form of single crystals, powders or Langmuir-Blodgett (LB) films, seem to be very promising materials for molecular electronics. In this respect, it is interesting to understand the factors which govern a degree of localization for physical processes in these systems. Inspired by experiments with irradiation of CT molecular crystals [1], which demonstrated that rather small molecular clusters are responsible for optical characteristics, we undertook a systematic study [2] of CT salts where molecules form quasi-isolated clusters of two, three and four entities. The cluster size is determined by the mode of overlap of adjacent molecules: the transfer integral between molecules forming a cluster should be much larger than the one between molecules from neighbouring clusters. Electronic correlations, the electron-molecular vibration (EMV) coupling, and the interaction with nearby counter-ions were shown to be of the same order of magnitude. This fact ensures that it is possible to modify the physical properties via the choice of constituents, stoichiometry, or external conditions.

Since 1962, when the synthesis of the first CT crystal based on tetracyanoquinodimethane (TCNQ) molecule was reported [3], a great number of salts with a variety of physical characteristics have been obtained (see e.g. [4,5]). However, the synthesis of CT molecular crystals is a difficult (and also costly) process and a typical size of single crystals is usually in the submillimeter range. So, the prospect of employing the Langmuir-Blodgett technique in order to produce ultrathin films with large surfaces in shorter time and at cheaper cost looks very attractive. Although this approach has been proposed several times in the past, the first successful attempt was reported only in 1985 by a French group from Sacley [6]. Since then, LB films based either on molecular conductors or on conducting polymers have been built up in many ways. One such possibility, the so called homodoping strategy, comes directly from supermolecular engineering and is based on the *in situ* mixing of two derivatives of the same electroactive molecule (in differently charged states) within the same layer. In this case, a conductive Langmuir film is formed already on the water surface and then transferred to the substrate. Contrary to the crystalline CT compounds, one can obtain via the homodoping strategy LB films with stoichiometry as an externally adjustable parameter. The latter possibility is very important for practical applications because a review of the optical properties of CT molecular crystals reveals their high sensitivity to stoichiometry [7].

It may be noticed that the optical properties of CT crystalline salts and LB films share major common features [8]. A typical absorption spectrum exhibits a broad (~ 2000 cm^{-1}) band of electronic excitations corresponding to a charge transfer between molecules and a number of vibrational bands in infrared. These bands with halfwidths about 100 cm^{-1} are at the wavenumbers close to those of totally symmetric a_g vibrational modes of an isolated molecule. The a_g-modes are normally inactive in infrared and their indirect excitation has been shown [9] to be possible via electron-intramolecular vibration coupling [10]. A relationship between d.c. conductivities and the wavenumbers of the electronic CT band is apparent. In order to allow a quantitative comparison, Richard *et al.* selected a series of bulk crystalline salts [11]. They established a linear relation between log $\sigma_{d.c.}$ and the wavenumber of the CT band: a compound with higher d.c. conductivity has a smaller optical energy gap. It means that compounds with more delocalized charge carriers have the electronic CT band in their absorption spectra at a lower wavenumber. In terms of the cluster approach, this situation corresponds to the decrease of CT excitation energy as the size of a cluster increases. The fact that the spectra of CT crystals and of LB films are so similar suggests that the optical response is of the same nature determined by short-range interactions and that the cluster approach based on the supremacy of the localized interactions is the most adequate method.

Our paper is organized in the following way. In Section 2, we consider in detail the simplest case of dimerized compounds, briefly discuss trimers and tetramers, and present the results of the general formalism for arbitrary cluster. The specific features of the LB films are evaluated in Section 3 and the main results and perspectives are summarized in Section 4.

2. Cluster approach to the optics of low-dimensional molecular crystals

Low-dimensional organic conductors and semiconductors are a particular class of molecular crystals in which charge transfer plays a dominant role. In the majority of these crystals, generally flat acceptor and/or donor molecules are packed face-to-face and form linear stacks with high density of radical (unpaired) electrons as a result of the charge transfer between donors and acceptors. During the last thirty years, a variety of charge-transfer crystalline compounds have been synthesized with a wide spectrum of physical properties. In the organic conductors, the molecules inside the stack are uniformly spaced; this means that both the distance and the mode of overlap are the same. The optical properties of organic metals are discussed in a recent review [12]. However, in the majority of quasi-one-dimensional organic solids the molecules are arranged to form linear clusters of 2, 3, 4 or even 5 molecules; these crystals exhibit semiconducting properties. The characteristic feature of these compounds is the coexistence of several equally important interactions. For example, the bandwidth calculated in the tight-binding scheme ($4t$) is approximately equal to the energy of the on-site Coulomb repulsion (U) between the two radical electrons occupying the same MO. Thus, usual crystal band theory fails and it is necessary to describe the electron-electron correlation in an explicit way. Secondly, the electron-molecular vibration coupling cannot be ignored, since it manifests itself so clearly in infrared spectra. Also, variations of the MO energies of non-equivalent sites along the stack are often of the same order of magnitude.

In the cluster approach it is possible to take into account all of these interactions explicitly assuming that the optical properties may be calculated as a superposition of the optical responses of isolated n-mers with an arbitrary, but constant number of radical electrons. These electrons interact with the totally symmetric intramolecular vibrations and an externally applied electric field **E**. The corresponding Hamiltonian has the form [13] ($\hbar = 1$):

$$H = H_e + H_v + \sum_{\alpha,i} g_\alpha Q_{\alpha i} n_i - \mathbf{p}\cdot\mathbf{E}, \tag{1}$$

where

$$H_v = \sum_{\alpha,i} \frac{\omega_\alpha}{4}\left(\Pi_{\alpha i}^2 + Q_{\alpha i}^2\right) \tag{2}$$

is the Hamiltonian of the totally symmetric (a_g) internal modes of vibration of the monomers. $Q_{\alpha i}$ denotes the dimensionless normal mode coordinate of the symmetric vibration α of the monomer i with the frequency ω_α and the canonically conjugated momentum $\Pi_{\alpha i}$. Linear EMV coupling is described explicitly by the third term in Eq. (1), where n_i is the occupation number operator for site i and g_α denotes the EMV coupling constant. Of course, if the molecules are not identical the index i should be added to ω_α and g_α. It follows from the symmetry arguments [14] that if the MO occupied by the radical electron is non-degenerate, the linear EMV coupling is allowed only for the totally symmetric a_g intramolecular vibrations. The last term in Eq. (1) describes the interaction of the radical electrons with an externally applied electric field **E**. The operator **p** denotes the electric dipole moment induced by **E**

In Eq. (1), H_e is the Hamiltonian of the radical electrons in the n-mer in the *absence* of vibronic coupling, at the equilibrium values of $Q_{\alpha i}$. As discussed in the Introduction, the electronic correlations are to be treated explicitly. The most convenient way [15] to do this is to employ the Hubbard Hamiltonian in its extended version [16], which includes the Coulomb repulsion of radical electrons residing in nearest neighbour sites, V:

$$H_e = \sum_i \varepsilon_i n_i - \sum_{i,\sigma} t_{i,i+1}\left(c_{i,\sigma}^+ c_{i+1,\sigma} + c_{i+1,\sigma}^+ c_{i,\sigma}\right) + \frac{U}{2}\sum_{i,\sigma} n_{i,\sigma} n_{i,-\sigma} + \sum_i V_{i,i+1} n_i n_{i+1}. \tag{3}$$

Here ε_i is the energy of MO occupied by the radical electron, $c_{i\sigma}^+$ and $c_{i\sigma}$ denote site fermion creation and destruction operators, σ labels the electronic spin, and t is the hopping or transfer integral between adjacent molecules.

The frequency dependent conductivity $\sigma(\omega)$ in the framework of linear response theory has been initially calculated [17] for dimerized simple salts with complete charge transfer and then extended for more sophisticated systems (see review [2]).

2.1. Dimer model

Dimerized compounds dominate in the family of low-dimensional CT crystals and attract a lot of attention. Already in 1977, Bozio et al. [18] showed that the anti-symmetric mode of dimeric internal molecular vibrations is active in infrared. Rice calculated [17] the frequency-dependent complex conductivity of an assembly of isolated molecular dimers each hosting two radical electrons as follows:

$$\sigma(\omega) = -\,\mathrm{i}\omega N \frac{e^2a^2}{4} \frac{1}{\dfrac{1}{\chi(\omega)} - D(\omega)} . \tag{4}$$

where a is the interdimer distance and N denotes the number of dimers per unit volume. Also:

$$D(\omega) = \sum_{\alpha} \frac{g_{\alpha}^{2}\,\omega_{\alpha}}{\omega_{\alpha}^{2} - \omega^{2} - \mathrm{i}\,\omega\,\gamma_{\alpha}} \tag{5}$$

and

$$\chi(\omega) = \frac{2\,M^2\,\omega_{\mathrm{CT}}}{\omega_{\mathrm{CT}}^{2} - \omega^{2} - \mathrm{i}\omega\Gamma} \tag{6}$$

is the reduced charge-transfer electronic polarizability in the absence of vibronic coupling, where M denotes the matrix element of the charge difference operator $(n_1 - n_2)$ for the electronic excitation with the energy

$$\omega_{\mathrm{CT}} = \frac{U}{2} + \sqrt{\frac{U^2}{4} + 4t^2} . \tag{7}$$

In Eqs. (5) and (6), γ_{α} and Γ are the phenomenological damping factors for the a_g-vibration and the CT excitation, respectively.

In the case of one radical electron per dimer, an interesting phenomenon, specific for molecular solids, was investigated [13] in detail. Let the site energies be slightly different: $\varepsilon_1 = \varepsilon - \Delta_c$, $\varepsilon_2 = \varepsilon + \Delta_c$ (for example due to an asymmetric arrangement of counter-ions with respect to two moieties in the dimer). As a result of such inequivalence of the monomers, the charge density is greater on the monomer with the energy $\varepsilon - \Delta_c$. Now, with the introduction of the EMV coupling, the internal bond lengths in both molecules readjust to this new unsymmetrical charge distribution, leading to a further displacement of the charge density and to an increase of the energy difference between the two sites: $\varepsilon_1 = \varepsilon - (\Delta_c + \Delta_v)$, $\varepsilon_2 = \varepsilon + (\Delta_c + \Delta_v)$. This process of lowering the electron energy is counterbalanced by the elastic restoring forces associated with molecular distortion. The equilibrium value of Δ_v, Δ_v^0 , has been found [13] to satisfy the relation

$$\Delta_v^0 = \frac{E_p\left(\Delta_c + \Delta_v^0\right)}{\sqrt{t^2 + \left(\Delta_c + \Delta_v^0\right)^2}}, \qquad \text{where} \qquad E_p = \sum_{\alpha} \frac{g_{\alpha}^{2}}{\omega_{\alpha}} \tag{8}$$

is the small polaron binding energy. Note, that the initial asymmetry of the dimer is not obligatory: even for $\Delta_c = 0$ we have $\Delta_v \neq 0$ (spontaneous breaking of the dimer symmetry) if the EMV coupling is strong enough: $E_p > t$. We wish to stress the importance of such readjustment of molecular geometry driven by a perspective of some energy gain: the same idea has been later used [19] to account for specific features of the neutral-ionic phase transition in CT molecular compounds. It can be shown that this effect can be neglected in the case of two electrons per dimer, because a strong Coulomb repulsion of two electrons on one site favours a homogeneous distribution of the charge density.

As for the optical properties of dimerized compounds with one radical electron per dimer, Eqs. (4) to (6) are still valid, but the matrix element M is different and the electronic CT excitation energy has a lower (compared to the case of two electrons per dimer in simple salts) value:

$$\omega_{CT} = 2\sqrt{t^2 + \left(\Delta_c + \Delta_v^0\right)^2}\,. \tag{9}$$

In the usual physics of classical three-dimensional semiconductors, the processes involving non-equilibrium charge carriers (e.g., recombination, diffusion, luminescence) have been extensively studied both theoretically and experimentally. In principle, quasi-one-dimensional molecular crystals present an even more interesting object of investigation of these phenomena due to crystal anisotropy and the specific role of electronic correlations and EMV coupling. The main difficulty of observing any effects with non-equilibrium charge carriers in quasi-one-dimensional molecular crystals is the rather high "background" electronic density $\sim 10^{21}$ cm^{-3}. Therefore, to be able to produce anything detectable, one needs a laser or any other source of external influence with a very short pulse in order not to melt the crystal. First theoretical models [20] examine the changes of the optical properties of dimerized quasi-one-dimensional molecular crystals caused by photo-excited charge carriers or by an injection of electrons. For the dimerized compounds with the stoichiometry providing two radical electrons per dimer it has been shown that the generation of photo-excited charge carriers by external illumination with the energy (7) leads to the creation of an additional electronic CT band at a lower wavenumber and a corresponding shift of vibronic bands. In the dimerized compounds with one radical electron per dimer, an injection of additional electrons creates dimers with two electrons and, consequently, the absorbance exhibits two CT bands given by (7) and (9). Also, due to the EMV coupling, both types of dimers produce absorption in the range of the totally symmetric intramolecular vibrations which could be observed as a doublet fine structure of vibrational bands in infrared spectrum. The intensity ratio of the doublet components depends on the density of the injected charge carriers. Some experimental evidence for the forecasted changes has been found in the case of irradiated crystals, in which additional electrons are created, but for practical applications some reversible methods of generation of non-equilibrium charge carriers should be investigated.

2.2. Trimers and tetramers

In the case of two electrons per trimer, it has been found [21] that there are two allowed CT excitation transitions. The first transition at a lower wavenumber is practically independent of U and corresponds to the "A" band of Torrance [22], while the second is almost proportional to U and refers to the "B" band. The complex conductivity can still be written as Eq. (4), only the reduced electronic polarizability $\chi(\omega)$ has now two terms with different sets of M and ω_{CT}. However, with parameter values typical for the trimerized ion-radical salts, the higher frequency excitation has a much lower oscillator strength and can be disregarded.

For tetramers, there are more combinations of intramolecular a_g-modes which can be activated via the EMV coupling and we have four allowed electronic transitions in the case of two radical electrons per tetramer [23]. The complex conductivity can still be written analytically, although the expressions are rather cumbersome. An important new feature is a doublet fine structure of vibrational bands, which arises from a non-homogeneous equilibrium charge distribution among internal and external molecules of the tetramer.

2.3. General formalism

Recently [8] we have proposed the use of linear response theory for several variables to describe the optical properties of arbitrary molecular clusters. Molecules included in the cluster may be different in chemical composition and structure and their relative positions are not constrained. The microscopic model used to describe the cluster is defined by the Hamiltonian (1), where

$$\mathbf{p} = \sum_{i}^{N_s} e\left(n_i - \overline{n_i}\right) \mathbf{a}_i \,. \tag{10}$$

Vector $\mathbf{a}_i$ denotes the site position, N_s is the total number of sites included in the cluster, $\overline{n_i}$ is the average charge density per site, and the rest of the notation is the same as used previously with obvious extensions allowing, for example, different frequencies $\omega_{\alpha i}$ for different sites, as follows from the equilibrium charge density distribution defined by the ground state of the electronic Hamiltonian H_e (3).

It is convenient for considering different polarizations to present the projection of the total electric moment $\mathbf{p}$ along the vector of external field $\mathbf{E}$ in the form

$$p = \sum^{N_s} p_i \, n_i \,. \tag{11}$$

The complex conductivity for an assembly of isolated clusters was found [8] to be

$$\sigma(\omega) = -\,\mathrm{i}\,\omega N \left(\mathbf{p}, \left[\mathbf{I} - \mathbf{X}\cdot\mathrm{diag}\mathbf{D}\right]^{-1} \cdot \mathbf{X} \cdot \mathbf{p}\right), \tag{12}$$

where $\mathbf{p}$ and $\mathbf{D}$ are N_s-component vectors defined by Eq. (11) and by

$$D_i = \sum_{\alpha} \frac{2 g_{\alpha i}^2 \omega_{\alpha i}}{\omega_{\alpha i}^2 - \omega^2 - \mathrm{i}\omega\,\gamma_{\alpha i}} \,. \tag{13}$$

diag $\mathbf{D}$ is the standard notation for the diagonal matrix with elements D_1, D_2,..., and $\mathbf{I}$ and $\mathbf{X}$ denote unitary and charge transfer polarizabilities

$$\chi_{ij}(\omega) = \sum_{\beta} \frac{\langle 1 \mid n_i \mid \beta\rangle \langle \beta \mid n_j \mid 1\rangle \, 2\omega_{\beta 1}}{\omega_{\beta 1}^2 - \omega^2 - \mathrm{i}\omega\,\Gamma_\beta} = \chi_{ji}(\omega) \tag{14}$$

matrices, respectively. The energies $\omega_{\beta 1} = E_\beta - E_1$; E_β and $|\beta\rangle$ are the exact eigenvalues and eigenfunctions of the electronic Hamiltonian (3). $\beta = 1$ labels the ground state. The optical properties are calculated according to the following steps:

(i) on the basis of the crystal structure data, determine the cluster size, the site positions, and the number of clusters per unit volume, and estimate the transfer integrals from the mode of overlap of neighbouring molecules and intermolecular distances and ε_i from the positions of counter-ions;

(ii) evaluate U and V for the molecules in question;

(iii) find the eigenvalues and eigenfuncions of the Hamiltonian (3);

(iv) calculate the matrix elements of $\mathbf{X}$,

(v) using the eigenfunctions from step (iii), find the equilibrium charge distribution in the cluster and employing the values of unperturbed vibrational frequencies for each molecule corresponding to its charge via linear interpolation of Raman data for neutral and singly charged molecules;

(vi) calculate the optical conductivity from Eq. (12) and then any other optical characteristics using standard relations.

Figure 1 shows the calculated spectra of the real part of the complex conductivity (12) for clusters composed of one, two, three, and four dimers in the case of $n/N_S = 0.5$ and infinite on-site Coulomb repulsion energy U. We note that for the same ratio n/N_S, the lowest CT band shifts to a lower wavenumber as the cluster size grows. This is consistent with the experimentally found [11] correlation between the positions of the CT band and the d.c. conductivity value.

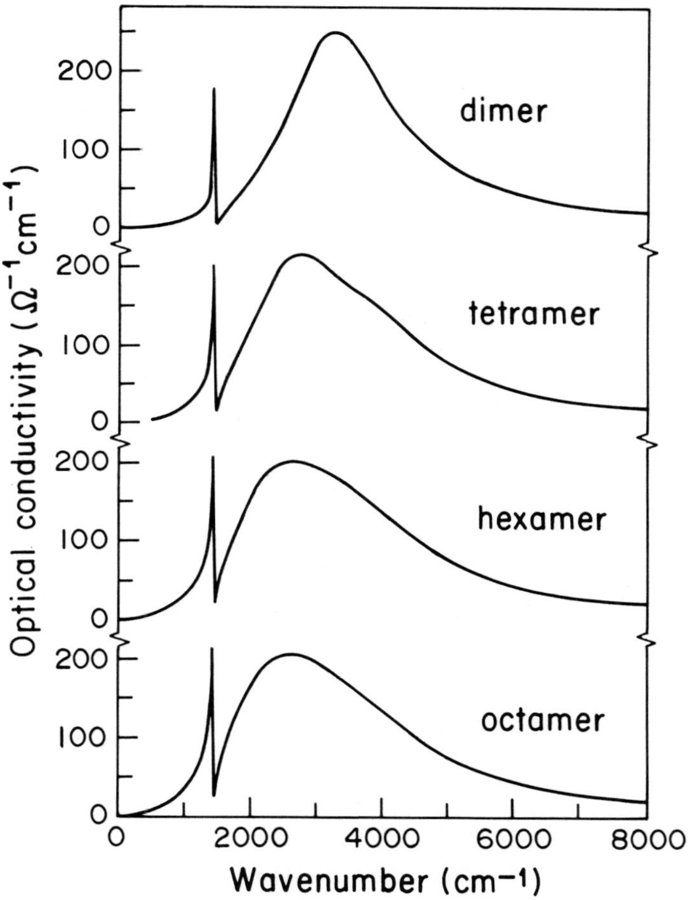

Fig. 1. Calculated conductivity of an assembly of linear superclusters composed of dimers with one electron per dimer. The parameters in Eqs. (3), (12) to (14) are: $t = 0.2$ eV, $t' = 0.08$ eV, $V = 0$, $\varepsilon_i = 0$, $\Gamma = 2000$ cm^{-1}, $\omega_\alpha = 1455$ cm^{-1}, $g_\alpha = 300$ cm^{-1}, $\gamma_\alpha = 10$ cm^{-1}.

In Table 1, we present the calculated positions of the electronic CT bands for parameters typical for tetracyanoquinodimethane salts: $t = 0.2$ eV, $U = 1.1$ eV.

Table 1. Positions of CT electronic excitation bands for different clusters: general expressions and the approximate wavenumbers of the CT bands obtained from numerical solution of the eigenvalue problem for the Hamiltonian (3).

n	Ns	ω_{CT}	ω_{CT}/cm^{-1}
1	2	Eq. (9)	3000
1	3	$\sqrt{2}\,t$	2300
1	4	t, $(1+\sqrt{5})\,t$	1500 5000
2	3	E_1, E_2	2300 11000
2	4	E_1, E_2, E_3, E_4	2200, 3500, 10000, 12000
2	2	Eq. (7)	10000
4	4	$E_1, E_2, ...$	8500, 13000,...

3. Modifications for LB films

The similarity of the optical properties of LB films and charge transfer salts based on the same molecular species has already been noted from the early absorption measurements [24]. This similarity signifies that the principal interactions are the same. For example, in the absorption of the LB film obtained by mixing in the same layer octadecyl (C_{18}) TCNQ and dimethyloctadecylsulfonium$^+$TCNQ$^-$ (ODS-TCNQ) in 1:1 composition shown in Fig. 2 (a), we observe a broad band of electronic CT excitations and a number of vibrational bands with Fano aniresonance shape, typical for excitation of a_g modes via the EMV coupling. The same features are present in molecular crystals with TCNQ although, strictly speaking, the symmetry of the TCNQ molecule is lowered by the attachment of an alkyl chain. If we consider the role of different atomic displacements in different a_g vibrations modes of TCNQ (see, e.g. [2]), we note that most of these modes do not involve any motion of hydrogen atoms of TCNQ and consequently are not sensitive to a substitution of one of the hydrogen atoms by the $C_{18}H_{37}$ chain. Clearly, it is not true for the mode ω_5, which is mainly related to C-H bond bending. Indeed, we observe new features in the corresponding part of the absorption spectrum. The band in the range 1050-1200 cm^{-1} is broader than in crystalline compounds and several peaks are observed superimposed on this broad band. We note that in the same wavenumber range several features in the Raman spectrum of the neutral form $C_{18}TCNQ^0$ have been attributed [8] to excitations in the C_{18} chain. We proposed [25] that vibrational modes of the C_{18} chain normally active in Raman spectroscopy are activated as a result of the coupling to a totally symmetric mode ω_5 of the TCNQ molecule, which is itself activated due to the EMV coupling mechanism, therefore involving a two-step coupling process.

Mathematically we can account for this effect by adding the term akin to Eq. (2) describing C_{18} chain vibrations with the normal coordinate $q_{\beta i}$ corresponding to the β-th mode with frequency ν_β of the C_{18} chain attached to the molecule i. The coupling itself is expressed by the term

$$\sum_{\alpha,\beta,i} k_{\alpha\beta}\, Q_{\alpha i}\, q_{\beta i}. \qquad (15)$$

where $k_{\alpha\beta}$ denote the coupling constant. The final expression for the complex conductivity of the assembly of independent clusters has the same matematical form as Eq. (4), except the function $D(\omega)$ differs from that given in Eq. (5) by an additional term in the denominator responsible for the description of the two-step coupling:

$$D(\omega) = \sum_{\alpha} \frac{g_{\alpha}^{2}\,\omega_{\alpha}}{\omega_{\alpha}^{2} - \omega^{2} - i\,\omega\,\gamma_{\alpha} - \sum_{\beta} \dfrac{k_{\alpha\beta}^{2}\,\omega_{\alpha}\,\nu_{\beta}}{\nu_{\beta}^{2} - \omega^{2} - i\,\omega\,\gamma_{\beta}}} \qquad (16)$$

Let us use this theory to discuss the absorption spectra [24] of $[\text{ODS-TCNQ}]_{\xi}{:}[C_{18}\text{TCNQ}]_{1-\xi}$ LB films for different values of ξ. We prefer to use this notation to define a stoichiometry of LB films instead of usual chemical formulas because in the homodoping preparation of a layer, it is not clear in advance what kind of aggregates will be built on the water surface and then transferred to the substrate. In fact, we intend to use spectroscopy to obtain this information.

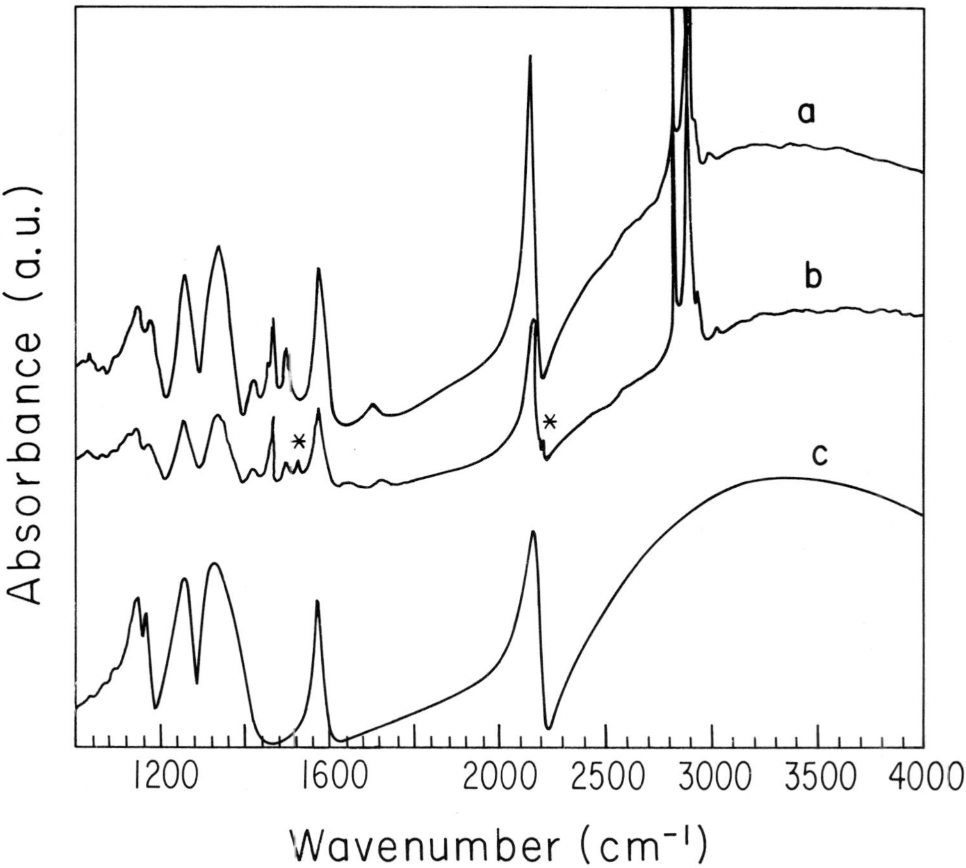

Fig. 2. Absorption spectra of $[\text{ODS-TCNQ}]_{\xi}{:}[C_{18}\text{TCNQ}]_{1-\xi}$ for (a) $\xi = 0.5$ and (b) $\xi = 0.25$. Curve (c) is the absorbance calculated from Eqs. (4), (6), (9) and (15). Adapted from Ref. [25].

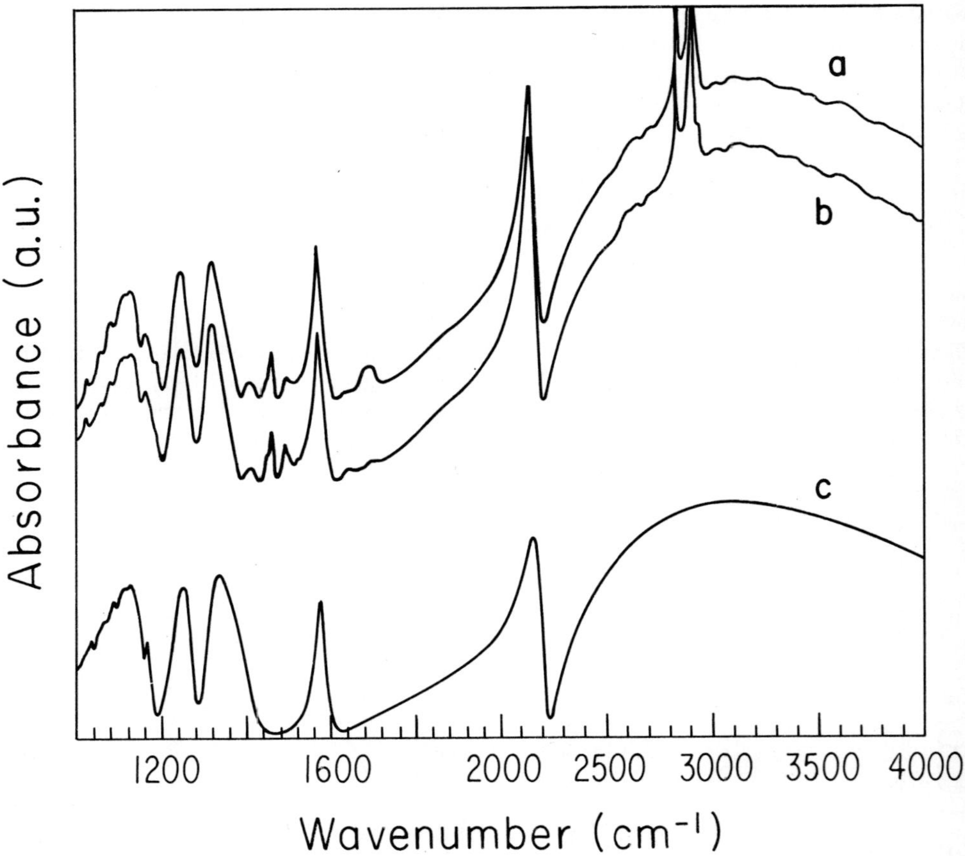

Fig. 3. Absorption spectra of $[\text{ODS-TCNQ}]_{\xi}{:}[\text{C}_{18}\text{TCNQ}]_{1-\xi}$ for (a) $\xi = 0.33$ and (b) $\xi = 0.285$. Curve (c) is the absorbance calculated from Eqs. (4), (6), (9) and (15). Adapted from Ref. [25].

For $\xi = 0.66$, the wavenumber of the electronic CT band at 8700 cm^{-1} points out the presence of quasi-isolated dimers ("B" band of Torrance [22]) as observed in a number of simple TCNQ ion-radical crystalline salts [26]. The observed CT band position corresponds well to that given by Eq. (7) for the values of $U = 1$ eV and $t = 0.2$ eV, typical for salts of TCNQ [5].

For $\xi = 0.5$ and 0.25, the electronic CT band shifts to a much lower energy of 3800 cm^{-1}, the value we expect for quasi-isolated dimers with a single electron (see Eq. (9)). In the case of $\xi = 0.25$, the absolute values for the absorbance with the same number of layers equals half that of the film with $\xi = 0.5$, but the spectra (a) and (b) in Fig. 2 are almost identical. It signifies that in both cases dimers with one electron are present, but simply the number of such dimers for $\xi = 0.5$ is two times larger than in the case of $\xi = 0.25$, just because in the latter case the number of radical electrons available is two times smaller. This assumption is supported by the presence of bands at 2220 and 1530 cm^{-1} which are signatures of neutral TCNQ molecules. These bands marked by an asterisk are observed only for $\xi = 0.25$, because this stoichiometry gives us only one electron per four molecules and it is shared by two molecules forming a dimer, the remaining two TCNQ molecules being neutral.

For $\xi = 0.33$, we may expect a formation of either dimers or trimers. However, the latter model gives EMV coupling constants very different from the commonly accepted values for crystalline salts and from the values obtained for films with $\xi = 0.5$ and 0.25, while the dimer model accords well with the expected values. Consequently we believe that the "optically active" clusters are dimers with one electron. The same kind of clusters are organized in the case of $\xi = 0.285$, where the intensity and frequency dependence of the absorbance are very close to those of $\xi = 0.33$, as shown in Fig. 3. It seems that formation of dimers is preferable in the case of LB films and probably is connected with the presence of a film-water interface which makes it difficult to ensure optimal mode of overlap between molecules for larger clusters [25].

We stress that even though ξ is a key parameter governing optical as well as electrical properties of the LB films [24], it is not the "average" charge per TCNQ site. Indeed, the best theoretical fits of the experimental absorption data are obtained for dimers with two electrons (average TCNQ charge is -1) for $\xi = 0.66$, and for dimers with a single electron (average TCNQ charge is -0.5) for $\xi = 0.5$, 0.33, 0.285 and 0.25.

4. Spectroscopy as a tool for studying molecular organization

In principle, X-ray investigations allow one to determine atomic positions and the relative orientation of constituent molecules. Unfortunately, this direct method cannot be applied in the case of powdered samples or LB films, and in this situation spectroscopic studies may help. Already early works on the optical properties of low-dimensional molecular CT crystals were used to obtain the information of whether we have a complete charge transfer in simple salts [26]. Charge transfer excitation occurs at a lower energy than any other electronic excitation, such as (for example) an intramolecular one. For an ionic salt we observe the CT band around 10000 cm^{-1}, labelled as the "B-band" in Torrance's classification [22]. In the case of a mixed-valence compound we detect a new absorption band, called the "A-band", at a lower wavenumber than the previous one. These CT bands can be interpreted as local elementary excitations. It is easy to see that, for a completely ionic system, this elementary excitation is associated with the creation of a doubly occupied site, whereas for a partially ionic one we just have to take account of the electrostatic interaction between two first-neighbour molecules. In a zero-order approximation, which neglects completely electron transfer interactions, the energies needed to produce the charge transfer are equal to the short-range Coulomb correlation values for the on-site repulsion energy (U) and that between electrons on two adjacent sites (V). These parameters (U and V) are those introduced in the extended Hubbard model

Along with the energy of the electronic CT band, some important information may be obtained from the positions and intensities of vibrational bands related to the linear electron-molecular vibration coupling. As demonstrated by Rice et al. [13], an intermolecular charge transfer polarized along the stack of dimers or any multimers induces (via linear EMV coupling) vibrations which are normally infrared inactive because they are intrinsically a_g modes. These vibrational modes are activated only if the translational lattice symmetry is broken initially or following a structural phase transition. For example, the absence of any a_g vibrational bands is considered a firm indication of the regular stacking [12], and a doublet structure of these bands points to a tetramerized structure [27,28].

On the basis of the calculations of the complex conductivity for different clusters, we can also try to get more details about the way molecules are organized. We note qualitatively that as the size of the cluster grows the charge transfer band moves to a lower wavenumber and its form more and more deviates from a simple Lotentzian. So, a decomposition of an experimental CT band in a number of Lorentzian bands and the use of Table 1 give us an indication of the size of the cluster and estimates of fundamental parameters (transfer integrals and electronic correlations). At the second stage, we can calculate the charge distribution in the cluster and obtain the frequencies of molecular vibrations (they are usually assumed to be linearly dependent on the charge [29]). Now we can verify our model by comparing the calculated wavenumbers with the experimentally measured vibrational bands in the Raman spectrum. Finally, from the positions of the vibrational band in the infrared spectrum, we can estimate (see [8]) the values of the EMV coupling constants from the shift of the band in the infrared with respect to the unperturbed wavenumber (measured by Raman spectroscopy). Additional support for the model of molecular organization may be found by examining a possible fine structure of vibronic bands as predicted in the case of a non-uniform equilibrium charge distribution [30].

To conclude, the absorption spectra of CT crystals and LB films can provide information about molecular organization. The next step would be a controlled construction of supermolecular aggregates sufficiently sensitive to photo-, thermo-, or voltage-induced charge transfer.

References

[1] K. Kamaras, K. Holczer, A. Janossy, Infrared Spectra of the Neutron Irradiated Quasi-One-Dimensional Charge Transfer Salts TEA(TCNQ)$_2$ and Qn(TCNQ)$_2$, *Phys. Status Solidi (b)*, Vol. 102, 1980, pp. 467-474.

[2] V.M. Yartsev, R. Świetlik, Infrared Properties of Quasi-One-Dimensional Organic Semiconductors, *Reviews of Solid State Science*, Vol. 4, 1990, pp.69-117.

[3] L.R. Melby, R.J. Harder, B. Welber, F.B. Kaufman, P.E. Seiden, Substituted Quinodimethans. II. Anion-Radical Derivatives and Complexes of 7,7,8,8-Tetracyanoquinodimethan, *J. Chem. Soc*, Vol. 84, 1962, pp. 3374-3387.

[4] J.R. Ferraro, J.M. Williams. *Introduction to Synthetic Electrical Conductors*, Academic Press, New York, 1987.

[5] A. Graja. *Low-Dimensional Organic Conductors*, World Scientific, Singapore, 1992.

[6] A. Ruaudel-Teixier, M. Vandevyver, A. Barraud, Novel Conducting LB Films, *Molec. Cryst. Liq. Cryst.*, Vol. 120, 1985, pp. 319-322.

[7] A. Graja, Optical Properties, *Organic Conductors. Fundamentals and Applications*, Marcel Dekker, New York, 1994, pp. 229-267.

[8] P. Delhaes, V.M. Yartsev, Electronic and Spectroscopic Properties of Conducting Langmuir-Blodgett Films, *Spectroscopy of New Materials*, John Wiley & Sons, Chichester, 1993, pp. 199-289.

[9] G.R. Andersen, J.P. Devlin, Electron Oscillation Effects in the Vibrational Spectra of Tetracyanoquinodimethane Ion Radical Salts, *J. Phys. Chem.*, Vol. 79, 1975, pp. 1100-1102.

[10] E.E. Ferguson, F.A. Matsen, Enhancement of Infrared Absorption Bands of Charge Transfer Complexes, *J. Chem Phys*, Vol. 29, 1958, pp. 105-107.

[11] J. Richard, P. Delhaes, M. Vandevyver, Electronic and Spectroscopic Properties of Conducting Langmuir-Blodgett Films Based on Semi-amphiphilic TCNQ Salts: a Comprehensive Comparison with Bulk Crystalline Molecular Conductors: a Review, *New J. Chem.*, Vol. 15, 1991, pp. 137-147.

[12] R. Bozio, C. Pecile, Charge Transfer Crystals and Molecular Conductors, *Spectroscopy of Advanced Materials*, John Wiley & Sons, Chichester, 1991, pp. 1-86.

[13] M.J. Rice, V.M. Yartsev, C.S. Jacobsen, Investigation of the Nature of the Unpaired Electron States in the Organic Semiconductor MEM(TCNQ)$_2$, *Phys. Rev. B*, Vol. 21, 1980, pp. 3437-3446.

[14] N.O. Lipari, C.B. Duke, R. Bozio, A. Girlando, C. Pecile, A. Padva, Electron-Molecular Vibration Coupling in 7,7,8,8-Tetracyano-p-quinodimethane (TCNQ), *Chem. Phys. Lett.*, Vol. 44, 1976, pp.236-240.

[15] P.J. Strebel, Z.G. Soos, Theory of Charge Transfer in Aromatic Donor-Acceptor Crystals, *J. Chem. Phys.*, Vol. 53, 1970, pp. 4077-4090.

[16] J. Hubbard, Generalized Wigner Lattices in One Dimension and Some Applications to Tetracyanoquinodimethane (TCNQ) Salts, *Phys. Rev. B*, Vol. 17, 1978, pp. 494-505.

[17] M.J. Rice, Towards the Experimental Determination of the Fundamental Microscopic Parameters of Organic Ion-Radical Compounds, *Solid State Commun.*, Vol. 31, 1979, pp. 93-98.

[18] R. Bozio, A. Girlando, C. Pecile, Correlation between Infrared Spectra and Magnetic and Optical Properties of Potassium Chloranil. Effects of Phase Transition and Solvation Processes, *Chem. Phys.*, Vol. 21, 1977, pp. 257-263.

[19] S. Matsuzaki, V.M. Yartsev, Charge Distribution in Mixed-Valence Trimer, *Solid State Commun.*, Vol. 89, 1994, pp. 941-944.

[20] V.M. Yartsev, Optical Properties of Quasi-One-Dimensional Molecular Crystals with Non-Equilibrium Charge Carriers, *J. Appl. Spectrosc.*, Vol. 45, 1986, pp. 744-748.

[21] V.M. Yartsev, Electron-Molecular Vibration Coupling in Trimerized Organic Ion-Radical Semiconductors, *Phys. Status Solidi (b)*, Vol. 112, 1982, pp. 279-287.

[22] J.B. Torrance, The Difference between Metallic and Insulating Salts of Tetracyanoquinodimethane (TCNQ): How to Design an Organic Metal, *Acc. Chem. Res.* Vol. 12, 1979, pp. 79-86.

[23] V.M. Yartsev, Charge Transfer and Electron-Molecular Vibration Coupling in Tetramerized Quasi-1D Semiconductors, *Phys. Status Solidi (b)*, Vol. 126, 1984, pp. 501-510.

[24] A. Ruaudel-Teixier, M. Vandevyver, M. Roulliay, J.-P. Bourgoin, A. Barraud, M. Lequan, R.M. Lequan, A New Strategy for Building Conducting Langmuir-Blodgett Films, *J. Phys. D: Appl. Phys.*, Vol. 23, 1990, pp. 987-990.

[25] V.M. Yartsev, J.-P. Bourgoin, P. Delhaes, M. Vandevyver, A. Barraud, Spectroscopic Properties of Conducting Langmuir-Blodgett Films Obtained by the "Homodoping Strategy", *J. Chem. Phys*, Vol. 99, 1993, pp. 3092-3099.

[26] D.B. Tanner, C.S. Jacobsen, A.A. Bright, A.J. Heeger, Infrared Studies of the Energy Gap and Electron-Phonon Interaction in Potassium-Tetracyanoquinodimethane (K-TCNQ), *Phys. Rev. B*, Vol. 16, 1977, pp. 3283-3290.

[27] R. Świetlik, V.M. Yartsev, Infrared Spectra of Quasi-One-Dimensional Semiconductors with Tetramerized Stacks, *Molec. Cryst. Lyq. Cryst*, Vol. 230, 1993, pp. 375-380.

[28] R. Świetlik, Infrared Spectra of Tetramerized Organic Semiconductors PrPht$(TCNQ)_2$ and PrQuin$(TCNQ)_2$ - the Problem of a_g Pnonon Mode Splitting, *Synth. Metals*, Vol. 74, 1995, pp. 115-122.

[29] S. Matsuzaki, R. Kuwata, K. Toyoda, Raman Spectra of Conducting TCNQ Salts; Estimation of the Degree of Charge Transfer from Vibrational Frequencies, *Solid State Commun.*, Vol. 33, 1980, pp. 403-405.

[30] V.M. Yartsev, M.J. Rice, Conductivity Spectrum Fine Structure of One-Dimensional Organic Semiconductors, *Phys. Stat. Solidi (b)*, Vol. 100, 1980, pp. K97-K100.

π-d interaction in phthalocyanine conductors

Kyuya Yakushi, Toshihiro Hiejima and Hideo Yamakado

Institute for Molecular Science and Graduate University for Advanced Studies
Myodaiji, Okazaki, Aichi 444, Japan

Abstract

Partially oxidized phthalocyanine conductors such as $CoPc(AsF_6)_{0.5}$ and $NiPc(AsF_6)_{0.5}$ consist of one-dimensional double chain comprising a transition-metal and organic macrocycle chains. Each of these chains generates a narrow d-band and a wide π-band. The d-band is located near the Fermi level of the π-conduction-band. The interplay between these π- and d-bands will be introduced. The first examples of the interplay is the influence of the magnetic moments of unpaired d-electrons at the Co site to the itinerant π-holes on Pc chain. The exchange coupling with these magnetic moments suppresses the coherent motion of π-holes in $CoPc(AsF_6)_{0.5}$. This conclusion was derived from the comparative study of the isostructural conductors, non-magnetic $NiPc(AsF_6)_{0.5}$ and magnetic $CoPc(AsF_6)_{0.5}$. Another example is the transfer of holes from π- to d-band which is induced by high pressure. This *charge transfer* occurs at 0.5 GPa and continues up to 6 GPa in $NiPc(AsF_6)_{0.5}$.. The same phenomenon was observed in $CoPc(AsF_6)_{0.5}$ as well. A metal-nonmetal transition is induced by the pressure at which the charge transfer begins in $NiPc(AsF_6)_{0.5}$.

1 Introduction

One-dimensional conductor based on a metallo-phthalocyanine (MPc) are the double-chain system which involves two conductive pathways in the same molecular column. The double-chain system is organized by the metal and macrocycle chains, consequently comprising two one-dimensional bands. One of these energy band is a wide π-band derived from the π-orbital (HOMO) of the macrocycle, and another is a narrow d-band from the $3d_{z^2}$-orbital of the central metal ion. Several groups has studied phthalocyanine conductors using several central metal ions (Co, Ni, Cu, Pt), macrocycle ligands (phthalocyanine, tetrabenzophorphyrine), and counter anions (I_3^-, BF_4^-, ClO_4^-, AsF_6^-, SbF_6^-), and has found the same double-chain structure in many phthalocyanine conductors.[1] This narrow d-band is closed to the Fermi level of the π-band.[2] Furthermore, they are not hybridized in the same molecule, or at the Γ point in the Brillouin zone because of the symmetry restriction. It is a delicate and interesting problem to which band the holes are doped by the counter anions. Since the bandwidth of the π- and d-bands are quite different, the solid property strongly depends upon the character of the doped holes.

Where are the holes are doped, π-band or d-band , in other words, what part is oxidized, metal or Pc,? This problem has been discussed from the beginning of the study of phthalocyanine conductors. The holes are doped into the π-band in almost all phthalocyanine conductors except in CoPcI. Interestingly, the π-holes are produced in the isostructural $CoPc(AsF_6)_{0.5}$.[3] In the first section we will discuss about the different properties of $CoPc(AsF_6)_{0.5}$ from those of $NiPc(AsF_6)_{0.5}$, and where this difference comes from. The lattice constant along the stacking axis, in other words, the metal-to-metal distance, is shortest in CoPcI. If this quantity is the parameter to distinguish the oxidized

Springer Proceedings in Physics, Vol. 81
Materials and Measurements in Molecular Electronics
Editors: K. Kajimura · S. Kuroda

part, the application of high pressure to $CoPc(AsF_6)_{0.5}$ will shorten the lattice constants and it might control the character of the holes. In the second section we will discuss about this problem.

2 Comparative study of two isostructural conductors, $NiPc(AsF_6)_{0.5}$ and $CoPc(AsF_6)_{0.5}$

2.1 Crystal structure

The crystal of $CoPc(AsF_6)_{0.5}$ belongs to a tetragonal system, the lattice parameters being $a = 14.234$, $c = 6.296$ Å, and Z = 2.[2] The molecules on a 4-fold axis are stacked exactly in the metal-over-metal overlapping mode. The neighboring molecules are equi-distantly arranged and rotated about 40 ° with each other. The repeating unit along the c-axis is two CoPc molecules. The different point from the following $NiPc(AsF_6)_{0.5}$ is the arrangement of AsF_6^-, which is displaced by 0.6 Å from the molecular sheet of CoPc and has not a three-dimensional order.

The crystal of $NiPc(AsF_6)_{0.5}$ belongs to an orthorhombic system, the lattice parameters being $a = 14.015$, $b = 28.485$, $c = 6.466$ Å, and $Z = 4$.[4] The unit cell contains two molecular columns, and has a pseudo-tetragonal symmetry. In this molecular column NiPc is stacked nearly in the metal-over-metal overlapping mode. NiPc is alternately displaced by -0.21 Å to the direction of b-axis from the stacking axis (0.5, 0.25, z), so that they are stacked in a zigzag way. The repeating unit along the c-axis is two NiPc molecules.

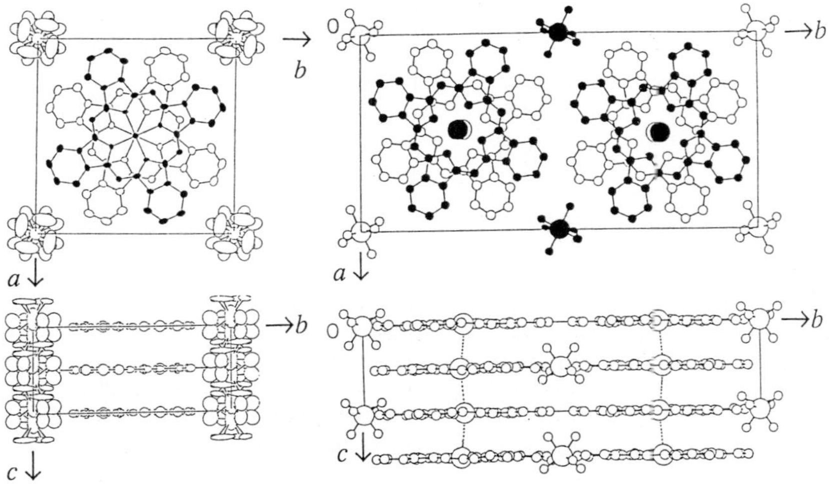

Figure 1:Crystal Structures of $CoPc(AsF_6)_{0.5}$ and $NiPc(AsF_6)_{0.5}$

2.2 Electrical resistivity and thermopower

The left panel of Fig. 2 shows the comparison of the high-frequency (9.4 GHz) electrical resistivity measured by a cavity perturbation method. The resistivity of $NiPc(AsF_6)_{0.5}$ decreases on lowering the temperature down to 40 K. In contrast to this metallic behavior of $NiPc(AsF_6)_{0.5}$, the room-temperature resistivity of $CoPc(AsF_6)_{0.5}$ is one order of magnitude higher and the resistivity increases on lowering the temperature in all temperature range from room temperature. As shown here, the electrical properties of these isostructural compounds are very different from each other.

The thermopowers of $NiPc(AsF_6)_{0.5}$ and $CoPc(AsF_6)_{0.5}$ are shown in the right panel of Fig. 2. Correspondingly to the metallic behavior of resistivity, the thermopower of $NiPc(AsF_6)_{0.5}$ also shows a typical metallic behavior: the absolute value of thermopower is small and linearly decreases on lowering the temperature. On the other hand, the thermopower of $CoPc(AsF_6)_{0.5}$ is twice as large as the room-temperature value of $NiPc(AsF_6)_{0.5}$ and is almost temperature independent. Interestingly, this constant value, 50 μV/K, is close to $(k_B / |e|) ln2$ (= 60 μV/K) which is theoretically expected in the one-dimensional Hubbard model of a large-U limit.

These behaviors of the electrical resistivity and thermopower are similar to those of $(TMTTF)_2X$[5] and $Ag(DMe\text{-}DCNQI)_2$[6], where Coulomb interaction is responsible for the bandgap. The corresponding values of the temperature-independent region are +36 μV/K for $(TMTTF)_2BF_4$[5] and -60 μV/K for $Ag(DMe\text{-}DCNQI)_2$ [7]. The turn to the negative sign at low temperature is explained by the presence of the n-type impurity levels, as in the case of $(TMTTF)_2X$.[5] The large-U behavior itself is not unusual in organic conductors. Due to the narrow bandwidth and on-site Coulomb repulsion, almost all one-dimensional organic conductors show such properties. However, when we compare with other metallic compounds in phthalocyanine family such as NiPcI, H_2PcI, or $NiPc(AsF_6)_{0.5}$, the conductivity and thermopower of $CoPc(AsF_6)_{0.5}$ are very unusual, because they have the same molecular size and shape, similar nature of HOMO, and almost the same crystal structure.

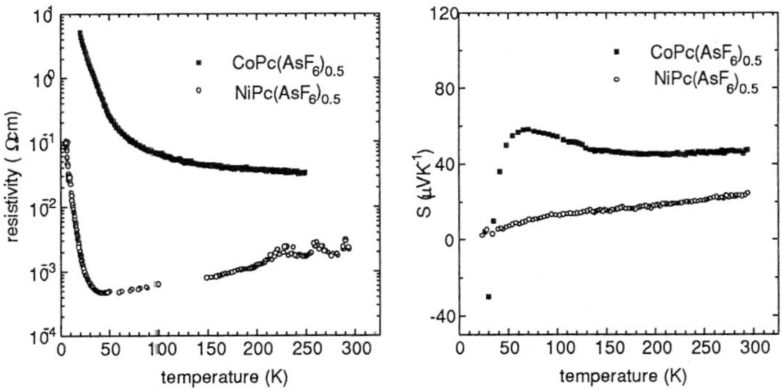

Figure 2: Microwave conductivity and thermoelectric power

2.3 Reflection spectrum

To know the density of state near the Fermi level, the optical spectrum is a good probe. However, the spectral region in this study is restricted down to 1000 cm^{-1} due to the small size of the crystal (2-3 mm in length and 50 μm in width). Figure 3 shows the polarized reflection spectra of $CoPc(AsF_6)_{0.5}$ and $NiPc(AsF_6)_{0.5}$. Since the one-dimensional band is produced along the c-axis, the optical transition near the Fermi level appears in the //c polarized spectrum. The optical transition in the ⊥ c spectrum corresponds to the interband transition with a character of an intramolecular local excitation. The //c reflectance spectrum of $CoPc(AsF_6)_{0.5}$ is quite different from a typical Drude-like curve which is observed in $NiPc(AsF_6)_{0.5}$. The reflectivity reaches 50 % at the highest and the reflectivity minimum near the plasma edge is not low enough. As shown in Fig. 4, the optical conductivity σ(ω) of $CoPc(AsF_6)_{0.5}$ calculated by the Kramers-Kronig transformation has a peak at

2300 cm^{-1}. This conductivity spectrum suggests the existence of a energy gap or pseudogap. It is difficult to know the absorption edge from the reflectance spectrum, since the spectral region is restricted down to 1000 cm^{-1}. The density of state at Fermi level may remain and make a pseudogap as shown in the right panel of Fig. 4, since the thermopower is not semiconductor-like but shows a

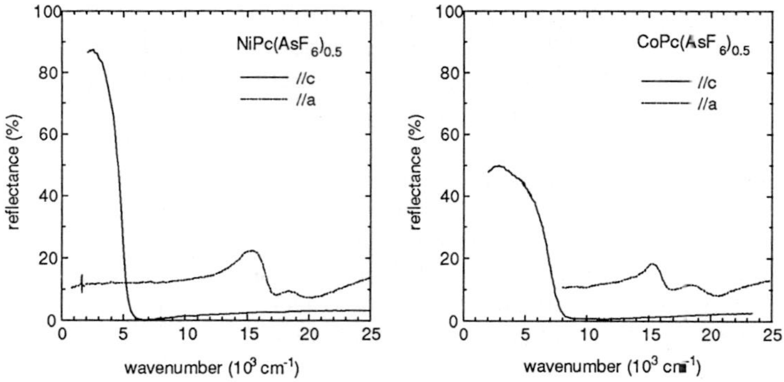

Figure 3: Polarized reflection spectra of CoPc(AsF$_6$)$_{0.5}$ and NiPc(AsF$_6$)$_{0.5}$

localized nature. To clarify the origin of this energy gap, we examined a possibility of the pseudo Peierls gap caused by the fluctuation of the lattice distortion. So far this kind of fluctuation of the lattice dimerization has been observed in many organic conductors such as TTF-TCNQ[8],(TMTSF)$_2$X[9],(TMTTF)$_2$X[9,10], (BEDT-TTF)$_2$X[11], M(DMe-CNQI)$_2$[12], *etc.*

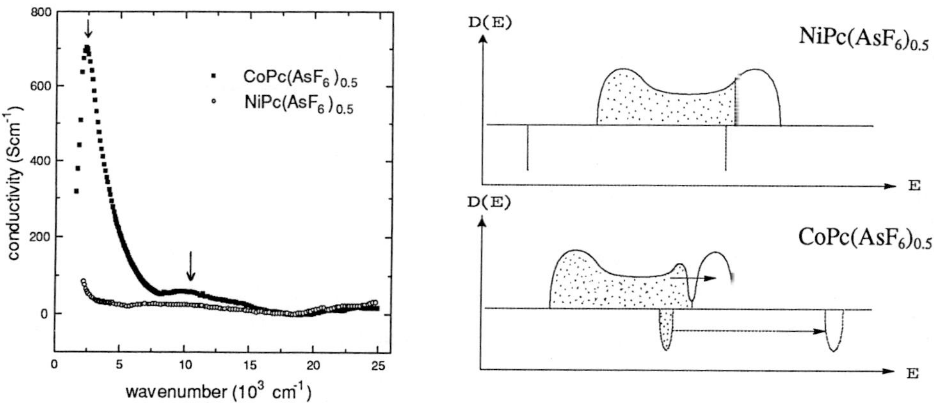

Figure 4: Conductivity spectra and schematic density of state of CoPc(AsF$_6$)$_{0.5}$ and NiPc(AsF$_6$)$_{0.5}$

In these cases, several strong vibronic modes (a_g mode) appear and increase their intensity on lowering the temperature. We measured the low-temperature absorption spectra of the powdered samples of NiPc(AsF$_6$)$_{0.5}$ and CoPc(AsF$_6$)$_{0.5}$ in the infrared region. However, no such vibronic modes were found in both compounds down to 11 K. This result shows that the glide-plane symmetry which requires the equi-distant stacking of CoPc and NiPc is not broken down to 11 K. This means that the one-dimensional lattice of CoPc(AsF$_6$)$_{0.5}$ and NiPc(AsF$_6$)$_{0.5}$ is very stable against the 2k$_F$ lattice distortion ($\lambda_F = 2c$). Thus the energy gap of CoPc(AsF$_6$)$_{0.5}$ is not ascribable to the Peierls gap. The upturn of the resistivity of NiPc(AsF$_6$)$_{0.5}$ at 40 K is also not attributed to the Peierls transition. The optical spectrum of $CoPc(AsF_6)_{0.5}$ resembles those of $(TMTTF)_2X$ and $Ag(DMe\text{-}DCNQI)_2$. The character of the 2300 cm^{-1} absorption band is, therefore, similar to the 2100 cm^{-1} band of $(TMTTF)_2X$[9,10] and 3000 cm^{-1} band of $Ag(DMe\text{-}DCNQI)_2$[12], which are assigned to the optical transition across the Hubbard gap. The resemblance of the thermopower of $CoPc(AsF_6)_{0.5}$ to that of $(TMTTF)_2X$ also suggest the Hubbard gap in $CoPc(AsF_6)_{0.5}$.

Another point which should be noticed in $CoPc(AsF_6)_{0.5}$ is the small hump at 10 000 cm^{-1} in $\sigma(\omega)$ of the //c spectrum of Fig. 4. This hump cannot be observed in the spectra of isostructural $NiPc(AsF_6)_{0.5}$. The electronic configurations of the central metals of MPc are denoted as $(3d)^7$ for Co^{2+}, $(3d)^8$ for Ni^{2+}. Since the singly occupied $3d_{z^2}$ orbital of Co^{2+} is elongated to the neighboring $3d_{z^2}$-orbital in $CoPc(AsF_6)_{0.5}$, a transfer of charge will occur between adjacent Co^{2+} ions through the overlap of $3d_{z^2}$-orbitals. We therefore attribute this hump to the charge-transfer transition such as $Co^{2+}Co^{2+} \rightarrow Co^{1+}Co^{3+}$. Since Co^{2+} in CoPc is aligned along the c-axis and Co^{2+} is expected to have a large on-site Coulomb energy, the $3d_{z^2}$-orbital forms a one-dimensional Hubbard band, which splits into two due to the strong Coulomb repulsion. This optical transition is equivalent to the interband transition across this Hubbard gap as shown in the right panel of Fig. 4. From the excitation energy (~1.3 eV) and intensity (oscillator strength: ~0.09) of this optical transition, we roughly estimated the effective on-site Coulomb energy U_{eff} as ~1.5 eV and the transfer integral $t_{M\text{-}M}$ as ~0.15 eV using the one-dimensional Hubbard model.[13] Since the transfer integral is an order of magnitude smaller than the on-site Coulomb energy, the electrons on the Co^{2+} chain is rather localized and thus have a spin degree of freedom. These spins on Co^{2+} are antiferromagnetically coupled through the super-exchange interaction.

2.4 Magnetic susceptibility

Figure 5 shows the temperature dependence of the static magnetic susceptibilities of $CoPc(AsF_6)_{0.5}$ and $NiPc(AsF_6)_{0.5}$. The diamagnetic contribution coming from the core electrons is subtracted using the experimental values of NiPc and AsF_6^-. There is a big difference in the absolute value of the paramagnetic susceptibilities of these two compounds. The small and almost temperature independent $\chi_p(T)$ of metallic $NiPc(AsF_6)_{0.5}$ is attributable to the Pauli paramagnetism. We think that the large paramagnetic susceptibility of $CoPc(AsF_6)_{0.5}$ mainly comes from the localized and antiferromagnetically interacting unpaired electrons on Co. To discuss the magnitude of the magnetic susceptibility, we tentatively divided this curve into two components: the Curie-Weiss component and temperature-independent component. The least-squares fitting by the equation,

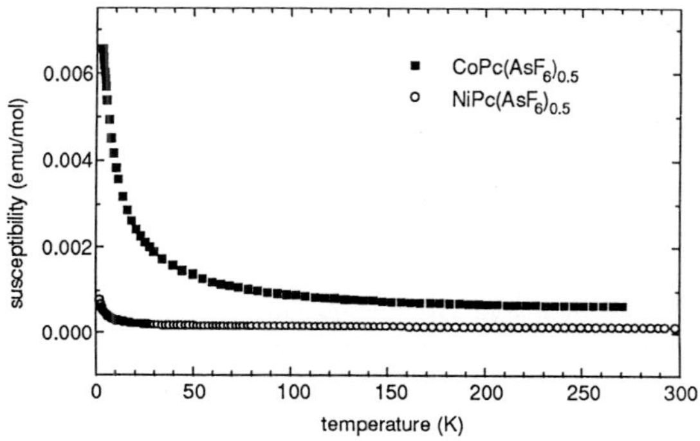

Figure 5: Magnetic susceptibilities of $CoPc(AsF_6)_{0.5}$ and $NiPc(AsF_6)_{0.5}$

$$\chi_p(T) = \chi_c + C/(T - \theta)$$

gives the following parameters: $C = 0.052 \pm 0.005$ emu mol^{-1} K, $\theta = -5.2 \pm 0.1$ K, $\chi_c = (4.3 \pm 0.1) \cdot 10^{-4}$ emu mol^{-1}, where $C = Ng^2 m_B^2 S(S+1)/3k_B$, θ, and χ_c are the Curie constant, Weiss temperature, and constant paramagnetic term, respectively. Assuming the g-value of 2.0023 and S=1/2, 10 % of CoPc contributes to the Curie-Weiss component. If the localized paramagnetic center has a higher spin state (S=1 or 3/2), the spin concentration becomes more small. From this Curie constant we can conclude that the magnetic moment on Co^{2+} is not free but suppressed by an antiferromagnetic interaction. The interpretation of the nearly constant term, which is slowly increasing upon temperature, is not simple, since it comes from the magnetic moment of Co^{2+} coupled with π-holes. It should be interpreted by the one-dimensional ladder-like spin system. Let us compare the magnitude of almost temperature-independent term with other organic conductors. The value of χ_c ($4.3 \cdot 10^{-4}$ emu mol^{-1}) resembles the large-U organic conductors, in which the corresponding value in the region of 100-300 K falls in the range of $5\text{-}6 \cdot 10^{-4}$ emu mol^{-1} for $(TMTTF)_2X$ (X=ClO_4, PF_6, and NO_3)[14] and $(MeX\text{-}DCNQI)_2Ag$ (X = Me, Cl, and Br)[15].

2.5 Exchange interaction

We pointed out that the π-holes of $CoPc(AsF_6)_{0.5}$ behave like those of the large-U organic conductors. The thermopower suggests the localized nature of the π-holes, and the temperature-independent term of the magnetic susceptibility is significantly larger than the ordinary Pauli paramagnetism of metallic phthalocyanine salts. The electrical resistivity shows nonmetallic temperature dependence. The optical spectrum polarized along the conducting axis shows an energy gap at Fermi level. The relation between $NiPc(AsF_6)_{0.5}$ and $CoPc(AsF_6)_{0.5}$ is similar to that of the isostructural materials, $(TMTSF)_2X$ and $(TMTTF)_2X$. In $(TMTSF)_2X$, the localized nature of the charge carrier is more weakened and shows metallic behavior. The crucial point to distinguish them is the presence of the interchain interaction (t~0.02 eV) in $(TMTSF)_2X$. The situation is almost the same in $(DMe\text{-}DCNQI)_2M$ (M=Ag and Cu), where the interchain interaction through the hybridization of the π-orbital of DMe-DCNQI and Cu 3d-orbital plays an important role in the metallic nature of $(DMe\text{-}DCNQI)_2Cu$.

In MPc, the difference seems to be in the inside of molecule. The configurations of the highest occupied d-orbitals are $[3d_{z^2}(a_g)]^2$ for NiPc, $[3d_{z^2}(a_g)]^1$ for CoPc. So the localized unpaired electron at the metal seems to be coupled strongly with the π-hole. To see the effect of exchange coupling, CuPcI is the suggestive compound, which contains a localized unpaired electron on Cu2+ as an electronic configuration of $[3d_{x^2-y^2}(b_{1g})]^1$. In this compound, the π-holes on Pc chain are almost magnetically decoupled with the localized spins on Cu^{2+}.[16] In this weak coupling case, the π-holes are more delocalized, so that the electrical resistivity, thermopower and optical spectrum of CuPcI more resemble NiPcI rather than $CoPc(AsF_6)_{0.5}$. These results suggests that the coupling parameter $J_{M\text{-}L}$ of CoPc salts seems to be much larger than that of CuPcI. Mishima have made a simple model to describe the ground state of $CoPc(AsF_6)_{0.5}$ and CuPcI, taking account of only $t_{M\text{-}M}$, $t_{L\text{-}L}$, and $J_{M\text{-}L}$. In this model $J_{M\text{-}L}$ produces a coupled spin system between π- and d-electrons, making a SDW state. According to Mishima's numerical calculation, the bandgap is formed by $J_{M\text{-}L}$ when $J_{M\text{-}L}/t_{M\text{-}M}$ overcomes a critical value which depends upon $t_{L\text{-}L}$.[17] This model may qualitatively explain the solid-state properties of $CoPc(AsF_6)_{0.5}$.

The strong coupling in CoPc is suggested by the spin density distribution determined by a spin polarized neutron diffraction study.[18] According to G.A.Williams *et al.*, the spin populations in CoPc are $3d_{xy}^{0.4(10)}$, $3d_{xz,yz}^{0.17(10)}$, $3d_{z^2}^{0.79(12)}$, $3d_{x^2-y^2}^{-0.21(10)}$, $4s^{-0.14(6)}$ for Co orbital and total population on macrocycle N and C atoms is -0.17(5) spin. Although the local spin is populated mainly in $3d_{z^2}$ orbital, it is also distributed in not only other 3d or 4s orbitals of metal but also the ligand orbital through spin polarization effects. Especially the local spins on the macrocyclic ligand induced by spin polarization effect would more strongly interact with π-holes.

3 Pressure-induced charge transfer

3.1 $NiPc(AsF_6)_{0.5}$: metal-nonmetal phase transition

Figure 6 shows the absorption spectra of $NiPc(AsF_6)_{0.5}$ and neutral NiPc along with the schematic band structure. The main absorption band, Q-band, corresponds to an interband transition from the HOMO (a_{1u}) band to the LUMO (e_g) band, both of which are mainly made by the π-orbital of macrocycle. In $NiPc(AsF_6)_{0.5}$ two additional absorptions (A and B) appear at 3250 cm^{-1} (A) and 18500 cm^{-1}(B). As shown in the schematic band structure, the HOMO band of $NiPc(AsF_6)_{0.5}$ has a vacant state due to the partial oxidation or π-hole doping. The absorption band B has been assigned to the interband transition from the lower filled band (e_g) to the HOMO band (a_{1u}).[19] Thus, this optical transition is observable only when the HOMO band is not completely filled. As will be discussed later, the band A is assigned to the absorption of the plasmon of the conduction electron.

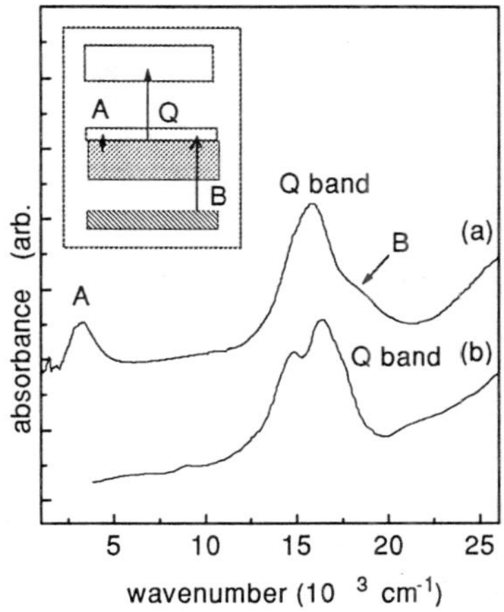

Fig. 6. Absorption spectra of the powdered samples of (a) $NiPc(AsF_6)_{0.5}$ and (b) NiPc. The inset shows the schematic diagram of the band structure of $NiPc(AsF_6)_{0.5}$.

The left panel of Fig. 7 shows the pressure dependence of the interband transitions. Upon applying pressure, the band B decreased its intensity at 0.5 GPa and disappeared above 1.0 GPa. Since the band B appears only when the HOMO band is incompletely filled, this result suggests that the HOMO-band is gradually filled from other filled energy band due to the effect of pressure. To examine the suggestion that the π-conduction band is filled by pressure and to know where the charge comes from, we conducted the pressure-dependent infrared absorption spectrum. The valence of the macrocycle has been characterized by the intensity of some diagnostic bands.[3] The vibrational mode characteristic of the oxidized macrocycle (Pc(1-)) is the 1356 cm^{-1} band, which is not observable in neutral Ni(2+)Pc(2-), whereas the 1534 cm^{-1} band is the characteristic vibration of the neutral phthalocyanine (Pc(2-)).[3] The right panel of Fig. 7 shows the pressure dependence of the infrared spectra in the range of 1300-1700 cm^{-1}. The intensities of the 1356 cm^{-1} band begins to decrease above 0.5 GPa, and the 1534 cm^{-1} band increases above 1.4 GPa. On the other hand, all other peaks shift monotonously to the high-wavenumber side, and they do not change their relative intensities under high pressure up to 4.0 GPa. Any additional peak was not observed up to 4.0 GPa. Furthermore, the 700 cm^{-1} band of AsF_6^-, which is incorporated as a counter anion, does not change under high pressure up to 4.0 GPa. The pressure dependence of this infrared spectrum was reversible for the compression and decompression processes as well as the visible spectrum.

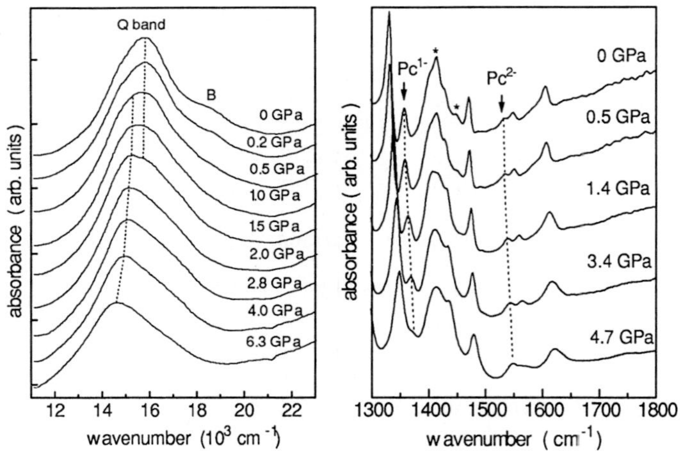

Figure 7: Pressure dependence of the interband transition and infrared modes.

The pressure dependence of the 1356 cm^{-1} and 1534 cm^{-1} bands indicates that the macrocycle approaches to the valence of Pc(2-), which is consistent with the pressure-induced charge transfer suggested by the band B under high pressure. No splitting or no additional peak at high pressure means that the high symmetry (D_{4h}) of the phthalocyanine molecule is maintained and also that the lattice distortion such as a Peierls distortion is not induced in the process of pressure-induced charge transfer. The behavior of the 700 cm^{-1} band of AsF_6^- excludes the possibility that the negative charge comes from the AsF_6^- anion. The infrared modes as well as the band B indicate that the valence of Pc(1.5-) at ambient pressure changes to Pc(2-) at high pressure, in other words, the π-hole begins to decrease under high pressure above *ca.* 0.5 GPa. Since the counter anion, AsF_6^-, shows no anomaly, the most reasonable source of the negative charge is the central metal. We call this phenomenon the pressure-induced d-π charge transfer. Under high pressure, therefore, the central metal must be a mixed-valent state. The ESR or magnetic susceptibility at high pressure may give a crucial information about the valence of the central metal. We consider that the formal valence is expressed as Ni(2.5+)Pc(2-) $(AsF_6(1-))_{0.5}$ at 4.7 GPa, since the 1356 cm $^{-1}$ band almost vanishes at this pressure.

Since the π-hole concentration appears to decrease upon applying high pressure, the pressure should make a significant influence on the plasmon which appears in the infrared region. The longitudinal mode such as a plasmon is observable when the particle size of the powdered sample is smaller than the wavelength of the light. According to the effective medium theory,[20] the absorption spectrum of a plasmon for a one-dimensional metal is given by the following equation,

$$\alpha(\omega) = \frac{3\pi\varepsilon_\alpha^{3/2}}{c} \frac{\varepsilon_{2z}(\omega)\omega}{(\varepsilon_{1z}(\omega)+2\varepsilon_\alpha)^2+\varepsilon_{2z}^2(\omega)}$$

, where $\alpha(\omega)$ is the extinction coefficient, $\varepsilon_{1z}(\omega)$ and $\varepsilon_{2z}(\omega)$ the real and imaginary parts of the ω-dependent dielectric constant of the *z*-component (conducting axis) of the material, ε_α the dielectric constant of the pressure medium, *c* the light velocity. The following approximations are assumed in this equation: each particle has a spherical shape (depolarization factor, P=1/3) and the concentration of the particles in the pressure cell is very low (filling factor, f=0). Based on this equation, we simulated the absorption spectrum using the Drude-type dielectric function. From the analysis of the polarized reflectance spectra of the single crystal shown in Fig. 3, the Drude parameters are obtained as $\omega_p = 7170\ cm^{-1}$ for plasma frequency, $1/\tau = 450\ cm^{-1}$ for the relaxation rate, and $\varepsilon_c = 2.18$ for the dielectric constant coming from the high frequency region. The dielectric constant of the pressure medium was approximated by the square of the refractive index of solid paraffin ($n_0 = 1.433$ at 589 nm).[21] The simulated spectrum showed an excellent agreement[22]. From this results, the 3250 cm^{-1} absorption is safely assigned to the absorption band of the plasma oscillation. The peak position and intensity of the absorption band of plasmon is characterized by the plasma frequency (ω_p). The peak position (ω_0) is given by $\omega_0 = \omega_p(\varepsilon_c + 2\varepsilon_\alpha)^{-1/2}$. From the numerical evaluation the integrated intensity is nearly proportional to ω_p^2, which is proportional to the π-hole concentration. Thus, the intensity and the peak position are not independent with each other. For example, this absorption should show a red shift and weaken the intensity when the concentration of the carrier decreases.

Figure 8 shows the pressure dependence of the plasmon absorption. The intensity of this absorption decreased upon pressure as we expected, since the π-hole concentration decreases on applying pressure. Contrary to our expectation, however, the peak position was almost independent of pressure. This behavior is incompatible with the simple picture that the concentration of the π-hole gradually decreases on applying pressure. For the more quantitative discussion, we estimated the relative intensities and peak positions of these spectra by fitting a Lorentz function to the observed spectrum. The pressure dependence of the intensity and peak position (solid circle) are shown in the right panel of Fig. 8. The intensity increases up to 0.6 GPa, abruptly decreases above 0.6 GPa, and finally vanishes above 6.1 GPa. On the other hand, the peak position is almost pressure independent. The open triangle in Fig. 8 is the pressure-dependent peak position calculated by the experimentally observed intensity via the effective medium theory, when we assume that only the band filling factor of the metallic π-band changes upon pressure. The open triangle in Fig. 8 is the pressure-dependent intensity calculated by the same way as the peak position. From the clear discrepancy shown in Fig.8, the experimentally observed pressure- dependence above 0.6 GPa is

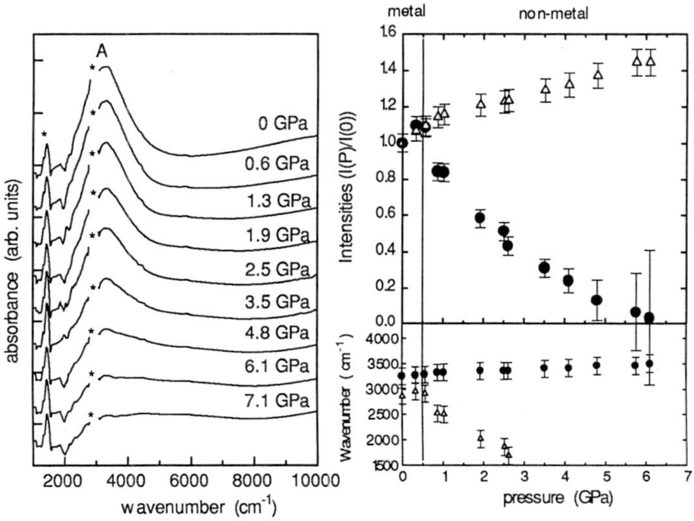

Figure 8: Pressure dependence of the plasmon absorption.

inconsistent with the plasmon absorption based on a simple band-filling picture. This result strongly suggests the phase change from metal to nonmetal at *ca.* 0.6 GPa as shown by the vertical line in the right panel of Fig. 8. The evolution of the energy gap seems to push the peak position to the high wavenumber side. Finally, we note that a new weak band appeared at 6000 cm^{-1} above 1 GPa. The intensity of this absorption slightly increases upon pressure. As will be discussed later, this extremely weak absorption band is assigned to the CT band between the central metal ions (Ni(2+)-Ni(3+)).

There is a boundary in the electronic state of $NiPc(AsF_6)_{0.5}$ at *ca.* 0.5 GPa, as we pointed out in the preceding section. In the low-pressure metallic region, where the d-π charge transfer does not occur, both the intensity and peak position of the plasmon absorption increases, which corresponds about 3 % (at 0.3 GPa) increase of the plasma frequency. The plasma frequency for a quarter-filled one-dimensional tight-binding band is given by the following equation,

$$\omega_p = \sqrt{\frac{4tNe^2d^2 \sin(\pi / 4)}{\varepsilon_0 \pi \hbar^2}}$$

, where t is the transfer integral, N the carrier concentration, d the distance between neighboring molecules, ε_0 the dielectric constant of vacuum. On raising the pressure, t and N increase, while d decreases. The pressure dependent experiment of the plasma edge has been done in TTF-TCNQ.[23] The increase of ω_p was attributed to the increase of the transfer integral, t, in TTF-TCNQ. Since the compressibility is not known yet and t is more sensitive to the pressure than a volume contraction,[32] we tentatively consider that the increase of ω_p comes from the increase of the transfer integral in $NiPc(AsF_6)_{0.5}$ as well.

In the high-pressure non-metallic region, where the charge transfer begins, the intensity of this absorption begins to decrease but the peak position is almost constant. If we treat the optical transition by the following Lorentz function,

$$\varepsilon_z(\omega) = \varepsilon_c + \frac{\Omega_p^2}{\omega_g^2 - \omega^2 - i\omega\Gamma}$$

the peak position of the absorption of the powdered sample is given by

$$\omega_0 = \omega_g \sqrt{\frac{\varepsilon_s + 2\varepsilon_\alpha}{\varepsilon_c + 2\varepsilon_\alpha}}$$

, where ε_s the dielectric constant at $\omega=0$ is given by $\varepsilon_s=\varepsilon_c+(\Omega_p/\omega_g)^2 > \varepsilon_c$.[20] At the pressure just after the charge transfer occurs, Ω_p is approximated to be ω_p. If $\omega_p >> \omega_g$, $\varepsilon_s \approx (\omega_p/\omega_g)^2 >> \varepsilon_\alpha$, so ω_0 is nearly equal to $\omega_p(\varepsilon_c+2\varepsilon_\alpha)^{-1/2}$, which is equal to the peak position of the plasmon. For this relation, we can surmise that $\omega_g \approx 0$ around 0.5 GPa. According as the pressure increases, ε_s should decrease due to the decrease of Ω_p which determines the intensity. Since ω_0 is almost pressure independent experimentally, the ω_g seems to increase upon pressure. At 6.1 GPa the absorption band almost vanishes, so Ω_p is nearly equal to 0 at this pressure. In this case $\varepsilon_s \gg \varepsilon_c$, so ω_g is estimated to be *ca.* 0.4 eV from the peak position ω_0 at 6.1 GPa. Although this estimation is very rough, this experimental result suggests that ω_g increases on increasing the degree of charge transfer. The d-π charge transfer is the interchange of holes between the π- and d-bands. In this case plasmon absorption of d-holes should be observed instead of π-plasmon. However, the absorption in this region almost disappears, and no additional absorption is observed in the infrared region down to 400 cm^{-1}. Instead of this plasmon absorption a very weak absorption shows up at 6000 cm^{-1}. This weak absorption is interpreted as the intermolecular charge-transfer transition between Ni(3+) and Ni(2+). The small intensity means that the induced holes in Ni chain are localized. The metal-nonmetal transition suggested by the optical spectrum was confirmed by the pressure dependence of the electrical resistivity at room temperature as shown in Fig. 9. The resistance of the sample slowly decreases like an ordinary organic conductor and increases around 0.5 GPa. This resistance increase returns to the initial value showing a large hysteresis when the pressure is released. The temperature dependences of the resistance below (0.3 GPa) and above (1.0 GPa) this phase transition pressure showed a metallic and nonmetallic behavior around room temperature.

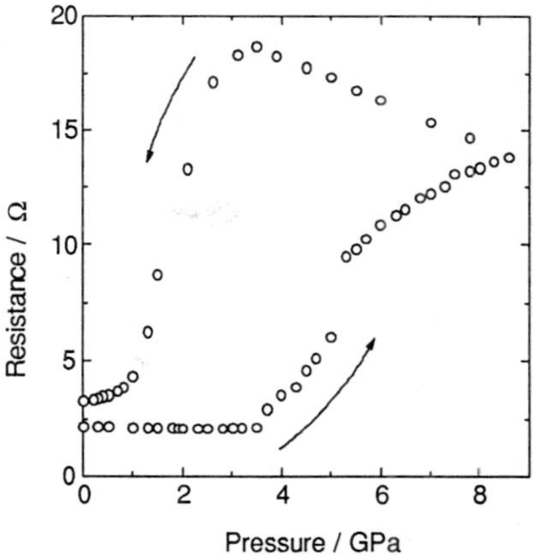

Figure 9: Pressure dependence of the resistance of $NiPc(AsF_6)_{0.5}$ at room

3.2 $CoPc(AsF_6)_{0.5}$

The same phenomenon was observed in $CoPc(AsF_6)_{0.5}$ as well under high pressure as shown in Fig. 10. The interband transition starts to decrease above 1.1 GPa and then vanishes above 4.4 GPa. Finally this spectrum resembles the visible spectra of the neutral Pc, so the π-band of the a_{1u} symmetry seems to be completely filled at 4.4 GPa. The infrared spectrum at high pressure also shows the similar behavior as in $NiPc(AsF_6)_{0.5}$.

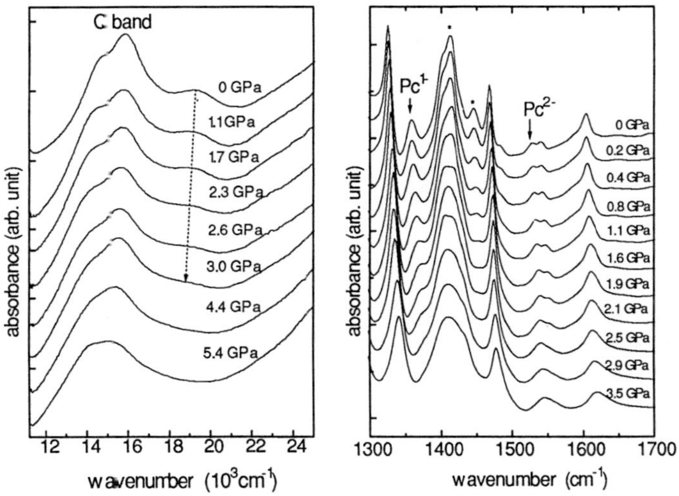

Figure: 10 Pressure dependences of the interband transition and infrared modes.

As shown in the right panel of Fig. 10, the characteristic mode of the oxidized ring (Pc(1-)), at 1358 cm^{-1} starts to decrease at 1.1 GPa and almost vanishes at 3.5 GPa, and characteristic mode of the unoxidized ring (Pc(2-)) at 1525 cm^{-1} increases above 0.8 GPa. These results indicates again that the valence of the macrocycle changes from Pc(1.5-) to Pc(2-). The left panel of Fig. 11 shows the pressure dependence of the charge-transfer band. We estimated the peak position and relative intensity by fitting the Lorentz function. The result is shown in the right panel of Fig. 11. The charge-transfer band continuously shifts to the high wavenumber side up to 7 GPa. Comparing with the plasmon absorption of $NiPc(AsF_6)_{0.5}$, this band shows a large blue shift. On the other hand, the intensity first increases, reaches the maximum at 1.0 GPa, and decreases down to half of the ambient-pressure spectrum at 6.2 GPa. This absorption still remains even at 7 GPa which is the highest hydrostatic pressure attainable in our pressure cell, although the interband transition and 1356 cm^{-1} vibrational band almost vanish at 4.4 GPa and 3.5 GPa, respectively. The lattice contraction enhances the transfer integral, t, between the neighboring macrocycles. The increase of the intensity in the pressure region of 0-1.1 GPa can be explained by this enhancement of the transfer integral. As shown in the preceding section, the interband transition and characteristic vibrational modes clearly shows that the d-π charge transfer is induced by high pressure in this compound. The pressure (1 GPa), at which the intensity of the charge-transfer band begins to decrease, exactly coincides with the pressure where the infrared modes and the interband transition start to change. Thus, the decrease of intensity is related to this charge-transfer phenomenon.

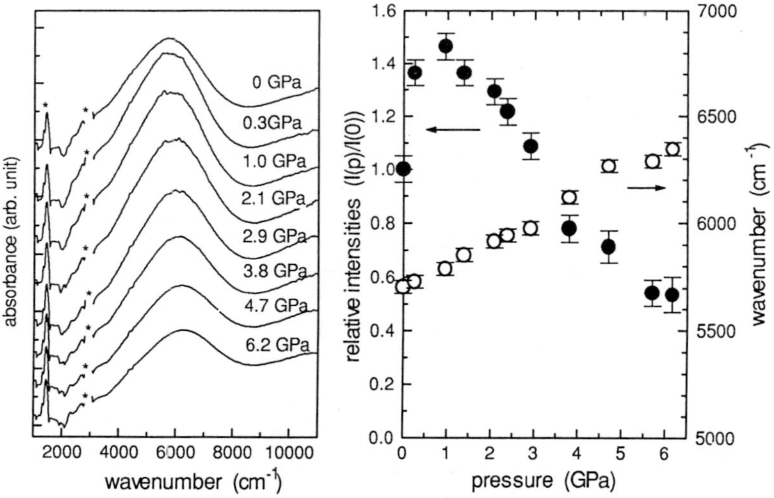

Figure: 11 Pressure dependence of the charge-transfer band.

The charge-transfer band does not vanish at 6 GPa but leaves half of the intensity at 6000 cm^{-1}, although the interband transition and infrared modes suggest the complete charge transfer at this pressure. As we described before, a very weak absorption band showed up at the same region in $NiPc(AsF_6)_{0.5}$, which was assigned to the charge-transfer transition between Ni(2+) and Ni(3+). We therefore ascribed this remaining band to the charge-transfer band between Co(2+) and Co(3+). This band is much stronger than that of $NiPc(AsF_6)_{0.5}$. The strong intensity in $CoPc(AsF_6)_{0.5}$ is attributed to the larger transfer integral between the neighboring Co $3d_{z^2}$ orbitals. In this case the d-holes are more delocalized than that of $NiPc(AsF_6)_{0.5}$ but still does not make a metallic band. $CoPc(AsF_6)_{0.5}$ has an exact metal-over-metal overlap, which is more advantageous to the overlap of the adjacent $3d_{z^2}$-orbital than the zigzag stacking mode in $NiPc(AsF_6)_{0.5}$. If we consider that the near-infrared absorption band consists of two charge-transfer transitions, the CT transition in a macrocycle chain and the CT transition in a Co chain, these CT transitions gradually changes their weight upon pressure. It is an interesting result that the Co $3d_{z^2}$-orbital is more extended to the z-direction than the Ni $3d_{z^2}$-orbital.

Acknowledgment

We acknowledge to Dr. K. Imaeda of IMS, Dr. Takahashi and Prof. Mouri of ISSP of Tokyo University for their assistance of the pressure dependent resistivity measurements.

References

[1] H. Yamakado, T. Ida, A. Ugawa, K. Yakushi, K. Awaga, Y. Maruyama, K. Imaeda, and H. Inokuchi, *Synth. Met.* **62**, 169 (1994), and references cited in this paper.

[2] F. W. Kutzler and D. E. Ellis, *J. Chem. Phys.* **84**, 1033 (1986).

[3] K. Yakushi, H. Yamakado, T. Ida, and A. Ugawa, *Solid State Commun.* **78**, 919 (1991).

[4] K. Yakushi, H. Yamakado, M. Yoshitake, N. Kosugi, H. Kuroda, T. Sugano, M. Kinoshita, A. Kawamoto, and J. Tanaka, *Bull. Chem. Soc. Jpn.* **62**, 687 (1989).

[5] K. Mortensen, E. M. Conwell, and J. M. Fabre, *Phys. Rev. B* **28**, 5856 (1983), and C. Coulon, P. Delhaes, S. Frandrois, R. Lagnier, E. Bonjour, and J. M. Fabre, *J. Phys.* **43**, 1059 (1982).

[6] J. U. von Schutz, M. Bair, H. J. Gross, U. Langohr, H.-P. Werner, H. C. Wolf, D. Schmeisser, K. Graf, W. Gopel, P. Erk, H. Meixner, and S. Hunig, *Synth. Met.* **27**, B249 (1988).

[7] T. Mori, H. Inokuchi, a. Kobayashi, R. Kato, and H. Kobayashi, *Phys. Rev. B*, **38**, 5913 (1988).

[8] H. Basista, D. A. Bonn, T. Timusk, J. Voit, D. Jerome, and K. Bechgaard, *Phys. Rev. B* 42, 4088 (1990).

[9] C. S. Jacobsen, D. B. Tanner, and K. Bechgaard, *Phys. Rev. B* **28**, 7019 (1983).

[10] K. Yakushi, S. Aratani, K. Kikuchi, H. Tajima, H. Kuroda, *Bull. Chem. Soc. Jpn.* **59**, 363 (1986).

[11] H. Kuroda, K. Yakushi, H. Tajima, A. Ugawa, M. Tamura, Y. Okawa, A. Kobayahi, R. Kato, and H. Kobayashi, *Synth. Met.* **27**, A491 (1988), and K. Kornelson, J. E. Eldrige, H. H. Wang, and J. M. Williams, *Phys. Rev. B* **44**, 5235 (1991).

[12] K. Yakushi, A. Ugawa, G. Ojima, T. Ida, H. Tajima, H. Kuroda, A. Kobayashi, R. Kato, and H. Kobayashi, *Mol. Cryst. Liq. Cryst.* **181**, 217 (1990).

[13] M.Yoshitake, K.Yakushi, H.Kuroda, A.Kobayashi, R.Kato, and H.Kobayashi, *Bull. Chem. Soc. Jpn.* **61**, 1115 (1988).

[14] C. Coulon, P. Delhaes, S. Frandrois, R. Lagnier, E. Bonjour, and J. M. Fabre, *J. Phys.* **43**, 1059 (1982).

[15] H.-P. Werner, J. U. von Shutz, H. C. Wolf, R. Kremer, M. Gehrke, A. Aumuller, P. Erk, and S. Hunig, *Solid State Commun.* **65**, 809 (1988).

[16] M.Y.Ogawa, J.Martinsen, S.M.Palmer, J.L.Stanton, J.Tanaka, R.L.Greene, B.M.Hoffman, and J.A.Ibers, *J. Am. Chem. Soc.* **109**, 1115 (1987).

[17] A.Mishima, *Synth. Met.* **55-57**, 1815 (1993).

[18] G. A. Willams, B. N. Figgis, and R. Mason, *J. Chem. Soc. Dalton* 734 (1981).

[19] K. Yakushi, T. Ida, A. Ugawa, H. Yamakado, H. Ishii and H. Kuroda, *J. Phys. Chem.*, **95** 7637 (1991).

[20] G. C. Papavassiliou, *Prog. Solid St. Chem.* **12**, 185(1979).

[21] *"Handbook of Chemistry Vol. II"*, ed. by K. Hata, Chemical Society of Japan, Tokyo, (1984), p556.
[22] T. Hiejima and K. Yakushi, *J. Chem. Phys.* **103**, 3950 (1995).
[23] B. Welber, P. E. Seiden and P.M. Grant, *Phys. Rev. B,* **18** 2692 (1978).

Low-Dimensional Electronic Systems of Molecular Conductors under High Magnetic Fields

Seiichi Kagoshima

Department of Pure and Applied Sciences, University of Tokyo, Komaba 3-8-1, Meguro, Tokyo 153, Japan

Abstract

Studies of novel electronic states of low-dimensional molecular conductors are briefly reviewed and magnetic field effects to them are discussed. Magnetic field effects to electron spins are discussed in $(R_1R_2DCNQI)_2Cu$. Effects to electron orbital motion are discussed in the quasi one-dimensional conductor $(TMTSF)_2X$ and in the quasi two-dimensional conductor $(BEDT\text{-}TTF)_2XHg(SCN)_4$. Electronic states in an anomalous state of $(BEDT\text{-}TTF)_2KHg(SCN)_4$ are investigated by magnetoresistance measurements, especially the angle dependent magnetoresistance oscillations.

1 Introduction

Low-dimensionality is one of the most characteristic properties of molecular conductors. It comes from the shape of constituent molecules far from spherical symmetry. TMTSF and BEDT-TTF molecules, which are most popular ones to make molecular conductors and superconductors, usually form columnar or sheet-like structures in crystal because of their planar shape. In those conductors one obtains quasi one- or two-dimensional conduction electron systems.

Various electronic properties characteristic of low-dimensional electronic systems have been found in molecular conductors [1]: The Peierls instability of one-dimensional system causes a metal-insulator transition forming charge-density (CDW) or spin-density waves (SDW), the latter of which are made when the electron-electron Coulomb interaction is not negligible. The Coulomb correlation is enhanced in low-dimensional systems. When it is large in a low-dimensional half-filled band, the system easily undergoes the Mott transition. This is considered to be responsible for metal-nonmetal transitions found frequently in molecular conductors.

Magnetic fields play several roles in studying electronic properties of low-dimensional molecular conductors: First, it gives the Zeeman energy to the system leading to some kinds of phase transitions due to reduction of the free energy of spins parallel to the magnetic field. Second, magnetic fields modify the orbital motion of electrons by exerting the Lorentz force. Electrons lose the freedom of translational motion perpendicular to the magnetic fields. This provides a clue to investigate electronic structures and their dynamics and, moreover, to control the electronic state. Under high magnetic fields even new electronic states are formed such as the Landau quantized state.

This article reviews shortly the magnetic field effects to spins and orbital motion of low-dimensional molecular conductors followed by discussion on novel electronic states of one of quasi two-dimensional conductors, $(BEDT\text{-}TTF)_2XHg(SCN)_4$, studied under high magnetic fields.

Springer Proceedings in Physics, Vol. 81
Materials and Measurements in Molecular Electronics
Editors: K. Kajimura · S. Kuroda

2 Magnetic field effects to low-dimensional electronic systems

2.1 Magnetic field effects to spins

The quasi one-dimensional conductor, $(R_1R_2DCNQI)_2Cu$ has columns of DCNQI molecules interconnected by Cu as shown in Fig. 1 [2,3]. It is metallic in the high-temperature-low-pressure regime and is insulating in the low-temperature-high-pressure one as shown in Fig. 2. XPS and infrared studies have verified that the Cu has the mixed valence state $Cu^{4/3+}$ in the metallic state [4,5]. Therefore, d-electrons of Cu are expected to take part in the electronic states at the Fermi level. A first principle band calculation has shown this intuitive expectation to be valid [6]. It is formed by one-dimensional π-like bands originating in DCNQI molecular orbitals and three-dimensional d-like ones from Cu. Electrical, magnetic and structural studies have shown that the insulating state has paramagnetic spins of Cu^{2+} and a three-fold superstructure parallel to the one-dimensional c-axis [7].

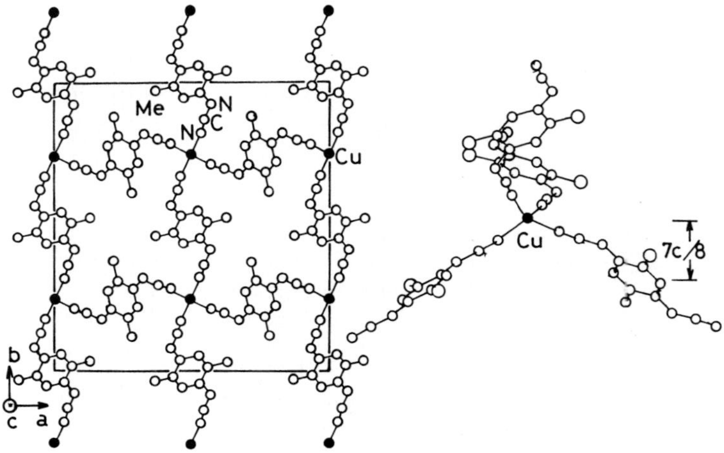

Fig. 1 Crystal structure of $(R_1R_2DCNQI)_2Cu$ in the case R_1 and R_2 are methyl groups.

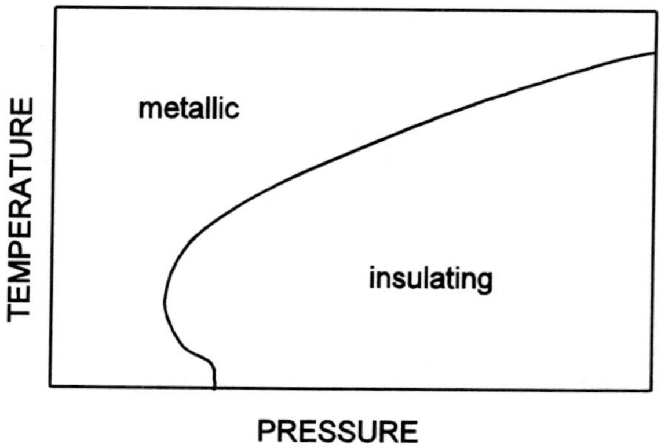

Fig. 2 Temperature-pressure phase diagram of $(R_1R_2DCNQI)_2Cu$.

The basic mechanism of the metal-insulator transition is ascribed to the Peierls instability in the π-like band and the Mott transition in the d-like one [8]. The Peierls instability is accompanied by the three-fold superstructure [9,10]. This superstructure makes the Mott transition easy to undergo because the Brillouin zone is reduced to 1/3 leading to the half-filled state of the d-like band.

One of novel properties of this material is the metal-insulator-metal reentrant transition observed in the narrow pressure range shown in Fig. 2. Electrical resistance measurements under high magnetic fields and heat capacity ones were made in order to clarify its origin [11,12]. The magnetoresistance is found to be quite normal unlike that observed in the heavy electron system. The electronic heat capacity was measured to be about 10 - 20 mJ/mol/K^2 at low temperatures. This is also normal as molecular conductors suggesting that the re-entered metallic state has basically the same properties as ordinary metallic states far from the reentrant regime.

Under high magnetic fields of the order of 10 T, however, we found a field induced transition from the metallic state to the insulating one. This transition occurs when the sample state is close to the insulating phase. It is found to be hysteretic characteristic of a first order phase transition. By analyzing experimental results we propose the following picture: The transition between the metallic and the insulating states is of first order. The magnetic field reduces the free energy of the insulating state through the net contribution of the Zeeman energy and the entropy terms. Therefore the metallic state is made unstable by high magnetic fields when the metallic system is close enough to the insulating regime. The above idea is consistent with theoretical studies from both phenomenological and microscopic view points [13,14].

2.2 Magnetic field effects to orbital motion

Examples of magnetic field effects to the orbital motion of electrons are found in quasi one-dimensional conductors $(TMTSF)_2X$ (X =ClO_4, PF_6 etc.) and quasi two-dimensional ones $(BEDT\text{-}TTF)_2XHg(SCN)_4$ (X=K, NH_4 etc.). At ambient pressure $(TMTSF)_2PF_6$, whose crystal structure is shown in Fig. 3, undergoes the Peierls transition at 12 K accompanied by SDW [15-18]. Under high pressures above about 6 kbar (0.6 GPa) it becomes superconducting below about 1 K. However, $(TMTSF)_2ClO_4$ remains metallic at ambient pressure and undergoes the superconducting transition at about 1 K. This difference between two materials is ascribed to the higher one-dimensionality of the former material than the latter. Therefore one may say that the Peierls instability is suppressed in $(TMTSF)_2ClO_4$ by the two- or three-dimensionality of the system.

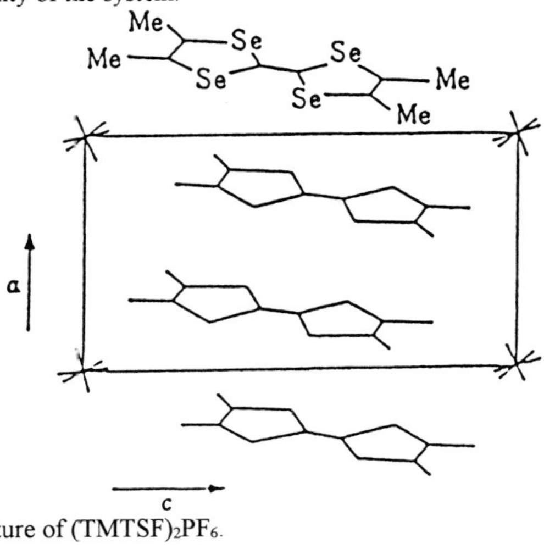

Fig. 3 Crystal structure of $(TMTSF)_2PF_6$.

It was found that application of high magnetic fields in the superconducting state restored SDW[19,20]. This suggests that the magnetic field lowers the effective system dimensionality. With further increasing magnetic fields a series of SDW subphases and a quantized Hall effect were found [21-23]. This phenomenon has been interpreted in terms of the so-called standard model [24-26]. It takes account of the interplay between the field-induced Peierls instability and the Landau quantization of two-dimensional electrons and holes in small pockets made by an imperfect nesting of the Fermi surface. It is considered that this interplay adjusts the nesting vector of CDW so as to keep the Fermi level just at the middle of two Landau levels. The former leads to the successive phase transition among SDW subphases and the latter to the associated quantized Hall effects.

The standard model seems to explain satisfactorily the observed phenomena. However, some problems are still remaining to be solved; nature of electronic state in the high field limit, mechanisms of sign change of the quantized Hall resistance, possible presence of infinitely many phases in the low temperature limit etc.

High magnetic fields cause another phenomenon originating from the orbital motion of electrons. When one rotates the direction of magnetic fields in the plane perpendicular to the one-dimensional a-axis of $(TMTSF)_2ClO_4$, the transverse resistance shows a series of local minima as a function of the angle between the magnetic field and a crystalline axis perpendicular to the one-dimensional axis as shown in Fig. 4 [27].

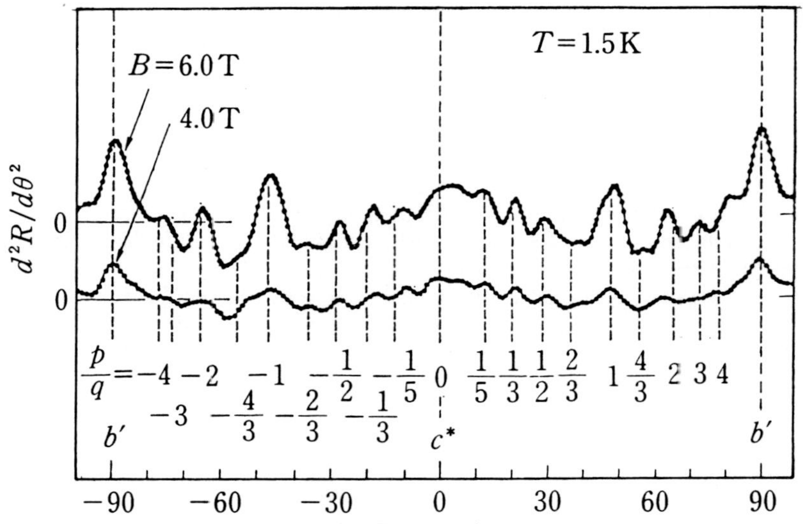

Fig. 4 Angle-dependent magnetoresistance oscillations of $(TMTSF)_2ClO_4$. The angle θ denotes that between the magnetic field and the c^*-axis.

This angle dependent magnetoresistance oscillation has been ascribed to a commensurability effect between the electron trajectory in momentum space and the reciprocal lattice [28]. Figure 5 depicts the trajectory on the Fermi surface. When the field direction is general, the trajectory reduced to the first Brillouin zone covers all over the warped Fermi surface. Transverse components of the Fermi velocity of electrons are averaged out leading to the net electron motion parallel to the one-dimensional axis.

When the field direction satisfies the relation $\tan\theta = pb'/qc'$, the trajectory leaves only a finite number of lines on the Fermi surface in the first Brillouin zone. Here p and q are arbitrary integers and b' and c' are projections of the lattice vector $\boldsymbol{b}$ and $\boldsymbol{c}$ onto the plane perpendicular to the one-dimensional a-axis,

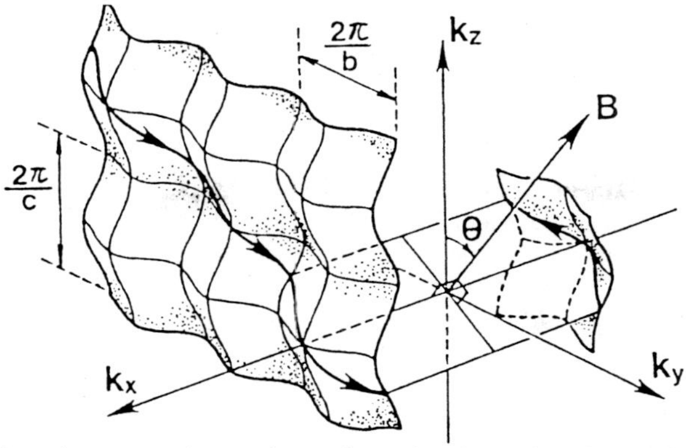

Fig. 5 Trajectory of an electron on the quasi one-dimensional Fermi surface under magnetic fields.

respectively. As the cancellation of the transverse components becomes imperfect, the trajectory of electrons in real space has the component perpendicular to the one-dimensional axis causing the local minima in the transverse resistance. A fully quantum mechanical calculation has given a result consistent with experimental results and this interpretation [29].

Also in quasi two-dimensional conductors the angle dependent magnetoresistance oscillation has been found with another interpretation : The resistance perpendicular to the two-dimensional plane shows a series of resistance maxima as a function of the angle between the magnetic field direction and the normal to the two-dimensional plane. This phenomenon was found in θ-(BEDT-TTF)$_2$I$_3$ by Kajita et al. and in (BEDT-TTF)$_2$ by Kartsovnik et al. independently [30,31].

A theoretical idea to explain this phenomenon was given by Yamaji [32]. When the field direction is general, the electron trajectory in real space is spiral-like leading to a finite resistance perpendicular to the two-dimensional plane. At magic angles given by the relation, $\tan\theta = (k_F/c^*)\ (n + 1/4)$ where k_F, c^* and n is the Fermi wave number, the reciprocal lattice vector and an arbitrary integer, the real space trajectory becomes closed as shown in Fig. 6.

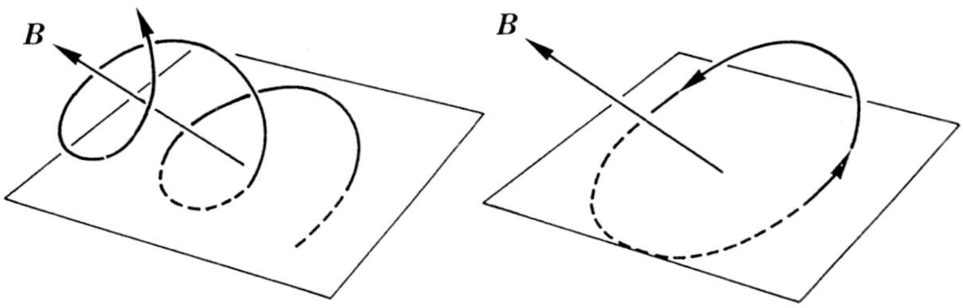

Fig. 6 Real space trajectories of an electron at the Fermi level under magnetic fields applied at general angles (left) and magic angles (right).

This makes the perpendicular resistance infinite. Quantum mechanical calculations of electronic energy spectra have shown that the band width at the Fermi level vanishes at the magic angles [33]. We consider this corresponds to the above naive picture of closed orbit in real space.

3 Anomalous electronic state of the quasi two-dimensional conductor $(BEDT\text{-}TTF)_2XHg(SCN)_4$ studied under high magnetic fields

The quasi two-dimensional conductor $(BEDT\text{-}TTF)_2XHg(SCN)_4$ has conducting layers of BEDT-TTF molecules sandwitched by anion layers of $XHg(SCN)_4$ as shown schematically in Fig. 7 [34]. When X is NH_4, it shows simple metallic properties at all temperatures and undergoes a superconducting transition at about 1.5 K [35,36]. The angle dependent magnetoresistance oscillations, analyzed by Yamaji's model above, suggest the presence of a quasi two-dimensional Fermi surface [37].

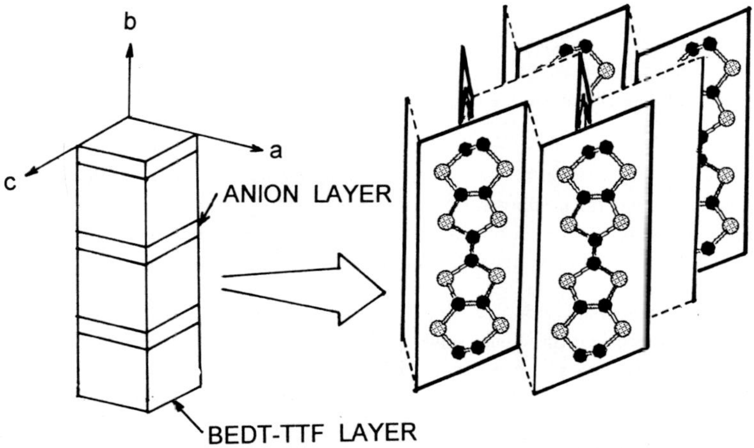

Fig. 7 A schematic view of the crystal structure of $(BEDT\text{-}TTF)_2XHg(SCN)_4$.

When X is K, however, the compound has an anomalous state, whose origin is unclear, in the low-temperature (<8 K)-low-pressure (<5 kbar)-low-magnetic-field (22 T) regime [38]. The magnetoresistance has a broad peak around 10 T accompanied by a kink-like structure at about 22 T as shown in Fig. 8 while the NH_4 compounds has no broad peak in magnetoresistance [39]. In addition the angle dependent magnetoresistance oscillations show a feature quite different from NH_4 compound and Yamaji's model.

To investigate possible origins of the anomalous state we measured angle dependent magnetoresistance oscillations under high magnetic fields and high pressures. Figure 9 shows the angle dependent magnetoresistance oscillations obtained by rotating a sample around two axes perpendicular to each other and to the magnetic field. One finds a series of resistance minima. Figure 10 shows a gnomonic projection of the angular position of resistance minima. It suggests the presence of a quasi one-dimensional Fermi surface rather than quasi two-dimensional one. However, the geometry of thus expected quasi one-dimensional Fermi surface is different from that of the band calculation [40].

Kartsovnik et al. ascribed the origin of the observed quasi one-dimensional Fermi surface to a possible nesting of the original one due to the Peierls instability as shown in Fig. 11 [41]. They showed that the reconstructed Fermi surface is in good agreement with the observed one.

This idea seems to be consistent with the possible antiferromagnetism suggested by the anisotropy of magnetic susceptibility measured by Sasaki et al [42]. Magnitude of magnetic moment is measured by μSR to be of the order of $10^{-3}\mu_B$ [43]. The Fermi surface nesting is considered to cause SDW rather than CDW.

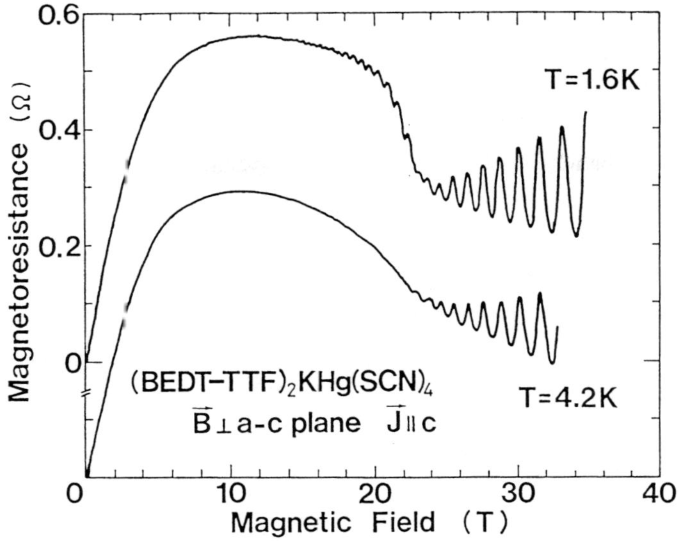

Fig. 8 Magnetoresistance of $(BEDT\text{-}TTF)_2KHg(SCN)_4$.

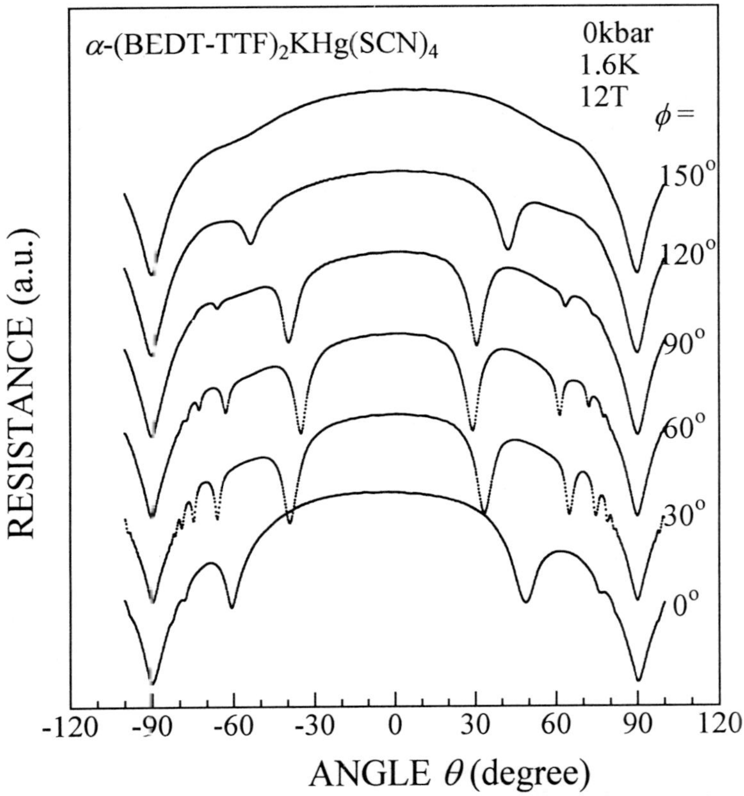

Fig. 9 Angle dependent magnetoresistance oscillations of $(BEDT\text{-}TTF)_2KHg(SCN)_4$ at ambient pressure. The angles θ and ϕ denote the polar angle measured from the normal of the two-dimensional plane and the azimuthal one from the a-axis, respectively.

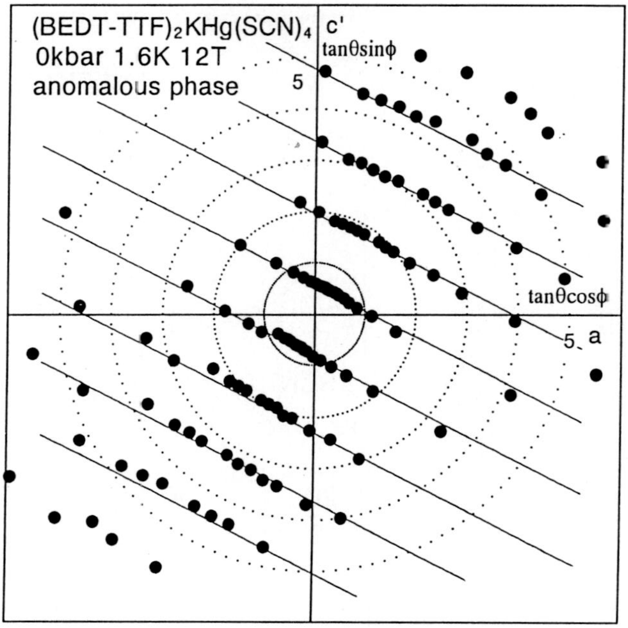

Fig. 10 A gnomonic projection of the angular position of resistance minima of $(BEDT\text{-}TTF)_2KHg(SCN)_4$ at ambient pressure under magnetic fields.

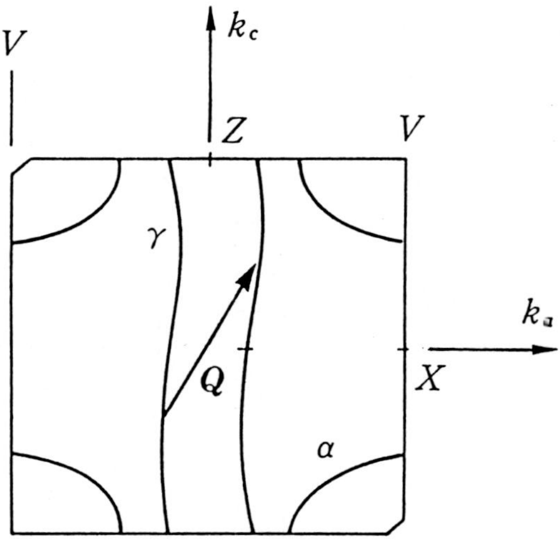

Fig. 11 Nesting of the quasi one-dimensional Fermi surface of $(BEDT\text{-}TTF)_2KHg(SCN)_4$.

Under high pressures above about 5 kbar (0.5 GPa) the angle dependent magnetoresistance shows the presence of both the quasi one- and two-dimensional Fermi surfaces in consistent with the band calculation as shown in Fig. 12.

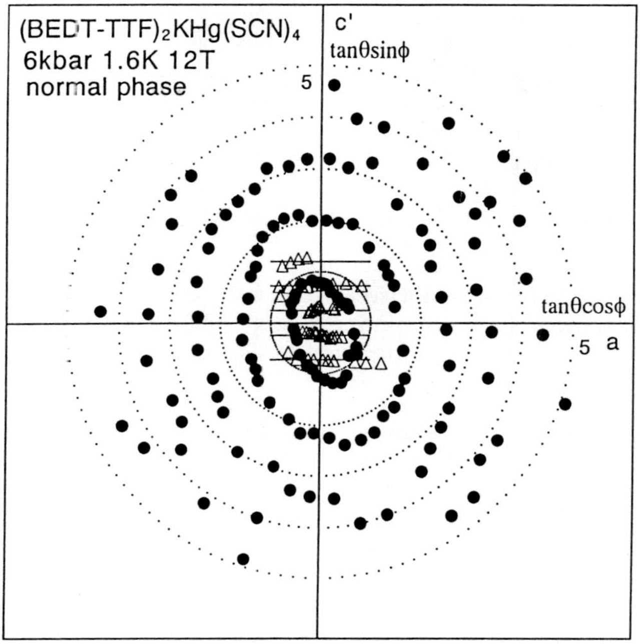

Fig. 12 A gnomonic projection of resistance minima (open triangles) and maxima (closed circles) of $(BEDT\text{-}TTF)_2KHg(SCN)_4$ at 6 kbar (0.6 GPa) under magnetic fields.

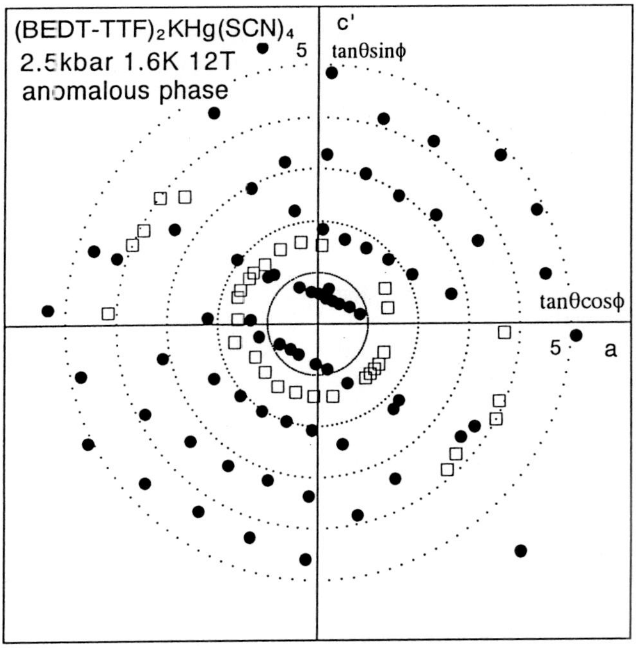

Fig. 13 A gnomonic projection of resistance minima of $(BEDT\text{-}TTF)_2KHg(SCN)_4$ at 2.5 kbar (0.25 GPa) under magnetic fields. Both the closed circles and the open squares denote the resistance minima.

In the intermediate pressure range of the anomalous state the angle dependent magnetoresistance shows predominantly the presence of the quasi one-dimensional Fermi surface made by the nesting. However, as shown in Fig. 13, another series of resistance minima was found. We verified that its angular position is really angle dependent rather than field-strength dependent although its magnitude increases with increasing the field strength. Figure 13 suggests that the possible origins for this resistance minima have some circular symmetry in the two-dimensional plane. However, the Fermi surface predicted by the band calculation cannot explain the presence of such circular symmetry. Therefore, the origin of this anomaly is remaining puzzling. Some unknown mechanisms are expected to work in the intermediate pressure range.

It has been proposed that more complicated phases are present in the anomalous state [44,45]. Actually magnetic susceptibility measurements have shown a hysteretic nature suggesting presence of first-order-like changes in the anomalous state. One can expect possible presence of more novel electronic states in the anomalous state.

Acknowledgements - This article is based on studies in collaboration with Mr. Y. Saito, Mr. N. Hanasaki, Dr. T. Osada, Profs. N. Miura, R. Kato and G. Saito. The author is grateful for discussion with Dr. H. Shinagawa, Profs. H. Kobayashi and Y. Iye. This work is partly supported by the Grant-in-Aid for Scientific Research from the Ministry of Education, Science and Culture and the grant from The Asahi Glass Foundation.

References

[1] S. Kagoshima, H. Nagasawa, T. Sambongi, *One-Dimensional Conductors,* Springer-Verlag, Berlin Heidelberg, 1988.

[2] A. Kobayashi, R. Kato, H. Kobayashi, T. Mori, H. Inokuchi, *Solid State Commun.,* Vol.64, 1987, 45.

[3] R. Kato, H. Kobayashi, A. Kobayashi, *J. Am. Chem. Soc.*, Vol.111, 1989, 5224.

[4] H. Kobayashi, A. Miyamoto, R. Kato, F. Sakai, A. Kobayashi, Y. Yamamoto, Y. Furukawa, T. Tasumi, T. Watanabe, *Phys. Rev. B*, Vol.47, 1993, 3495.

[5] I. H. Inoue, A. Kakizaki, H. Namatame, A. Fujimori, A. Kobayashi, R. Kato, H. Kobayashi, *Phys. Rev. B*, Vol.45, 1992, 5828.

[6] T. Miyazaki, K. Terakura, Y. Morikawa, T. Yamasaki, *Phys. Rev. Lett.*, Vol.74, 1995, 5104.

[7] See, for example, R. Kato, S. Aonuma, H. Sawa, *Synth. Met.*, Vol.70, 1995, 1071.

[8] H. Fukuyama, *J. Phys. Soc. Jpn.*, Vol.61, 1992, 3452.

[9] R. Moret, *Synth. Met.*, Vol.27, 1988, B301.

[10] H. Sawa, M. Tamura, S. Aonuma, R. Kato, M. Kinoshita, H. Kobayashi, *J. Phys. Soc. Jpn.*, Vol.62, 1993, 2224.

[11] T. Mori, H. Inokuchi, A. Kobayashi, R. Kato, H. Kobayashi, *Phys. Rev. B*, Vol.38, 1988, 5913.

[12] S. Kagoshima, Y. Saito, T. Hanasaki, N. Wada, H. Yano, R. Kato, N. Miura, H. Kobayashi, *Synth. Met.*, Vol.70, 1995, 1065.

[13] M. Nakano, M. Kato, K. Yamada, *Physica B*, Vol.186-188, 1993, 1077.

[14] T. Ogawa, Y. Suzumura, *Phys. Rev. B*, Vol.51, 1995, 10293.

[15] K. Bechgaard, C. S. Jacobsen, K. Mortensen, H. J. Pedersen, N. Thorup, *Solid State Commun.*, Vol.33, 1980, 1119.

[16] D. Jerome, A. Mazaud, M. Ribault, K. Bechgaard, *J. de Phys. Lett.*, Vol.41, 1980, L95.

[17] J. M. Delrieu, M. Roger, Z. Toffano, E. Wope Mbougue, R. Saint James, K. Bechgaard, *Physica B*, Vol.143, 1986, 412.

[18] T. Takahashi, Y. Maniwa, H. Kawamura, G. Saito, *Physica B*, Vol.143, 1986, 417.

[19] J. F. Kwak, *Mol. Cryst. Liq. Cryst.*, Vol.79, 1982, 111.

[20] H. Bando, K. Oshima, M. Suzuki, H. Kobayashi, G. Saito, *J. Phys. Soc. Jpn.*, Vol.51, 1982, 2711.

[21] M. Ribault, D. Jerome, J. Tuchendler, C. Weyl, K. Bechgaard, *J. de Phys. Lett.*, Vol.44, 1983, 953.
[22] K. Oshima, M. Suzuki, K. Kikuchi, H. Kuroda, I. Ikemoto, K. Kobayashi, *J. Phys. Soc. Jpn.*, Vol.53, 1984, 3295
[23] O. M. Chaikin, M. -Y. Choi, J. F. Kwak, J. S. Brooks, K. P. Martin, M. J. Naughton, E. M. Engler, R. L. Greene, *Phys. Rev. Lett.*, Vol.51, 1983, 2333.
[24] L. P. Gor'kov, A. G. Lebed', *J. de Phys. Lett.*, Vol.45, 1984, 433.
[25] G. Montambaux, M. Heritier, P. Lederer, *Phys. Rev. Lett.*, Vol.55, 1985, 2078.
[26] K. Yamaji, *J. Phys. Soc. Jpn.*, Vol.54, 1985, 1034.
[27] T. Osada, S. Kagoshima, N. Miura, *Physica B*, Vol.184, 1993, 481.
[28] T. Osada, A. Kawasumi, S. Kagoshima, N. Miura, G. Saito, *Phys. Rev. Lett.*, Vol.66, 1991, 1525.
[29] T. Osada, S. Kagoshima, N. Miura, *Phys. Rev. B*, Vol.46, 1992, 1812.
[30] M. V. Kartsovnik, P. A. Kononovich, V. N. Laukhin, I. F. Shchegolev, *Pis'ma Zh. Eksp. Theo.*, Vol.48, 1988, 498 (*JETP Lett.*, Vol.48, 1988, 541).
[31] K. Kajita, Y. Nishio, T. Takahashi, W. Sasaki, R. Kato, H. Kobayashi, Y. Iye, *Solid State Commun.*, Vol.70, 1989, 1189.
[32] K. Yamaji, *J. Phys. Soc. Jpn.*, Vol.58, 1989, 1520.
[33] T. Osada, R. Yagi, S. Kagoshima, N. Miura, M. Oshima, G. Saito, *Physics and Chemistry of Organic Superconductors*, Springer-Verlag, Berlin Heidelberg, 1990, 220.

[34] H. Mori, S. Tanaka, M. Oshima, G. Saito, T. Mori, Y. Maruyama, H. Inokuchi, *Bull, Chem. Soc. Jpn.*, Vol.63, 1990, 2183.
[35] H. H. Wang, K. D. Carlson, U. Geiser, W. K. Kwok, M. D. Vashon, J. E. Thompson, N. F. Larsen, G. D. McCabe, R. S. Hulscher, J. M. Williams, *Physica C*, Vol.166, 1990, 57.
[36] H. Mori, S. Tanaka, K. Oshima, M. Oshima, G. Saito, T. Mori, Y. Maruyama, H. Inokuchi, *Solid State Commun.*, Vol.74, 1990, 1261.
[37] T. Osada, A. Kawasumi, R. Yagi, S. Kagoshima, N. Miura, M. Oshima, H. Mori, T. Nakamura, G. Saito, *Solid State Commun.*, Vol.75, 1990, 901.
[38] T. Kouno, T. Osada, M. Hasumi, S. Kagoshima, N. Miura, M. Oshima, H. Mori, T. Nakamura, G. Saito, *Synth. Met.*, Vol.55-57, 1993, 2425.
[39] T. Osada, R. Yagi, A. Kawasumi, S. Kagoshima, N. Miura, M. Oshima, G. Saito, *Phys. Rev. B*, Vol.41, 1990, 5428.
[40] T. Mori, H. Inokuchi, H. Mori, S. Tanaka, M. Oshima, G. Saito, *J. Phys. Soc. Jpn.*, Vol.59, 1990, 2624.
[41] M. V. Kartsovnik, A. E. Kovalev, N. D. Kushch, *J. de Phys. I*, Vol.3, 1993, 1187.
[42] T. Sasaki, H. Sato, N. Toyota, *Synth. Met.*, Vol.41-43, 1991, 2211.
[43] F. L. Pratt, T. Sasaki, N. Toyota, K. Nagamine, *Phys. Rev. Lett.*, Vol.74, 1995, 3892.
[44] J. S. Brooks, S. J. Klepper, C. C. Agosta, M. Tokumoto, N. Kinoshita, Y. Tanaka, S. Uji, H. Aoki, A. S. Perel, G. J. Athas, X. Chen, D. A. Howe, H. Anzai, Physica B, Vol.184, 1993, 489.
[45] J. Caufield, J. Singleton, P. T. J. Hendriks, J. A. A. J. Perenboom, F. L. Pratt, M. Doporto, W. Hayes, M. Kurmoo, P. Day, *J. Phys. Cond. Matt.*, Vol.6, 1994, L155.

Part VI

Conjugated Electroluminescent Polymers

Electronic Processes Associated with Electroluminescence in Conjugated Polymers

D. R. Baigent[1], R. H. Friend[1] A. B. Holmes[2] and S. C. Moratti[2]

1. Cavendish Laboratory, Madingley Road, Cambridge CB3 0HE, UK
2. Melville Laboratory for Polymer Synthesis, Pembroke Street, Cambridge CB2 3RA, UK

Abstract

Processible conjugated polymers can be used to fabricate a range of thin-film diodes which can be designed to show good characteristics both as electroluminescent diodes and also as photoconductive diodes. We consider the present understanding of the operation of light-emitting diodes which use conjugated polymers for both charge transport and emission. We highlight the improvement to the electroluminescence efficiency that can be produced by the use of two polymer layers selected so that the heterojunction between the two layers is able to confine charge and thus bring about electron-hole capture to generate excitons at this interface. We present results on the photophysical properties of the cyano-substituted poly(phenylene vinylene)s which provide electron transport layers in these heterostructure devices; we find evidence for strong interchain interactions (excimer formation) for the neutral excited states in these polymers, with strong red-shifts of the luminescence, though preserving high photoluminescence efficiencies.

Keywords: Conjugated polymer, electroluminescence.

1 Introduction

Conjugated polymers offer the possibility of combining the desirable processing and structural properties of polymers together with the electronic functionality of a metal or semiconductor. In the context of semiconductor devices, the particular interest lies in the scope for fabrication of thin-film devices over large areas, which, with the exception of amorphous and polycrystalline silicon, has proved difficult to achieve with inorganic materials. The initial interest in polymer-based devices was in the field-effect transistor, FET, with the discovery in the late 1980's that the field-effect could be routinely achieved in devices made by the deposition of conjugated polymers onto insulator layers such as silicon dioxide [1-4]. This work has now been extended to the use of well-defined oligomers of thiophene, principally α-sexithiophene and its derivatives, which exhibit the best field-effect mobilities among organic FET's, of order 0.1 $cm^2/Vsec$ [5, 6].

Electroluminescence from conjugated polymers was first reported by the Cambridge group in 1990 [7] using poly(phenylene vinylene), PPV, as the single semiconductor layer between metallic electrodes, as is illustrated in figure 1. PPV has an energy gap between π and π^* states of about 2.5 eV, and produces luminescence in a band below this energy. PPV is an intractable material with a rigid-rod microcrystalline structure, so that it is infusible and insoluble in common solvents. This gives it excellent mechanical properties, with high elastic moduli and thermal stability to 400°C. These are excellent properties for a film of polymer once formed, but processing cannot be carried out directly with this material. PPV can however be obtained conveniently by in situ chemical conversion of films of a suitable precursor polymer which is itself processed from solution by spin-coating. Of these, the route most commonly used is the sulphonium precursor [8-12] which is conveniently processed from solution in methanol, and is converted to PPV by thermal treatment at temperatures of between 200 and 300°C. The alternative strategy for polymer processing is to attach flexible side-groups to the polymer chain so that the polymer is directly processible from solution. This is particularly convenient in that a single process step is required, but the trade-off is that the polymer film is softer and less stable thermally. The derivative of PPV that is most frequently used is poly(2-methoxy, 5-(2'-ethyl)hexyloxy-p-phenylene vinylene), MEH-PPV [13, 14].

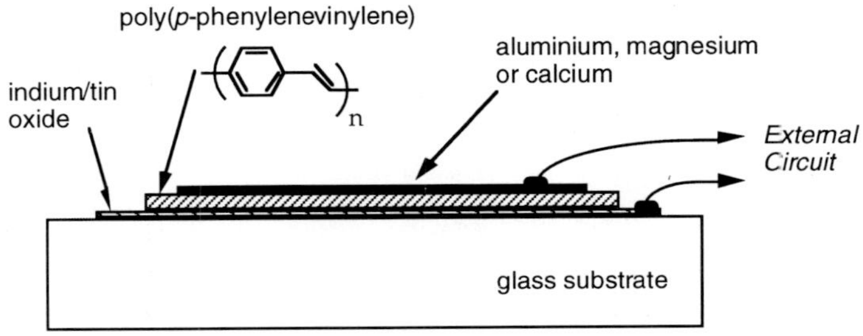

Fig. 1 Schematic structure of a polymer LED formed with a single layer of conjugated polymer.

LED operation is achieved when the diode is biased sufficiently to achieve injection of positive and negative charge carriers from opposite electrodes. Capture of oppositely-charged carriers within the region of the polymer layer can then result in formation of the singlet exciton which is generated by photoexcitation across the π-π* gap [15], and this can then decay radiatively, to produce the same emission spectrum as that produced by photoexcitation. The absorption and emission spectra for PPV are shown in figure 2. Note that the absorption rises rapidly above the onset of the π-π* threshold, and that the emission spectrum appears on the low-energy side of the absorption. Both absorption and emission spectra show broadening due to vibronic coupling, as is characteristic for optical transitions in molecular semiconductors where the excited state is a singlet exciton. Note that the similarity of the emission spectra produced by photoexcitation and by charge injection establishes that the excited state responsible for light generation in the LED is the same as that produced by photoexcitation.

The levels of efficiency of the first, simple LEDs based on PPV, which were fabricated with aluminium negative electrodes were relatively low, of order 0.01% photon generated within the device per electron injected [7]. External quantum efficiencies are strongly affected by the refractive indices of the layer that comprise the device, and the relationship between the two has been discussed by Greenham et al [16]. These values have risen rapidly over the past 4 years as improved understanding of the operation of these devices, aided in considerable measure by parallel developments made with sublimed molecular film devices [17-23], has allowed considerable optimisation of the device characteristics. The use of negative electrodes with lower work functions was shown by Braun and Heeger [14] to improve efficiency to as high as 1%, in devices made with ITO/MEH-PPV/Ca. The use of copolymers based on PPV with higher luminescence efficiencies also raised efficiencies to similar levels [24], and the use of heterostructure devices has brought efficiencies to 4% and above [25, 26].

In the sections which follow, we discuss some of the aspects of the device operation that control device operation. We do not consider in detail the control of colour of emission of these devices, but note that emission over the whole of the visible spectrum is possible by control of the polymer bandgap [24, 27]. Emission in the blue part of the spectrum is of particular interest, and there are now several reports of polymers and polymer blends which can be used to generate blue emission [28-33].

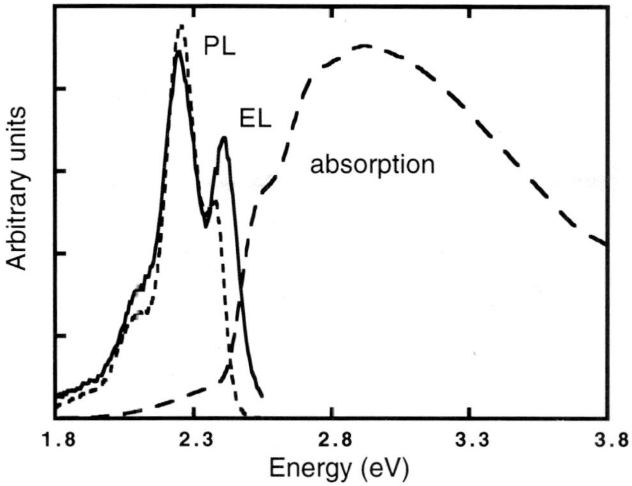

Fig. 2 Optical absorption and photoluminescence (PL) and electroluminescence (EL) emission spectra for PPV.

Light generation in a polymer LED requires a series of steps, which include charge injection, transport, electron-hole capture, and radiative decay of the exciton thus produced.

2 Charge Injection, Current Balancing

2.1 Charge injection

Injection of charge from most electrode materials requires that charges surmount or tunnel through a barrier at the interface. This is expected on examination of the positions of the electrode metal work functions and the positions of the π, highest occupied molecular orbitals, and π^*, lowest unoccupied molecular orbitals in the polymer. For the case of PPV, ITO provides a relatively good match for hole injection, though there is a barrier of around 0.2 eV [25]. However, electron injection is more difficult to achieve without the use of low work function, reactive metals such as calcium (for which the barrier as determined by the difference in work function of the metal and electron affinity of the polymer is of a similar magnitude).

Studies of the mechanism for charge injection into the polymer layer have been performed by several groups [14, 34-37]. The clearest evidence for the presence of a barrier for charge injection is provided by the finding that current density is dependent on the electric field across the device (i.e. it scales inversely with polymer layer thickness at constant applied bias) [34, 37]. Parker [37] concludes from the weak temperature dependence of the current-voltage characteristics that tunnelling is the dominant injection mechanism. He is also able to estimate the barrier heights, by analysis using the Fowler-Nordheim tunnelling model, and finds that these are consistent with the barriers expected from the differences in band edge positions in the polymer with respect to the metal work functions, both for electron and for hole injection.

Detailed understanding of the nature of the injection process will require a microscopic model for the chemistry at the polymer-metal interface, and considerable progress has been made on the nature of the polymer/cathode interface, experimentally by the Linköping group, and in terms of the quantum chemical modelling by the Mons group [38-40]. Photoemission measurements from clean polymer surfaces with low coverages of metals evaporated *in situ* reveal that alkali metals such as sodium [40] and the group II metal calcium [39] reductively dope the surface layer of PPV, to form a conductive charge-transfer complex. In the case of calcium it appears that the metal does not diffuse far into the PPV layer (it provides a relatively stable electrode, and long-lived devices [41]), so that the barrier for electron injection may in fact be between the calcium-doped surface layer of the polymer and the undoped polymer beneath it. In contrast, aluminium bonds covalently with the polymer, attacking the vinylic carbons to saturate the bonds at these positions on the chain, and thereby produce barriers for charge injection [38].

We comment finally on the use of chemically-doped conjugated polymers as injection electrodes. Several p-doped conjugated polymers show good environmental stability, including polypyrrole, polythiophene and polyaniline, and such materials have been used as hole-injecting electrodes. This was first reported by Hayashi et al. [42] using doped poly(3-methylthiophene) in combination with a sublimed film of perylene as emissive layer. For the conjugated polymers, this was first performed with polyaniline as hole-injecting electrode by Nakano et al [43], and subsequently by Yang et al [44], who demonstrate that doped polyaniline can show reasonable optical transparency and report lower barriers for hole injection in comparison with ITO. Pei et al [45] have recently reported that with the addition of a salt and ion-transporting material such as poly(ethyleneoxide) to the conjugated polymer, very low barriers for charge injection are found. They attribute this to the formation of p-doped regions in the polymer near the positive electrode and an n-doped region near the negative electrode, with the doping arising from electrochemical reactions which occur under drive conditions.

2.2 Heterostructure Devices

Injection and transport of holes from the positive electrode into the bulk of the polymer film must be matched by injection and transport of electrons from the opposite electrode. The use of two-layer structures to control injection rates of electrons and holes by introducing barriers for charge transport at the heterojunction between the two semiconductor layer is now well-established, both for devices made with sublimed molecular films [17, 18] and more recently with devices containing only conjugated polymers [26]. Most of the conjugated polymers which are readily available, including the PPVs and polythiophenes, are suitable as hole-injecting and transporting materials, and until recently it has been necessary to use electron-transporting layers made either as sublimed molecular films [46], or as blends of electron-transporting molecular semiconductors in polymer hosts [25, 47].

The synthesis of a family of solution-processible poly(cyanoterephthalylidene)s which are derivatives of PPV with nitrile groups attached to the vinylic carbons has however provided the materials necessary to complement the existing hole-transporting PPVs [26, 48, 49]. The polymers which have been studied in greatest detail are illustrated in figure 3. The alkoxy side groups are chosen to ensure that the polymer is soluble in convenient solvents, and the dialkoxy polymer (II) which has been reported previously [26, 48, 49] can be processed from solution in chloroform. Polymer (I) has asymmetric side chains (chosen to the same as MEH-PPV) and is also soluble in toluene. Both show similar optical properties to the polymers without the nitrile groups present, though in detail there are important differences which we discuss in section 5. The electron-withdrawing effect of the nitrile groups is calculated to increase the binding energies of both occupied π and unoccupied π^* states, whilst maintaining a similar π-π^* gap [50]. The increase in binding energy is measured experimentally by cyclic voltammetry to be about 0.5 eV [49], and these polymers are found to dope readily with electron donors such as sodium [51]. As mentioned in section 5, these polymers are found to be highly fluorescent, and therefore useful both for electron transport and as emissive layers.

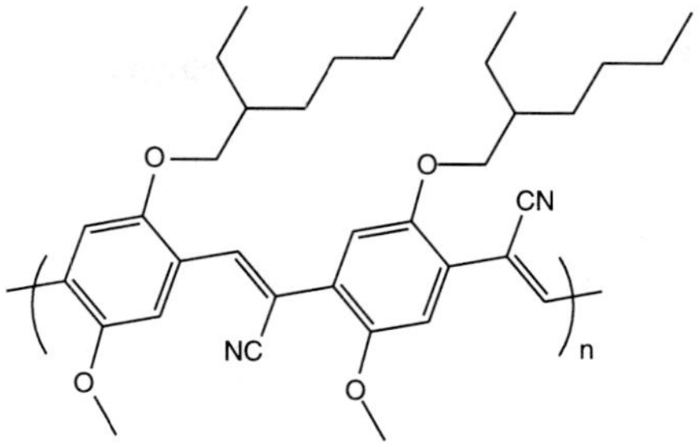

(I)

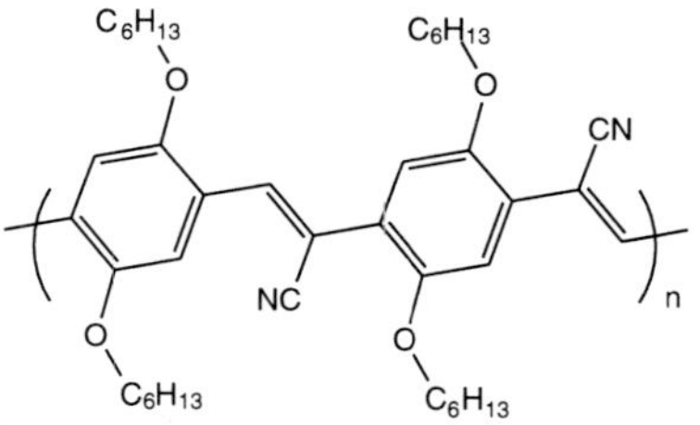

(II)

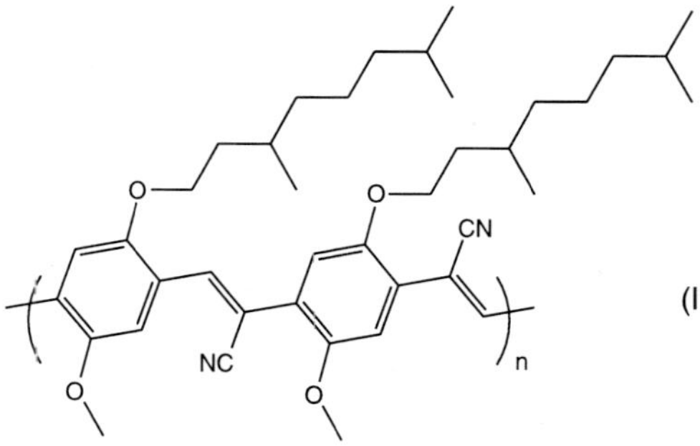

(III)

Fig. 3 Structures of some cyano-derivatives of PPV.

Fabrication of two-layer devices is as shown in figure 4. PPV is processed via the standard precursor polymer onto ITO on a glass substrate, and the nitrile (or cyano-) derivative can then be spin-coated directly on top of this intractable PPV layer. The final process step is to thermally evaporate top electrodes, for which purpose aluminium is found to be convenient. The band scheme for a two-layer device of this type is shown in figure 5. At the interface between the two polymers there are sizeable offsets in the energies of the π, highest occupied molecular orbitals (HOMOs) for the PPV and the CN-PPV and also between the π^*, lowest unoccupied molecular orbitals, (LUMOs) of the PPV and CN-PPV layers. Under forward bias, as illustrated in figure 5, injection of holes from the ITO into the HOMOs of the PPV layer results in transport of holes to the heterojunction, at which they are confined by the potential barrier which must be surmounted if they are to progress into the

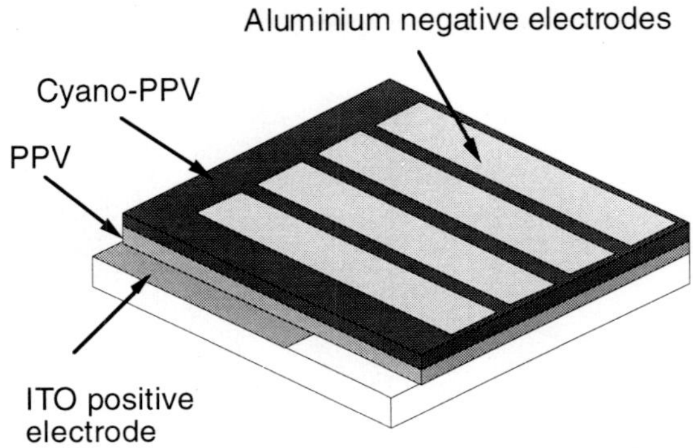

Fig. 4 Two-layer polymer LED.

CN-PPV layer. Similarly, electrons injected from the negative electrode are confined at the heterojunction by the potential step required to get transfer into the PPV LUMOs. This results in the setting up of space charge to either side of the heterojunction as indicated in figure 5. Tunnelling across one or other barrier will allow electron-hole capture and electroluminescence. We expect tunnelling across the lower of the two barriers, which should be that for holes. Note that it is now possible to get easy injection of electrons into the CN-PPV using higher work function metals, such as aluminium, in place of the calcium necessary for the single layer devices based on PPV.

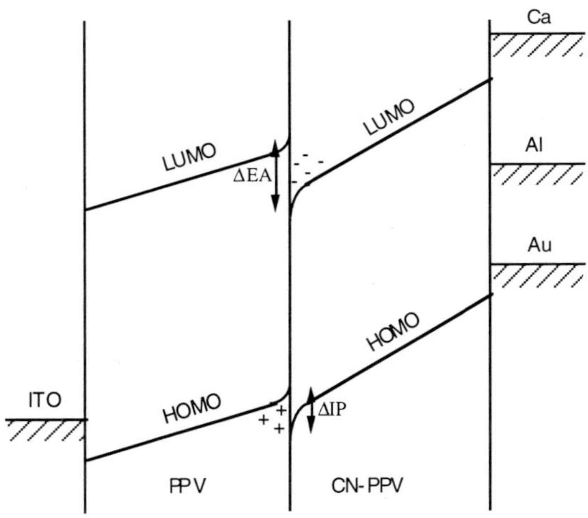

Fig. 5 Band scheme for a two-layer LED under forward bias. Positions of the Fermi energies for the electrode metals with respect to the π, highest occupied molecular orbitals (HOMOs) and the π^*, lowest unoccupied molecular orbitals (LUMOs) are illustrative.

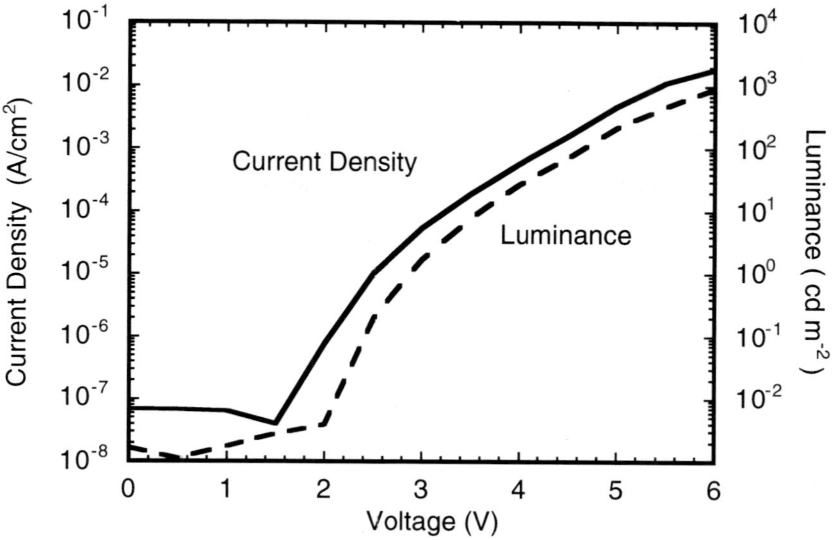

Fig. 6 Current and luminance versus forward bias for a heterostructure device of the type shown in figure 5, formed with PPV and the CN-PPV shown as (I) in figure 3. The polymer layers were of similar thicknesses and the total thickness of the polymer layers was about 75 nm [52].

Experimental results for a device made with PPV and the CN-PPV polymer (I) are shown in figure 6 [52]. The use of the toluene-soluble CN-PPV allowed more controllable fabrication than previously when polymer (II) was processed from chloroform, and the results shown in figure 6 are for a relatively thin device. Note that the turn-on voltage is close to 2 V and that the light output scales with the current through the device over many orders of magnitude of current. These devices exhibit a quantum efficiency a factor of two higher than that reported previously for similar devices made with polymer (II) [26], and using the procedures reported by Greenham et al [16] we measure an external quantum efficiency for light emitted in the forward direction of 2.5% and estimate an internal quantum efficiency in excess of 10%. With the peak output at 600 nm, the forward luminous intensity is now 3.5 cd/A at high current densities. Note that the luminous intensity reaches above 1 cd/m^2 at 3 V bias, and exceeds 100 cd/m^2 at 5 V bias.

The performance of these devices in terms of quantum efficiency is very high, and indicates that the matching of electron and hole currents injected from opposite electrodes is probably achieved.

3 Charge Transport

Carrier mobilities in polymers of this type are not well known, but there is good evidence that holes are considerably more mobile than electrons in the poly(phenylene vinylene)s, as is expected for polymers with relatively low ionisation potentials and electron affinities [34, 53]. It is therefore harder to get electrons to move into the bulk of the polymer film, and this may be responsible for poor device efficiency if excitons are generated too close to the negative electrode so that non-radiative decay becomes enhanced at the electrode [54]. Poor electron mobility, possibly resulting from charge trapping by included oxygen, is considered to play a major role in the operation of diodes of the structure as shown in figure 1 when operated in a photovoltaic or photoconductive mode [55].

We consider that the most significant problem in the context of charge injection and transport is to ensure that the currents of electrons and holes injected at opposite electrodes are balanced, so that there is not a preponderance of one charge type that passes current from one electrode to the other without encountering the opposite sign of carrier within the bulk of the polymer. There are two strategies for controlling this which have been reported. The first is to match the barriers for electron and hole injection by selection of the work functions of the electrode metals; as mentioned above, this has proved to be very successful when calcium has been used in place of aluminium. However, this suffers from the disadvantage that a reactive metal with low work function must be used (calcium has proved to be the favoured material) and this will constrain the application of these devices. The second method is to form a device with at least two semiconductor layers, making use of the heterojunction between them to confine electrons and holes travelling in opposite directions. This is the strategy that was developed for the sublimed molecular film devices [17, 18], and it has now been used successfully with the polymers [25, 26], as was discussed in section 2.2.

4 Electron-hole Recombination and Exciton Formation

The process of electron-hole capture in these devices is not well studied at present. However, in order to get efficient capture in these very thin structures (typically 100 nm total thickness of polymer layers), it is necessary that one or other charge carrier is of very low mobility so that the local charge density is sufficiently high to ensure that the other charge carrier will pass within a collision capture radius of at least one charge [34]. This is certainly enhanced in the heterostructure devices discussed below, where confinement at the heterojunction causes a build up in charge density.

Electron-hole capture is expected to produce excitons with spin wavefunctions in the triplet and singlet configurations in the ratio 3:1. For these polymeric semiconductors, there is firm evidence that the triplet exciton is strongly bound with respect to the singlet, indicating that at least the triplet (and probably also the singlet) excitons thus produced are spatially localised, to one polymer chain, and to a confined segment of the chain. That there is limited cross-over from triplet to singlet is evident in the very different lifetimes of the two species, with the singlet exciton decaying within typically 300 psecs [56], and the triplet exciton surviving for up to 1 msec at low temperatures [15]. In this model, we expect to lose 75% of the electron-hole pairs to triplet excitons which do not decay radiatively with high efficiency. Experiments to detect the formation of triplet excitons have been carried out in two ways. First, measurements of electroluminescence-detected electron-spin resonance have shown strong half-field resonances at the singlet emission energy [57]. This is considered to arise from creation of singlet states through processes such as triplet-triplet collisions. Secondly, the allowed optical transition between the lowest lying triplet and a higher lying triplet at 1.4 eV, which has been characterised in photoexcitation experiments [15, 58], can be measured directly under conditions of forward bias in a diode structure. Brown et al [59] consider that the strength of the observed absorption band in relation to the light emitted from radiative decay of singlet excitons is consistent with the 3:1 branching ratio expected within this simple model.

5 Radiative and Non-radiative Decay of Singlet Excitons

Radiative decay of the singlet exciton thus produced is required for light emission, and it is necessary to find conjugated polymers which show efficient luminescence in the solid state. Although measurement of photoluminescence efficiency for conjugated polymers in dilute solution have been known for some time [56, 60, 61], measurements for thin solid films of polymer are harder to make, and values obtained using an integrating sphere have only recently been reported [54, 62, 63]. Luminescence efficiency in the solid state tends to be lower than that measured for isolated molecules, as a result of exciton migration to quenching sites and also through interchain interactions which produce lower energy excited states which are not strongly radiatively coupled to the ground state [64]. An important source of quenching sites is provided by chemical doping, and many of the first conjugated polymers to become available were sufficiently doped so that luminescence yields were very low.

There has however been considerable progress made in improving luminescence yields in conjugated polymers, both through synthesis of higher purity polymers and also through the use of copolymers formed with segments of chain with different π-π* gaps [24, 27]. These are arranged so that excitons are trapped in lower gap regions of the chain, so that they are unable to move easily to quenching sites. Measured values of solid-state photoluminescence efficiency in the solid state for a range of poly(phenylenevinylene)s are now high. PPV prepared by the sulphonium precursor route in Cambridge shows an efficiency of 27% [54], and some of the soluble derivatives, such as the cyano-substituted polymers shown in figure 3 and copolymers [62, 63] show efficiencies of around 50%.

There has been considerable evidence that interchain interactions can significantly modify the energetics of exciton formation in conjugated polymers. For example, interchain interactions present in the ground state of some of ladder-poly(phenylene)s, give strong red-shifts of the luminescence, through the formation of 'dimer' or 'aggregate' states [65, 66]. For these materials, such interchain interactions seem to produce a reduction in quantum yield for luminescence, and the red-shift is undesirable. The situation for PPV is controversial at present. The Cambridge measurements of photoluminescence efficiency (27%), together with decay rate (0.32 nsec) are consistent with initial photogeneration of singlet, intra-chain excitons which then decay both radiatively (lifetime of order 1 nsec) and non-radiatively [54]. However, there is evidence for formation of interchain charge-separated states in samples investigated by Rothberg and co-workers [67], though this might be associated with partially oxidised polymer, for which carbonyl groups act as electron traps, thus facilitating charge separation [68].

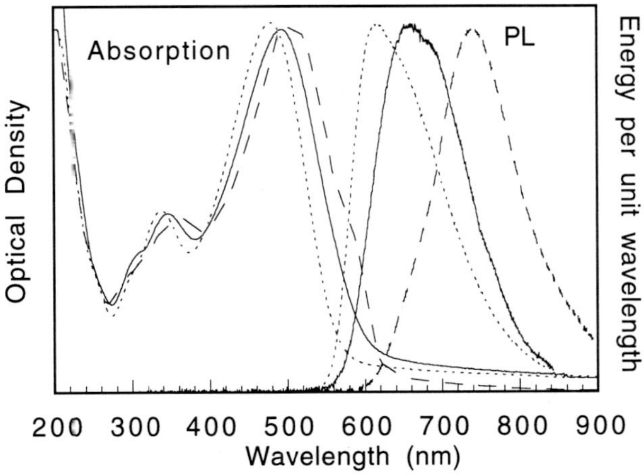

Fig.7 Optical absorption and photoluminescence, PL, spectra of solid films of the cyano-PPVs shown in figure 3: Polymer (I) (dotted line), Polymer (II) (solid line) and Polymer (III) (dashed line).

We have found that the behaviour of the cyano-PPVs shown in figure 3 is unusual, and indeed provides evidence for strong inter-chain interactions, although a high photoluminescence efficiency is still obtained. This is illustrated by the data for absorption and emission of solid films of these three cyano-PPV polymers shown in figure 7, and the data for emission from solutions of these polymers, which is contrasted with the solid film data in figure 8. We note that the Stokes' shift for the

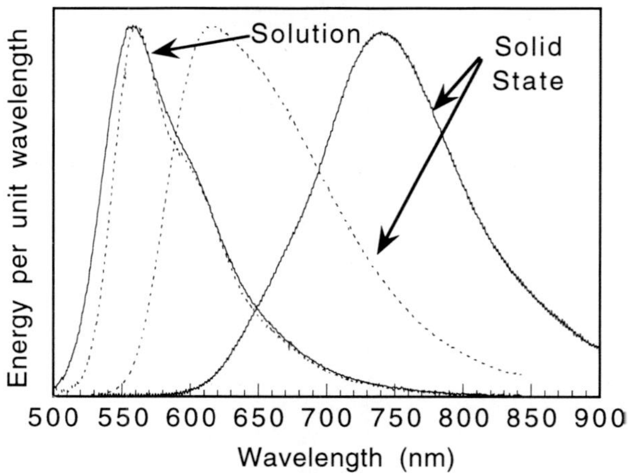

Fig. 8 Photoluminescence emission spectra of Polymer (I) (broken line) and Polymer (III) (solid line) in dilute solution and in the solid state.

solutions is small, comparable to that of PPV (figure 2), and similar for all three polymers. In contrast, there is a strong Stokes' shift for the solid films, and this is very different for the three polymers, so that the emission maxima range from near 600 nm for polymer (I) to near 750 nm for polymer (III). We consider that this red shift is due to the formation of an exciton which is delocalised over more than one chain (and thus described as an excimer), and we consider that the strong variation of the luminescence spectrum with selection of the alkyl side-chain results from different packing arrangements for these polymers, which control the strength of the interchain coupling. Further evidence to support this has been provided by Samuel et al [69] who report that the luminescence decay rate (0.8 nsec) in solutions of polymer (II) which show a photoluminescence efficiency of 52% suggest a radiative decay rate of 1.7 nsec, much as for PPV, but in contrast, the solid films of this polymer were measured to show a photoluminescence efficiency of 35%, and lifetime of 5.5 nsec, implying a radiative decay rate of 16 nsec. This factor of 10 increase in lifetime is considered to arise from the modified dipole matrix elements for radiative transitions of the inter-chain excited state.

Summary

Conjugated polymers now provide an important range of materials for use in electroluminescent devices. With improvements in synthesis and materials purity, there are now a range of materials with high luminescence efficiencies, comparable to those of the better sublimed molecular films. Improved understanding of the mechanisms for charge injection, transport, recombination, and the process of radiative and non-radiative decay of the excitons thus formed has also contributed to the performance enhancements of these devices.

Acknowledgements

We thank the Engineering and Physical Sciences Research Council, and the European Commission (ESPRIT programme 8013 LEDFOS) for support. We thank Dr. E. Staring (Philips, Eindhoven) for the supply of the monomer used for the synthesis of polymer (III), figure 3.

References

[1] H. Koezuka, A. Tsumara and T. Ando, Field-effect transistor with polythiophene thin film, *Synthetic Metals*, Vol.**18**, (1987), pp. 699-704 .

[2] A. Assadi, C. Svensson, M. Willander and O. Inganäs, Field-effect Mobility of Poly(3-hexyl thiophene), *Appl. Phys. Lett.*, Vol.**53**, (1988), pp. 195-197 .

[3] J. Paloheimo, H. Stubb, P. Yli-Lahti and P. Kuivalainen, Field Effect Conduction in Polyalkylthiophenes, *Synthetic Metals*, Vol.**41-43**, (1991), pp. 563-566 .

[4] J. H. Burroughes, C. A. Jones and R. H. Friend, Polymer Diodes and Transistors: New Semiconductor Device Physics, *Nature*, Vol.**335**, (1988), pp. 137-141 .

[5] F. Garnier, G. Horovitz, X. Peng and D. Fichou, An all-organic "soft" thin film transistor with very high carrier mobility, *Adv. Materials*, Vol.**2**, (1990), pp. 592-594 .

[6] F. Garnier, R. Hajlaoui, A. Yassar and P. Srivastava, All-Polymer Field-Effect Transistor Realised by Printing Techniques, *Science*, Vol.**265**, (1994), pp. 1684-1686 .

[7] J. H. Burroughes, D. D. C. Bradley, A. R. Brown, R. N. Marks, K. Mackay, R. H. Friend, P. L. Burn and A. B. Holmes, Light-Emitting Diodes Based on Conjugated Polymers, *Nature*, Vol.**347**, (1990), pp. 539-541 .

[8] R. A. Wessling and R. G. Zimmerman,*U.S. Patent*, (1968), 3401152 .

[9] R. A. Wessling and R. G. Zimmerman, *U.S. Patent*, (1972), 3706677 .

[10] I. Murase, T. Ohnishi, T. Noguchi and M. Hirooka, Highly Conducting Poly(p-phenylene vinylene) Prepared From a Sulphonium Salt, *Polymer Commun.*, Vol.**25**, (1984), pp. 327-329

[11] D. R. Gagnon, J. D. Capistran, F. E. Karasz and R. W. Lenz, Conductivity Anisotropy in Oriented Poly(p-phenylene vinylene), *Polymer Bulletin*, Vol.**12**, (1984), pp. 293-298 .

[12] P. L. Burn, D. D. C. Bradley, R. H. Friend, D. A. Halliday, A. B. Holmes, R. W. Jackson and A. M. Kraft, Precursor Route Chemistry and Electronic Properties of Poly(1,4-phenylenevinylene), Poly(2,5-dimethyl-1,4-phenylenevinylene), and Poly(2,5-dimethoxy-1,4-phenylenevinylene), *J. Chem. Soc. Perkin Trans 1*, (1992), pp. 3225-3231 .

[13] F. Wudl, P. M. Allemand, G. Srdanov, Z. Ni and D. McBranch, Polymers and an Unusual Molecular Crystal with Nonlinear Optical Properties, in *Materials for Nonlinear Optics: Chemical Perspectives* S. R. Marder, J. E. Sohn, G. D. Stucky, Eds. (American Chemical Society, Washington, 1991), vol. 455, pp. 683-686.

[14] D. Braun and A. J. Heeger, Visible light emission from semiconducting polymer diodes, *Appl. Phys. Lett.*, Vol.**58**, (1991), pp. 1982-1984 *erratum:* (1991) **59**, 878.

[15] N. F. Colaneri, D. D. C. Bradley, R. H. Friend, P. L. Burn, A. B. Holmes and C. W. Spangler, Photoexcited States in Poly(p-phenylene vinylene): Comparison with *trans,trans*-distyrylbenzene, a Model Oligomer, *Phys. Rev. B*, Vol.**42**, (1990), pp. 11671-11681 .

[16] N. C. Greenham, R. H. Friend and D. D. C. Bradley, Angular Dependence of the Emission From a Conjugated Polymer Light-Emitting Diode: Implications for Efficiency Calculations, *Adv. Mater.*, Vol.**6**, (1994), pp. 491-494 .

[17] C. W. Tang and S. A. VanSlyke, Organic Electroluminescent Diodes, *Appl. Phys. Lett.*, Vol.**51**, (1987), pp. 913-915 .

[18] C. W. Tang, S. A. VanSlyke and C. H. Chen, Electroluminescence of Doped Organic Thin Films, *J. Appl. Phys.*, Vol.**65**, (1989), pp. 3610-3616 .

[19] C. Adachi, S. Tokito, T. Tsutsui and S. Saito, Organic Electroluminescent Device with a three-layer Structure, *Jap. J. Appl. Phys.*, Vol.**27**, (1988), pp. 713-715 .

[20] C. Adachi, S. Tokito, T. Tsutsui and S. Saito, Electroluminescence in Organic Films with Three-Layer Structure, *Jap. J. Appl. Phys.*, Vol.**27**, (1988), pp. 269-271 .

[21] C. Adachi, T. Tsutsui and S. Saito, Organic electroluminescent device having a hole conductor as an emitting layer, *Appl. Phys. Lett.*, Vol.**55**, (1989), pp. 1489-1491 .

[22] C. Adachi, T. Tsutsui and S. Saito, Blue light-emitting organic electroluminescent devices, *Appl. Phys. Lett.*, Vol.**56**, (1990), pp. 799-801 .

[23] C. Adachi, T. Tsutsui and S. Saito, Confinement of charge carriers and molecular excitons within 5-nm-thick emitter layer in organic electroluminescent devices with a double heterostructure, *Appl. Phys. Lett.*, Vol.**57**, (1990), pp. 531-533 .

[24] P. L. Burn, A. B. Holmes, A. Kraft, D. D. C. Bradley, A. R. Brown, R. H. Friend and R. W. Gymer, Chemical Tuning of Electroluminescent Copolymers to Improve Emission Efficiencies and Allow Patterning, *Nature*, Vol.**356**, (1992), pp. 47-49 .

[25] A. R. Brown, J. H. Burroughes, N. Greenham, R. H. Friend, D. D. C. Bradley, P. L. Burn, A.Kraft and A. B. Holmes,Poly(p-phenylene vinylene) Light-Emitting Diodes: Enhanced Electroluminescence Efficiency Through Charge Carrier Confinement, *Appl. Phys. Lett.*, Vol.**61**, (1992), pp. 2793-2795 .

[26] N. C. Greenham, S. C. Moratti, D. D. C. Bradley, R. H. Friend and A. B. Holmes, Efficient Polymer-Based Light-Emitting Diodes Based on Polymers with High Electron Affinities, *Nature*, Vol.**365**, (1993), pp. 628-630 .

[27] P. L. Burn, A. Kraft, D. R. Baigent, D. D. C. Bradley, A. R. Brown, R. H. Friend, R. W. Gymer, A. B. Holmes and R. W. Jackson, Chemical Tuning of the Electronic Properties of Poly(p-phenylenevinylene)-Based Copolymers, *J. Am. Chem. Soc.*, Vol.**115**, (1993), pp. 10117-10124 .

[28] G. Grem, G. Leditzky, B. Ullrich and G. Leising, Realisation of a Blue Light Emitting Device Using Poly(para phenylene), *Adv. Materials*, Vol.**4**, (1992), pp. 36-37 .

[29] Y. Ohmori, M. Uchida, K. Muro and K. Yoshino, Blue Electroluminescent Diode Utilizing Poly(alkylfluorene), *Jap. J. Appl. Phys. Part 2 Lett.*, Vol.**30**, (1991)

[30] I. Sokolik, Z. Yang, F. E. Karsz and D. Morton, Blue-Light Electroluminescence From p-Phenylene-Based Copolymers, *J. Appl. Phys.*, Vol.**74**, (1993), pp. 3584-3586 .

[31] C. Zhang, H. von Seggern, K. Pakbaz, B. Kraabel, H. W. Schmidt and A. J. Heeger, Blue Electroluminescent Diodes Utilizing Blends of Poly(p-phenylene vinylene) in Poly(9-carbazole), *Synthetic Metals*, Vol.**62**, (1994), pp. 35-40 .

[32] J. F. Grüner, R. H. Friend, U. Scherf, J. Huber and A. B. Holmes, A High Efficiency Blue Light-Emitting Diode Based on Novel Step-Ladder Poly(para-phenylene)s, *Adv. Materials*, Vol.**6**, (1994), pp. 748-752 .

[33] W. Tachelet, S. Jacobs, H. Ndayikengurukiye, H. J. Geise and J. Grüner, Blue Electroluminescent Devices With High Quantum Efficiency From Alkoxy-Substituted Poly(para-phenylene vinylene)-trimers in a Polystyrene Matrix, *Appl. Phys. Lett.*, Vol.**64**, (1994), pp. 2364-2366 .

[34] A. R. Brown, N. C. Greenham, J. H. Burroughes, D. D. C. Bradley, R. H. Friend, P. L. Burn, A. Kraft and A. B. Holmes, Electroluminescence from Multilayer Conjugated Polymer Devices- Spatial Control of Exciton Formation and Emission, *Chem. Phys. Lett.*, Vol.**200**, (1992), pp. 46-54 .

[35] R. N. Marks, D. D. C. Bradley, R. W. Jackson, P. L. Burn and A. B. Holmes, Charge Injection and Transport in Poly(p-phenylene vinylene) Light Emitting Diodes, *Synthetic Metals*, Vol.**55-57**, (1993), pp. 4128-4133 .

[36] S. Karg, W. Riess, V. Dyakanov and M. Schwoerer, Electrical and Optical Characterisation of Poly-phenylene-vinylene Light Emitting Diodes, *Synthetic Metals*, Vol.**54**, (1993), pp. 427-433 .

[37] I. D. Parker, Carrier Injection and Device Characteristics in Polymer Light-Emitting Diodes, *J. Appl. Phys.*, Vol.**75**, (1994), pp. 1656-1666 .

[38] P. Dannetun, M. Lögdlund, M. Fahlman, M. Boman, S. Stafström, W. R. Salaneck, R. Lazzaroni, C. Fredriksson, J. L. Brédas, S. Graham, R. H. Friend, A. B. Holmes, R. Zamboni and C. Taliani, The Chemical and Electronic Structure of the Interface Between Aluminium and Conjugated Polymers or Molecules Interfaces, *Synthetic Metals*, Vol.**55-57**, (1993), pp. 212-217 .

[39] P. Dannetun, M. Fahlman, C. Fauquet, K. Kaerijama, Y. Sonoda, R. Lazzaroni, J. L. Brédas and W. R. Salaneck, Interface formation between calcium and poly(2,5 diheptyl-p-phenylenevinylene): Implications for light emitting diodes, *Synthetic Metals*, Vol.**67**, (1994), pp. 133-136 .

[40] M. Fahlman, D. Beljonne, M. Lögdlund, A. B. Holmes, R. H. Friend, J. L. Brédas and W. R. Salaneck, Experimental and Theoretical Studies of the Electronic Structure of Na-Doped Poly(para phenylene vinylene), *Chem. Phys. Lett.*, Vol.**214**, (1993), pp. 327-332 .

[41] F. Cacialli, R. H. Friend, S. C. Moratti and A. B. Holmes, Characterization of Properties of Polymeric Light Emitting Diodes over Extended Periods., *Synthetic Metals*, Vol.**67**, (1994), pp. 157-160 .

[42] S. Hayashi, H. Etoh and S. Saito, Electroluminescence of Perylene Films with a Conducting Polymer as an Anode, *Jap. Journ. Appl. Phys.*, Vol.**25**, (1986), pp. 773-775 .

[43] T. Nakano, S. Doi, T. Noguchi, T. Ohnishi and Y. Iyechika, Organic Electroluminescence Device, *European Patent Application*, (1991) Application Number 91301416.

[44] Y. Yang and A. J. Heeger, Polyaniline as a Transparent Electrode for Polymer Light-Emitting Diodes: Lower Operating Voltage and Higher Efficiency, *Appl. Phys. Lett.*, Vol.**64**, (1994), pp. 1245-1247 .

[45] Q. Pei, G. Yu, C. Zhang, Y. Yang and A. J. Heeger, Polymer Light-Emitting Electrochemical Cells, *Science*, Vol.**269**, (1995), pp. 1086-1089 .

[46] S. Doi, T. Nakano, T. Noguchi and T. Ohnishi, Al-q/PPV Light-Emitting Diodes, *Polymer Preprints, Jpn*, Vol.**40**, (1991), pp. 3594 .

[47] P. Burn, A. B. Holmes, A. Kraft, A. R. Brown, D. D. C. Bradley and R. H. Friend, Light emitting diodes based on conjugated polymers: Control of colour and efficiency, *Mat. Res. Soc. Symp. Proc.*, Vol.**247**, (1992), pp. 647-654 .

[48] S. C. Moratti, D. D. C. Bradley, R. H. Friend, N. C. Greenham and A. B. Holmes, Molecularly Engineered Polymer LEDs, *Mat. Res. Soc. Symp. Proc.*, Vol.**328**, (1994), pp. 371-376 .

[49] S. C. Moratti, D. D. C. Bradley, R. Cervini, R. H. Friend, N. C. Greenham and A. B. Holmes, Light-Emitting Polymer LEDs, *SPIE proceedings series*, Vol.**2144**, (1994), pp. 108-114 .

[50] J. L. Brédas and A. J. Heeger, Influence of Donor and Acceptor Substituents on the Electronic Characteristics of Poly(paraphenylene vinylene) and Poly(paraphenylene), *Chem. Phys. Lett.*, Vol.**217**, (1994), pp. 507-512 .

[51] P. Bröms, M. Fahlman, K. Z. Xing, W. R. Salaneck, P. Dannetun, J. Cornil, D. A. dos Santos, J. L. Brédas, S. Moratti, A. B. Holmes and R. H. Friend, Optical Absorption Studies of Sodium Doped Poly(cyanoterephthalylidene), *Synthetic Metals*, Vol.**67**, (1994), pp. 93-96 .

[52] D. R. Baigent, N. C. Greenham, J. Grüner, R. N. Marks, R. H. Friend, S. C. Moratti and A. B. Holmes, Light-Emitting Diodes Fabricated With Conjugated Polymers - Recent Progress, *Synthetic Metals*, Vol.**67**, (1994), pp. 3-10 .

[53] B. R. Hsieh, H. Antoniadis, M. A. Abkowitz and M. Stolka, Charge Transport Properties of Poly(p-phenylene vinylene) and its Sulfonium Precursor Polymers at Different Degrees of Conversion, *Polym. Preprints*, Vol.**33**, (1992), pp. 414 .

[54] N. C. Greenham, I. D. W. Samuel, G. R. Hayes, R. T. Phillips, Y. A. R. R. Kessener, S. C. Moratti, A. B. Holmes and R. H. Friend, Measurement of Absolute Photoluminescence Quantum Efficiencies in Conjugated Polymers, *Chem. Phys. Lett.*, Vol.**241**, (1995), pp. 89-96 .

[55] R. N. Marks, J. J. M. Halls, D. D. C. Bradley, R. H. Friend and A. B. Holmes, The Photovoltaic Response in Poly(p-phenylene vinylene) Thin Film Devices, *J. Phys. Condensed Matter*, Vol.**6**, (1994), pp. 1379-1394 .

[56] I. D. W. Samuel, B. Crystall, G. Rumbles, P. L. Burn, A. B. Holmes and R. H. Friend, The Efficiency and Time-Dependence Of Luminescence From Poly(*p*-phenylene vinylene) and Derivatives, *Chem. Phys. Lett.*, Vol.**213**, (1993), pp. 472-478 .

[57] L. S. Swanson, J. Shinar, A. R. Brown, D. D. C. Bradley, R. H. Friend, P. L. Burn, A. Kraft and A. B. Holmes, Electroluminescence detected magnetic resonance (ELMDR) study of poly(paraphenylene vinylene)-based diodes, *Phys. Rev. B*, Vol.**46**, (1992), pp. 15072-15077.

[58] X. Wei, B. C. Hess, Z. V. Vardeny and F. Wudl, Studies of Photoexcited States in Polyacetylene and Poly(paraphenylenevinylene) by Absorption Detected Magnetic Resonance: the Case of Neutral Photoexcitations, *Phys. Rev. Lett.*, Vol.**68**, (1992), pp. 666-669 .

[59] A. R. Brown, K. Pichler, N. C. Greenham, D. D. C. Bradley, R. H. Friend and A. B. Holmes, Optical Spectroscopy of Triplet Excitons and Charged Excitations in Poly(p-phenylenevinylene) Light-Emitting Diodes, *Chem. Phys. Lett.*, Vol.**210**, (1993), pp. 61-66 .

[60] L. Smilowitz, A. Hays, A. J. Heeger, G. Wang and J. E. Bowers, Time-Resolved Photoluminescence From Poly(2-methoxy, 5-(2'-ethyl-hexyloxy)-p-phenylene-vinylene): Solutions, gels, films, and blends, *J. Chem. Phys.*, Vol.**98**, (1993), pp. 6504-6509 .

[61] C. L. Gettinger, A. J. Heeger, J. M. Drake and D. J. Pine, A Photoluminescence Study of Poly(phenylene vinylene) Derivatives: The Effect of Intrinsic Persistence Length, *J. Chem. Phys.*, Vol.**101**, (1994), pp. 1673-1678 .

[62] D. Braun, E. G. J. Staring, R. C. J. E. Demandt, G. L. J. Rikken, Y. A. R. R. Kessener and A. H. J. Venhuizen, Photoluminescence And Electroluminescence Efficiency In Poly(dialkoxy-p-Phenylenevinylene), *Synth. Met.* , Vol.**66**, (1994), pp. 75-79.

[63] E. G. J. Staring, R. C. E. Demandt, D. Braun, G. L. J. Rikken, Y. A. R. R. Kessener, T. H. J. Venhuizen, H. Wynberg, W. ten Hoeve and K. J. Spoelstra, Photo- and Electroluminescence Efficiency in Soluble Poly(dialkoxy-p-phenylenevinylene), *Adv. Mater.*, Vol.**6**, (1994), pp. 934-937 .

[64] J. W. P. Hsu, M. Yan, T. M. Jedju, L. J. Rothberg and B. R. Hsieh., Assignment Of The Picosecond Photoinduced Absorption In Phenylene Vinylene Polymers, *Phys. Rev. B*, Vol.**49**, (1994), pp. 712-715 .

[65] A. Köhler, J. Grüner, R. H. Friend, U. Scherf and K. Müllen, Photovoltaic Measurements on Aggregates in Ladder-Type Poly(phenylene), *Chem. Phys. Lett.*, Vol.**243**, (1995), pp. 456-461 .

[66] U. Lemmer, S. Heun, R. F. Mahrt, U. Scherf, M. Hopmeier, U. Siegner, E. O. Göbel, K. Müllen and H. Bässler, Aggregate Fluorescence in Conjugated Polymers, *Chem. Phys. Lett.*, Vol.**240**, (1995), pp. 373-378 .

[67] M. Yan, L. J. Rothberg, F. Papadimitrakopolous, M. E. Galvin and T. M. Miller, Spatially Indirect Excitons as Primary Photoexcitations in Conjugated Polymers, *Phys. Rev. Lett.*, Vol.**72**, (1994), pp. 1104-1107 .

[68] M. Yan, L. J. Rothberg, F. Papadimitrakopoulos, M. E. Galvin and T. M. Miller, Defect Quenching of Conjugated Polymer Luminescence, *Phys. Rev. Lett.*, Vol.**73**, (1994), pp. 744-747 .

[69] I. D. W. Samuel, G. Rumbles and C. J. Collison, Efficient Inter-chain Photoluminescence in a High Electron Conjugated Polymer, *Phys. Rev.*, Vol.**B52**, (1995), pp. 11573-11572 .

Polymer LED Utilizing Poly(arylene vinylenes)

Toshihiro Ohnishi, Shuji Doi, Yoshihiko Tsuchida, and Takanobu Noguchi

Tsukuba Research Laboratory, Sumitomo Chemical Co. Ltd.,
6 Kitahara, Tsukuba, Ibaraki , 300-32 Japan

Abstract

Polymer light emitting diode, P-LED, has successfully been fabricated using highly luminous poly(arylene vinylene) derivatives. The P-LED showed a maximum luminance of 55,000cd/m^2. This high performance of the device was achieved by employing copolymers of arylene vinylene units as a light-emitting material and a multi-layer structure for the device.

Structural and energetical irregularity introduced into the conjugated polymers gave us a highly luminous polymer due to confinement of excitons. The irregularity can be formed by copolymerization of conjugated/non-conjugated segments, *m/p*-phenylene vinylenes and alkyl/alkoxy-substituted phenylene vinylenes. Among the copolymers, the *m/p*-phenylene vinylene copolymer gave a highly efficient P-LED device due to the balance of the exciton confinement and the charge transporting property .

1 Introduction

Organic and polymer light emitting diodes, O-LED and P-LED, have attracted much attention as an accessible flat panel display and have shown good progress in the last few years. Since bright and low-voltage driven O-LED devices were reported by Tang et al.[1], many light-emitting and charge transporting materials and their devices have been reported[2]. Some of the devices showed high luminance and long life time[3]. Catching up with O-LED, a P-LED device was reported in 1990[4] and then many kinds of conjugated polymers known as a conducting polymer were studied as a light-emitting material. P-LED devices using PPV derivatives[5,6], poly(*p*-phenylene)[7] and poly(3-alkyl- thiophene)[8] have been reported since 1990.

We have already reported soluble light-emitting poly(2,5-dialkoxy-*p*-phenylene vinylene), RO-PPV, and its side chain length dependence of fluorescent intensity[9]. Intensity of photo-luminescence, PL, of the RO-PPV film increased with the alkoxy chain length, which is attributable to the decrease of intermolecular interaction and confinement of excitons in the film.

High efficiency and long life time are essential factors for commercialization of O- and P-LED's. There have been many attempts made to improve device performance so far[10] . Device efficiency, η_d, is defined in the following equation[11].

$$\eta_d = \gamma \cdot \eta_{e\text{-}h} \cdot \phi_{fl}$$

where γ is the ratio of the number of minority carrier to that of major carrier, $\eta_{e\text{-}h}$ is the efficiency of electron-hole recombination in the device and ϕ_{fl} is the fluorescent efficiency of the light emitting material.

Springer Proceedings in Physics, Vol. 81
Materials and Measurements in Molecular Electronics
Editors: K. Kajimura · S. Kuroda

The value of γ is related to the injection process and depends on electrode materials. In P-LED, a calcium electrode is used to increase the electron injection efficiency due to its low work function[12, 13]. The value of $\eta_{e\text{-}h}$ is related to not only materials but also device structure. Multi-layer structure is commonly used to increase the value of $\eta_{e\text{-}h}$, generating the hole or/and electron accumulation near a light-emitting layer surface[1,14]. The maximum value of $\eta_{e\text{-}h}$ is 25% in case of only singlet exciton formation [11]. To improve ϕ_{fl}, we should use highly efficient materials or dye-doped materials[15].

After fundamental studies on soluble RO-PPV derivatives[9], we have focused on the synthesis of highly luminous polymers and fabrication of a highly efficient device to commercialize P-LED. Taking exciton confinement and charge transporting into account, we have copolymerized various kinds of arylene vinylene units to introduce structural and energetical irregularity into poly(arylene vinylene) .

In this paper we will discuss spectroscopic properties of the copolymers and the device performance using arylene vinylene copolymers.

2 Experimental

Poly(arylene vinylene) derivatives were prepared by Wittig reaction of diphosphonium compounds(I and V), and dialdehyde compounds(II, III,and IV) as shown in Fig.1. When I and II, I and III , I and IV, and V and II are used, soluble alternating copolymers of (A), (B), (C), and(D) are obtained respectively. The diphosphonium compounds were prepared using triphenylphosphine and 2,5-dioctyloxy(dioctyl)-*p*-xylylene dichloride, and polymerized by addition of dialdyhyde compounds and tert-BuOK as a catalyst in tert-BuOH at 90℃ for 7 hours to give a copolymer. Random copolymers having (A) and (B), (A) and (C) , and (A) and (D) (abbreviated as A/B, A/C and A/D respectively) were also synthesized in a similar manner to the copolymers using corresponding three monomers selected among (I), (II) , (III), (IV), and (V).

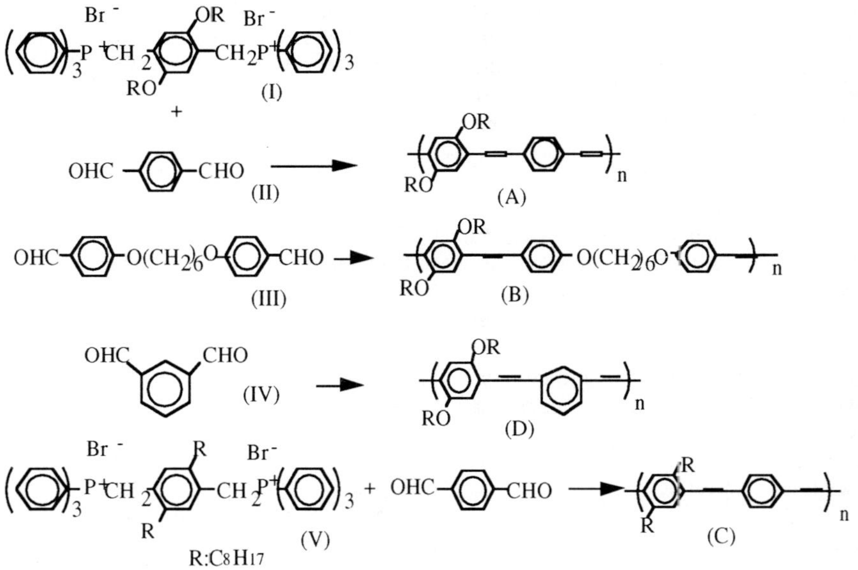

Figure 1 Synthetic route for alternating and random copolymers of arylene vinylenes

The copolymers were obtained by precipitation in methanol from reactant solution and purified by reprecipitation. Poly(2,5-dioctyloxy-p-phenylene vinylene), OO-PPV, used as a standard polymer was prepared by dehydrohalogenation of 2,5-dialkoxy-p-xylylene dibromide as reported previously[9].

The P-LED device consists of a transparent electrode, a light emitting polymer film, an electron-transporting or hole-blocking layer, and a negative electrode as shown in Fig.2. A buffer or additional layer[9] was used between the transparent electrode and the light- emitting polymer film in the case of three-layer structure.

Polymer films, 50 to 100 nm in thickness, were spin-cast from the toluene solution onto indium tin oxide, ITO, transparent electrode. A tris(8-quinolinol) aluminum, Alq_3, layer as an electron transporting layer was then deposited onto the polymer film followed by coevaporation of Li-Al or Mg-Ag alloy as a negative electrode under a vacuum of 10^{-6} Torr.

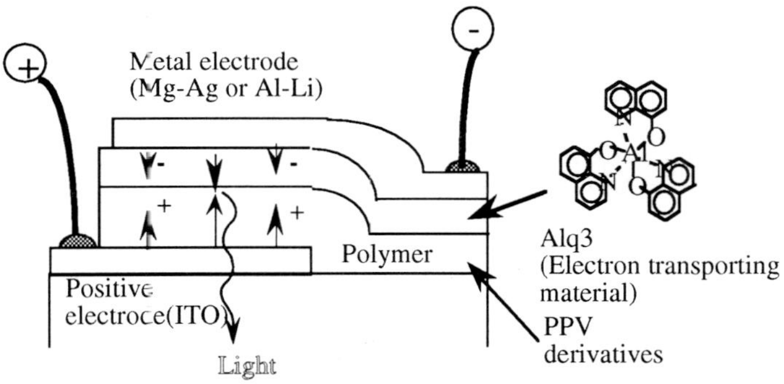

Figure 2 Schematic structure of P-PED

Absorption and luminescence spectra for thin films and device using copolymers were measured with a Shimadzu UV-365 spectrophotometer and a Hitachi 850 fluorescence spectrophotometer, respectively. Luminance and current-voltage curves were obtained with a TOPCON BM-8 luminance meter, a Keithley 990 digital multimeter and a Takasago GP050-2 DC voltage source which were controlled with a personal computer.

3 Results and Discussion

3.1 Spectroscopic properties of Copolymers

(a) Alternating copolymers
All copolymers are obtained in yellow or yellowish orange powder, are soluble in organic solvent such as toluene and chloroform and show strong fluorescence in both solution and film. Figure 3 shows the absorption spectra of the copolymers together with that of poly(2,5-dioctyloxy- *p*-phenylene vinylene), OO-PPV. The spectra of copolymers were blue-shifted compared with OO-PPV and band gaps or absorption edges of the copolymers (B) and (D) were larger than that of OO-PPV, while those of (A) and (C) were in between as shown in Table 1. This result indicates that the non-conjugated segment and the *m*-phenylene structure shorten conjugation length in the polymer due to low resonance effect and thus enlarge the band gap.

The fluorescent spectra of the copolymers were also blue-shifted similarly to the absorption spectra as shown in Fig. 4. Spectral properties of all the alternating copolymers are summarized in Table 1 together with OO-PPV. The fluorescence intensity was estimated from the signal intensity divided by the absorbance at excitation wavelength to give the relative quantum efficiency of fluorescence. The (B) copolymer having the ether group in the chain shows the strongest fluorescence in all the copolymers although it has the shortest conjugation length. The *m*-phenylene structure((D) copolymer) gives stronger fluorescence than *p*-phenylene structure((A) copolymer) because of less resonance effect. These results indicate that the fluorescence intensity depends on the degree of conjugation in the polymer chain. Since the (C) copolymer having the alkyl substituent shows the stronger fluorescence than the alkoxy substituted copolymer (A), the alkoxy substituent suppresses the fluorescence. In fact methoxy-substituted PPV shows weaker fluorescence than methyl-substituted PPV[16].

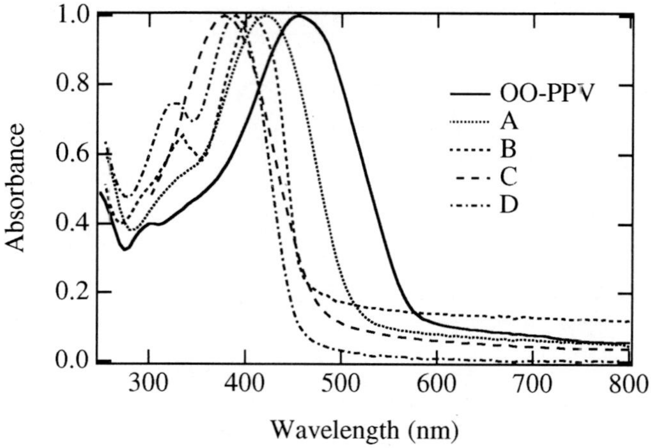

Figure 3 Absorption spectra of alternating copolymer of (A), (B), (C) and (D) together with that of OO-PPV in film.

Table 1 Optical properties of alternating copolymers in film

Polymer	repeating unit	Absorption peak(nm)	Band gap (eV)	Fluorescent peak(nm)	Fluorescent intensity (a.u.)
OO-PPV	OO-PV	455	2.12	582	3.6
Alternating	(A)	420	2.25	534	7.5
Ether type	(B)	405	2.66	504,528	50.5
R-PPV	(C)	375	2.53	496	9.8
m/p-phenylene	(D)	390	2.66	504	15.5

Band gap was estimated from wavelength of absorption edge

Fluorescent intensity=(area of fluorescent peak plotted against wave number) /(absorbance at excitation wavelength)

In these copolymers the substituent effect on the fluorescent intensity was observed similarly to that in RO-PPV reported previously[9]. The effect is explainable on the basis of exciton migration reported by Baesslar[17].

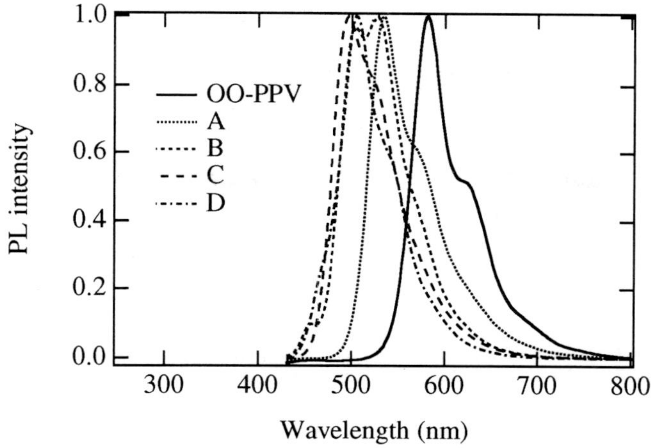

Figure 4 Fluorescent spectra of alternating copolymers of (A), (B), (C) and (D) together with that of OO-PPV

(b) Random copolymer

Figures 5 and 6 show the absorption and fluorescence spectra of the random copolymers consisting of (A) and (D) units. The absorption peak shifts proportionally to the ratio of (A) and (D), while the absorption edge shifts less than the peak. Similar spectral shifts were observed in the copolymers of A /B, and A/C.

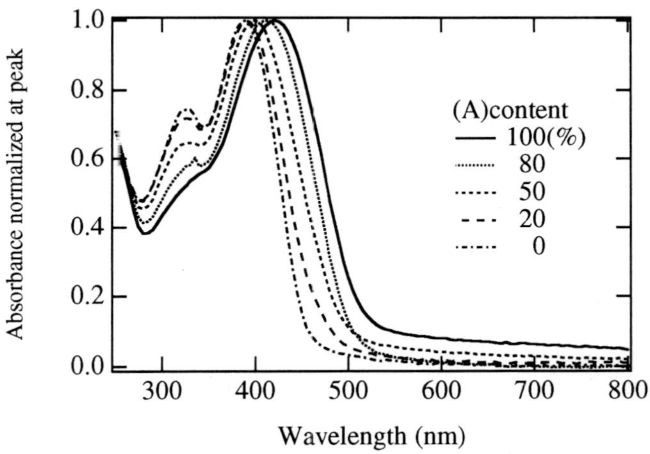

Figure 5 Absorption spectra of random copolymers consisting of (A) and (D) at various ratio.

The fluorescent spectra of random copolymers were, on the contrary, similar to that of (A) copolymer and were hardly shifted with increasing component of (D) unit as shown in Fig.6. Such small spectral shifts in fluorescence were also observed in A/B and A/C copolymers. This result indicates that there are segments having an energy level similar to that of (A) polymer which has the lowest band gap in the four polymers. Randomly copolymerizing (A) unit and the others generates the low energy segment consisting of a sequence of (A) in the polymer chain.

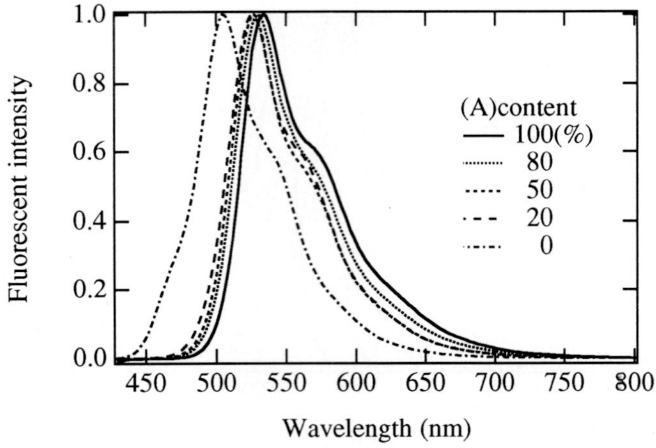

Figure 6 Fluorescent spectra of random copolymers consising of (A) and (D) at various ratio.

3.2 Quantum efficiency of PL

Figure 7 shows the relation between the fluorescent intensity in film form and the ratio of (A) unit in the polymers for A/B, A/C, and A/D. For the A/B and A/D copolymers, the fluorescent intensity increased with decreasing the ratio of (A) in the copolymers. Since wavelengths of fluorescent peak in the spectra are close to that of (A) polymer as shown in Fig. 6, the fluorescence is emitted from the lower energy segments consisting of the (A) unit which are incorporated with segments having higher energy. This means that the exciton generated by light absorption migrates to the lowest energy state and is confined efficiently to generate radiation. The exciton confinement thus takes place more easily in copolymers having less content of (A) and leads to give strong fluorescence. In the case of the A/C copolymers, the fluorescent intensity basically increased with decreasing the ratio of (A) unit except for the (C) polymer which shows smaller intensity than the copolymer having 10% of the (A) unit. This is

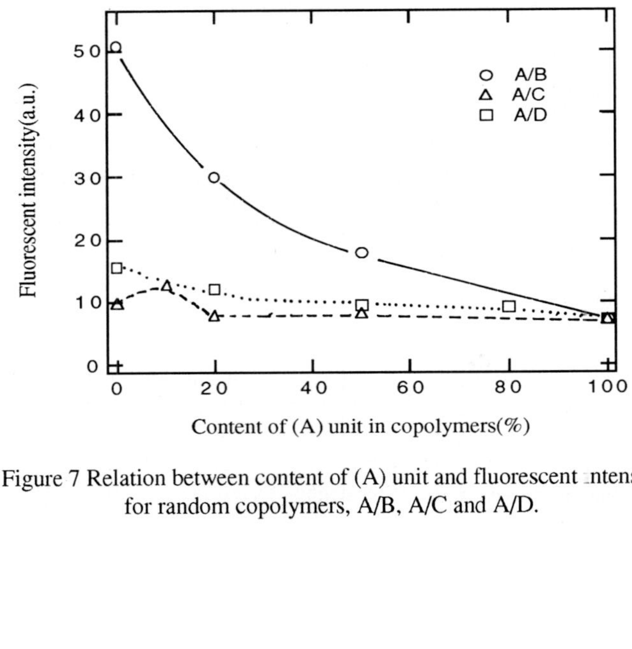

Figure 7 Relation between content of (A) unit and fluorescent intensity for random copolymers, A/B, A/C and A/D.

attributable to smaller confinement of the exciton in (C) copolymer than random copolymers because there is no energetical irregularity in the (C) copolymer.

3.3 Electroluminescence of LED using copolymers

(a) Multi-layer device

Alq_3 used most commonly in O-LED[1] possesses not only light-emitting but also electron-transporting properties. When Alq_3 is used as an electron-transporting material, its light-emitting property should be taken into account. We reported that P-LED consisting of Alq_3 and RO-PPV layers showed the emission from RO-PPV and high efficiency compared with a RO-PPV single-layer device and that Alq_3 was a good electron-transporting material for RO-PPV-based P- LED Using the copolymers, we found out that Alq_3 also acted as a good electron-transporting material and gave us highly luminous devices. Figure 8 shows the EL spectra of the devices using (A) and A/D copolymers together with the PL spectrum of Alq_3. The both EL spectra differ from the photo-luminescence PL, spectrum of Alq_3 and are in good agreement with PL spectra of copolymers shown in Figures 4 and 6.

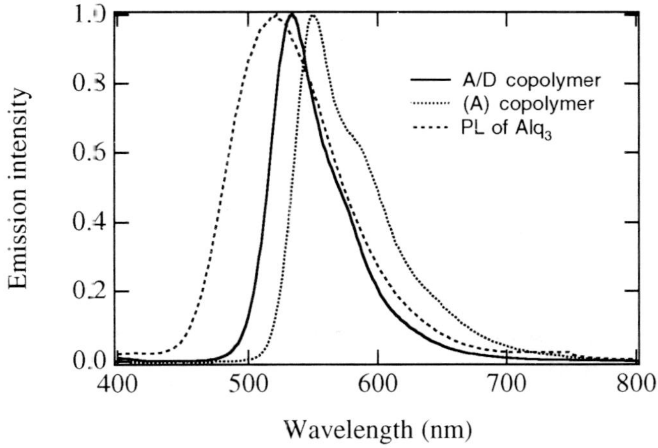

Figure 8 EL spectra of multi-layer P-LED using light emitting copolymer s and Alq_3 as a electron-transporting material together with PL spectrum of Alq_3

This result indicates that the hole-injection from copolymer to Alq_3 layer does not take place but the electron-injection from Alq_3 to copolymer layer does. From the electrochemically estimated energy levels of Alq_3 and copolymers, there are energy barrier about 0.3eV high at the polymer-Alq_3 interface for the hole injection and no barrier for the electron injection as shown in Fig. 9[18]. It is explainable that the electron-injection takes place more easily than the hole-injection and that the emission from the polymer is observed for multi-layer devices fabricated in this study using Alq_3.

(b)Device efficiency

The device efficiency depends on not only PL efficiency of the emitting polymer but also the efficiency of hole-electron recombination which is related to the charge transporting properties of materials used in the device and to the device structure as described above. Employing a multi-layer device[19] and a Mg-Ag or Li-Al alloy electrode[9], we have already improved the device efficiency but have to further improve the efficiency for the commercialization. We thus have studied on the relation between composition of the random copolymer and the device efficiency. The results were shown in Fig.10.

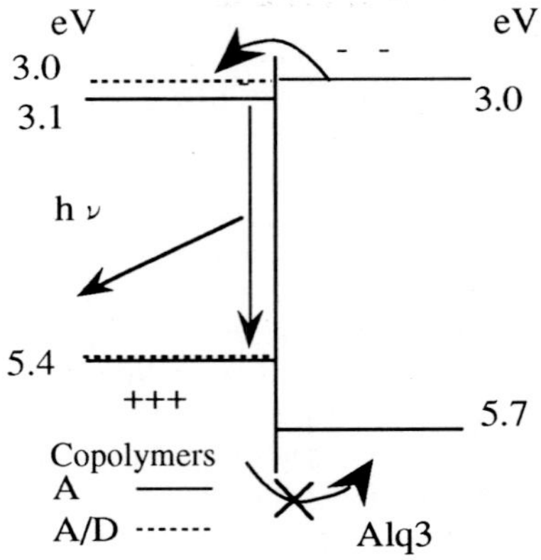

Figure 9 Schematic diagram of energy levels for the junction between the copolymers and Alq_3

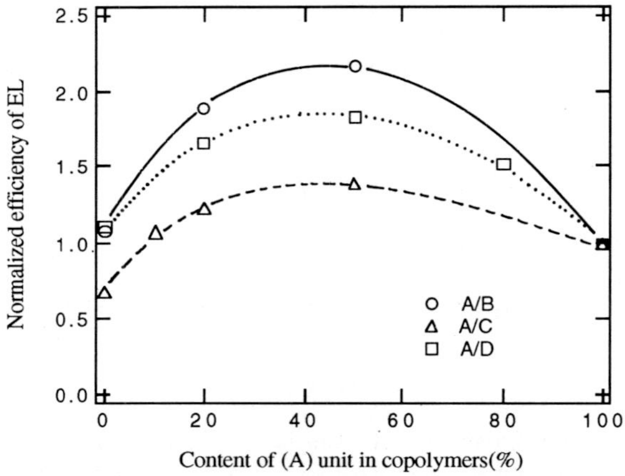

Figure 10 Relation between content of (A) unit and EL efficiency for random copolymers, A/B, A/C and A/D.
Device structures are a bi-layer device for A/B and a multi-layer device for A/C and A/D.

The device efficiency of the devices using all the random copolymers showed the maximum around 50% of (A) unit, while the fluorescent efficiency increased with decreasing the ratio of (A) as shown in Fig.6. The increase of the device efficiency in the region of large ratio of (A) unit is due to the increasing the fluorescent efficiency. The decrease in the small content of (A) unit is attributable to decreasing conductivity of the copolymer,because the structural and energetical irregularity in the copolymer increased.

These results suggest that the device efficiency depends on not only the fluorescent intensity but also on the charge carrier mobility of the copolymers.

Figure 11 shows the current density-luminance, I-L, curves of the P-LED devices using the random copolymers together with that for the (A) copolymer. The luminance of the devices increased almost linearly with the current-density for all devices. The maximum luminance reached 55,000 cd/m^2 for the A/D copolymer , 40,000cd/m^2 for A/C and (A) copolymers, and several thousands cd/m^2 for the A/B copolymer before break-down of the devices. This result implies that the multi-layer structure does not work well in the device using the A/B copolymer which shows the strongest PL efficiency. Since the A/B copolymer has a large band gap and a high LUMO(the lowest unoccupied molecular orbital) level compared with other copolymers, the energy barrier at the interface with Alq_3 is enlarged and consequently suppresses the electron injection process from Alq_3 to the polymer.

In the case of the single-layer device using RO-PPV, superlinearity in a I-L curve was observed, while no superlinearity was seen in these devices[9]. This is attributable to the difference in thermal properties between RO-PPV and copolymers. RO-PPV has higher glass transition temperature, Tg, around 80℃ than copolymers of which Tg values are around or below a room temperature. In spite of low Tg, the P-LED devices using copolymers showed good reproducibility in I-L curves and driving life time.
Since the slope of the I-L curve represents the device efficiency, the A/D polymer-based device having the steepest slope showed the highest efficiency in the four devices. The maximum device efficiency reached about 6 cd/A.

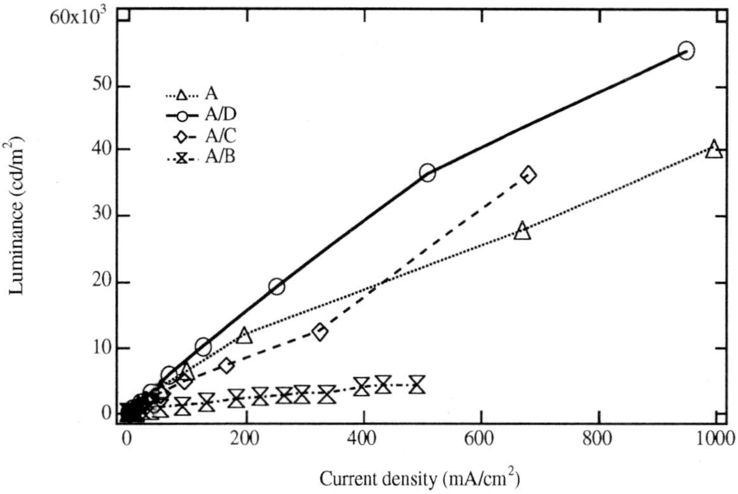

Figure 11 Luminance-current density curves of the P-LED's using (A) alternating copolymer and A/B, A/C, A/D random copolymers

Figure 12 shows current density - voltage, I-V, curves for the devices. All the devices showed on-set voltage around 5V at about 0.01mA/cm^2 which was near the measurement limitation of current density. All of the devices can be driven below 15V at high current density without break-down but showed the break-down over 20V. In the case of the A/B copolymer, the high current density was obtained although the luminance of the device is lower than those of the other devices. This indicates that the device using A/B copolymer has the lowest electron-hole recombination efficiency in four devices.

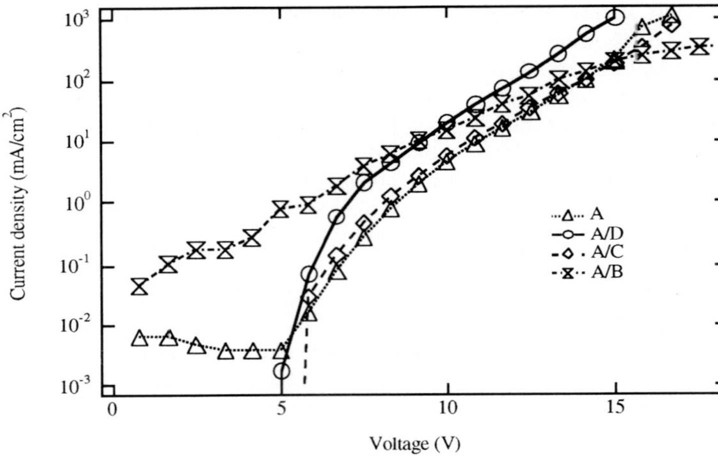

Figure 12 Current density- voltage curves of P-LED devices using (A) alternating copolymer and A/B, A/C and A/D random copolymers

4 Conclusion

The structural and energetical irregularity generated by copolymerization is effective to enhance fluorescent intensity of poly(arylene vinylene) derivatives and gives highly luminous P-LED devices. The *m*-phenylene structure induces the irregularity effectively and gives the good balance of the exciton confinement and the charge transporting property for high performance devices.

The multi-layer structure plays an important role in the P-LED using copolymers of arylene vinylenes to enhance the efficiency of hole-electron recombination. The optimized P-LED device using *m*- and *p*-phenylene vinylene copolymers showed high efficiency and the maximum luminance as high as 55,000cd/m^2.

References

[1]C.W.Tang and S.A.VanSlyke,*Applied Physics Letters* , Vol.51, 1987, pp913.

[2]C.Adachi, S.Tokito, T.Tsutsui, and S.Saito,*Japanese Journal of Applied Physics*, Vol.27, 1988, ppL269.

[3]Y. Hamada, T. Sano, K. Shibata and K.Kuroki, *Japanese Journal of Applied Physics*, Vol.34, 1998, ppL824.

[4]J.H.Burroughes, D.D.C.Bradley, A.R.Brown, R.N.Marks, K.Mackay, R.H. Friend, P.L.Burns and A.B.Holmes, *Nature*, Vol.347, 1990, pp539.

[5] T. Nakano, S. Doi, T. Noguchi, T. Ohnishi, and Y. Iyechika, EPC0443861A3 (1991)

[6]D.Braun and A.J.Heeger, *Applied Physics Letters,* Vol.58, 1991, pp1982.

[7]G.Grem, G.Leditzky, B.Ullrich and G.Leising, *Advanced Materials*, Vol.4, 1992, pp36.

[8]Y.Ohmori, M. Uchida, K. Muro and K.Yoshino, *Japanese Journal of Applied Physics*, Vol.30, 1991, ppL1938.

[9]S.Doi, M.Kuwabara, T.Noguchi, and T. Ohnishi, *Synthetic Metals*, Vol.55-5, 1993, pp4174.

[10]S.Saito, T. Tsutsui, M. Era, N. Takada, E. Aminaka and T. Wakimoto, *Molecular Crystal and Liquid Crystal,* Vol. 253, 1994, pp125-132.

[11]T. Tsutsui and S. Saito, NATO ASI Ser., Ser. E, Vol. 246, 1993, pp123-134.
[12]D.Braun and A.J. Heeger, *Journal of Electronic Materials*, Vol.20, 1991, pp945.
[13]P.L.Burn, A.B. Holmes, A.Kraft, A.R. Brown, D.D.C.Bradley, and R.H.Friend, *Material Research Society Symposium Proceedings*, Vol.247, 1992,pp647-654.
[14]N.C.Greenham, S.C.Moratti, D.D.C.Bradley, R.H.Friend and A.B. Holmes, *Nature*,Vol.365, 1993, pp628.
[15]C.W.Tang, S.A.VnSlyke and C.H.Chen, *Journal of Applied Physics*, Vol.65, 1989, pp3610-3616.
[16]Unpublished data
[17]U.Rauscher, L. Shuts, A.Rreiner and H.Baessler, *Journal Physics:Condensed Matter*, Vol.1, 1989, pp9751-9763.
[18]The energy level of HOMO(highest occupied molecular orbital) was calculated from the threshold potential of oxidation current in cyclic voltamogram and the potential of standard hydrogen electrode(4.5eV). The LUMO level was estimated from a threshold wavelength of an absorption spectrum.
[19]T. Ohnishi, T. Noguchi and S. Doi, *Japanese Patent* 5-247460(1993)

Electron-Nuclear Double-Resonance Spectroscopy of Polarons in Conjugated Polymers

S. Kuroda[1], K. Murata[1], Y. Shimoi[1], S. Abe[1], T. Noguchi[2] and T. Ohnishi[2]

1. Electrotechnical Laboratory, Tsukuba, Ibaraki 305, Japan
2. Tsukuba Research Laboratory, Sumitomo Chemical Co. Ltd., Tsukuba, Ibaraki 300-32, Japan

Abstract

Polarons are fundamental nonlinear excitations in conjugated polymers. Spin density distribution of polarons, or their spatial extension in other words, is one of the essential physical quantities that can be used to identify polarons. It is also known that the magnitude of electron correlation sensitively affects the detailed form of spin distribution. In this paper we discuss the electron nuclear double resonance (ENDOR) observation of the polaron in an electroluminescent polymer, poly(paraphenylene vinylene), PPV. Observed spin distribution extends over about 4 phenyl rings in half width and is well described by the theoretical calculation using PPP (Pariser-Parr-Pople) model in the case of finite electron correlation. We also present evidence for the photo-generation of polarons from the light-induced ESR spectra and their excitation profile.

1 Introduction

Since the discovery of metallic conductivity upon doping of polyacetylene films, many conjugated polymers have been synthesized to develop conducting polymers [1]. More recently, another important property of conjugated polymers has been found, that is, intense electroluminescence has been discovered in a polymer, poly(paraphenylene vinylene), PPV and its analogs [2-4]. These conjugated polymers, at the same time, attract much attention as one-dimensional electronic systems that can generate nonlinear excitations such as solitons, polarons and bipolarons [1, 5]. When the polymer has a degenerate ground state structure as in the case of *trans*-polyacetylene, solitons and polarons become primary excitations. On the other hand, for the polymers with nondegenerate ground state structures, solitons can not exist and polarons and bipolarons become primary charged excitations. These excitations give rise to characteristic absorptions in optical, magnetic and other spectroscopic measurements.

Among these spectroscopic methods, electron nuclear double resonance (ENDOR) spectroscopy of protons on the conjugated chain can directly measure the spin density distribution of paramagnetic excitations through the study of hyperfine coupling [6, 7]. This is because the spin density ρ on the carbon $p\pi$ orbital gives rise to a hyperfine coupling of magnitude $-\rho A$ with the proton bonded to the carbon atom. Here A is the hyperfine tensor characteristic of π–electron.

In the case of solitons in polyacetylene, ENDOR studies of oriented polymers have been successful in revealing the detailed spin distribution in the case of finite electron correlation [7]. The obtained spin distribution has the extension of about 18 carbon atoms in the full width at half maximum and is associated with negative spin sites, arising from electron correlation effect. The results have been used to estimate the magnitudes of Coulomb interactions in polyacetylene. Prior to the ENDOR analyses of the soliton spin distribution, the π–electron nature of the soliton has been confirmed from the studies of the anisotropy of ESR and ENDOR spectra of stretch-oriented samples, by noticing that π–electron shows characteristic anisotropy of g value and hyperfine

Springer Proceedings in Physics, Vol. 81
Materials and Measurements in Molecular Electronics
Editors: K. Kajimura · S. Kuroda

coupling in solids [6-8]. The results have justified the use of the theory of π–electron in subsequent ENDOR analyses. The use of stretched samples has been useful also in enhancing the spectral resolution. While the spin distribution of the soliton in polyacetylene has been established by the ENDOR studies by a few research groups [6, 7, 9-12], there had been no quantitative studies of polarons in conjugated polymers until recently.

In an electroluminescent polymer, PPV with non degenerate ground-state structure, excitations such as polarons, excitonic polarons and bipolarons have been attracting much attention in relation to their possible roles in the mechanism of electroluminescence. ENDOR spectroscopy has been successful in detecting the spatial extension of the π–electron spin defects in undoped PPV, using stretch-oriented samples [13, 14]. The half-width of spin distribution of several phenyl rings has been obtained. The resemblance between the dark and photo induced ESR signals strongly suggests that the observed spins are trapped polarons. More recently, it has been shown that the observed spin distribution is successfully reproduced by the theoretically calculated spin distribution of polarons in the case of finite electron correlation [15], which supports further the trapped polarons. Evidence has also been obtained for the photo-generation of polarons from the light-induced ESR spectra. Observed spectra are well reproduced by the theoretical spin distribution and their excitation profile shows a prominent increase with increasing photon energy above a threshold energy of about 3.2eV [16]. The latter behavior is similar to that of the photo current in PPV derivatives [17], indicating that the generated spins are charged species, that is, polarons. Thus the ENDOR studies of stretch-oriented PPV provide the first example of observing polaron extension in conjugated polymers. These results are reviewed in this paper.

2 ESR and ENDOR spectroscopy of π–electron in conjugated systems

ESR and ENDOR spectra of an unpaired π–electron interacting with nuclear spins in conjugated systems are well described by the following spin hamiltonian [18, 19].

$$\mathcal{H} = \mu_B \mathbf{S}\cdot\mathbf{g}\cdot\mathbf{H} - g_N\mu_N \sum_i \mathbf{I}_i\cdot\mathbf{H} + \sum_i \mathbf{S}\cdot\mathbf{A}_i\cdot\mathbf{I}_i. \quad (1)$$

The first and second terms of the right-hand side are the electronic and nuclear Zeeman terms, respectively. The third term shows hyperfine interactions. $\boldsymbol{g}$ and $\boldsymbol{A}_i$ show g tensor of the electron spin and the hyperfine tensor of i-th nuclear spin, respectively.

It is well known that unpaired π–electron shows characteristic g and hyperfine anisotropy [20]. Figure 1 shows the definition of the coordinate axes used in this article, where

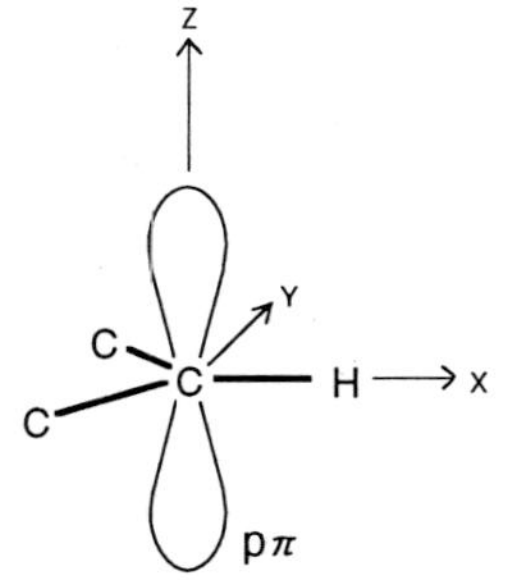

Fig. 1 pπ orbital of a carbon atom bonded to a proton in conjugated structure. Coordinate axes used in this paper are defined. These axes become principal axes of g and hyperfine tensors of a π-electron.

the x- and z-axes are taken parallel to the C-H bond axis and the pπ orbital axis, respectively. g value of a π–electron shows a nearly uniaxial anisotropy around the pπ orbital axis, with the z-axis component being close to the free electron g-value of g_0=2.0023 and the other two components being larger than g_0.

On the other hand, a finite spin density ρ on a carbon pπ orbital gives rise to a hyperfine coupling between the proton bonded to the carbon with the magnitude being given as -$\rho \boldsymbol{A}$. Here $\boldsymbol{A}$ is the hyperfine tensor of a π–electron due to a C-H proton. The principal tensor components are given as [6-8, 20],

$$A_{xx} = -(1-\alpha)A,\ A_{yy} = -(1+\alpha)A \ \text{ and } A_{zz} = -A. \tag{2}$$

Here A is so called McConnell's constant, having the magnitude of 56 ~ 84 MHz (or in the magnetic field unit, 20 ~ 30 gauss). α~0.5 represents the relative magnitude of anisotropic coupling. Among these tensor components A_{yy} has the largest absolute magnitude.

When the direction cosines of the external magnetic field to the hyperfine axes are given by ℓ_i, for a proton bonded to a particular carbon site with a spin density ρ, the ENDOR frequency is given as [6, 19],

$$\nu_{\pm} = \sqrt{\sum_i (\nu_p \pm \rho A_{ii}/2)^2 \ell_i^2}. \tag{3}$$

Here $\pm$ represent the two branches of the ENDOR frequency and ν_p is the free proton frequency. This equation reduces to the conventional form,

$$\nu_{\pm} = \left| \nu_p \pm \frac{1}{2}\rho A_{ii} \right|, \quad \text{for } i=x, y, z. \tag{4}$$

when the external field direction is parallel to one of the principal axes of the hyperfine tensor. The second term of the right hand side shows the frequency shift due to the hyperfine field and $\pm$ sign corresponds to up and down electron-spin orientation. The major advantage of ENDOR over ESR in determining hfc's is that for N nuclear spins of I=1/2 interacting with an electron spin, only N pairs of lines are observed for ENDOR through eqs. (3) or (4), while 2^N hyperfine lines result in ESR spectra that make ESR analyses much more complicated.

From the above discussions, it can be seen that the hyperfine coupling tensor of π–electron given in eq. (2) plays a crucial role in obtaining the spin density from the observed ENDOR spectra determined by eqs. (3) or (4). Therefore the verification of the π–electron character of spins becomes an important prerequisite when one is engaged in the ENDOR analyses of paramagnetic excitations associated with π–electron system in conjugated polymers. The verification is adequately done by using oriented polymers where the molecular axes are preferentially oriented even in non-single-crystalline materials, for example, by stretching [8].

In the case of polyacetylene, ESR and ENDOR spectra of stretch-oriented *trans* and *cis*-rich samples have shown clear anisotropy consistent with that of π–electron [6-8]. The ENDOR determination using stretch-oriented cis-rich polyacetylene has revealed the half-width of the spin distribution of 18 carbon sites and the ratio of the peak value of the negative spin density $\rho(1)$ to that of the positive density $\rho(0)$ of $\rho(1)/\rho(0)$~-0.44 [9-11]. The results have been theoretically reproduced with on-site Coulomb energy of several eV and nearest-neighbor energy of about half in magnitude of the former [21]. More details of ENDOR studies of polyacetylene are reviewed in Ref. 7.

Thus, the ENDOR spectroscopy of oriented polymer is useful in obtaining the spin density distribution of paramagnetic excitations. In the following sections, we describe the ENDOR spectroscopy of polarons in PPV, using stretch-oriented polymers.

3 ENDOR spectra of undoped PPV - evidence for trapped polarons

Figure 2 shows the chemical structure of PPV. Block letters from A to F show the eight carbon sites in one PPV monomer unit. The principal axes of proton hyperfine coupling of an unpaired π–electron, defined in Fig. 1, are also shown for two proton sites of inequivalent bond orientations. In undoped samples of PPV paramagnetic species exist with the spin concentration of about 1 spin/10^6 PPV monomer unit [13, 14, 22]. ESR spectra of these dark spins show clear anisotropy of π–electron on PPV chain in stretch-oriented samples, similar as in the case of polyacetylene. In particular, the line width anisotropy suggests the dominance of the unresolved proton hyperfine structures of a π–electron, which can arise from the interactions of the distributed π–electron spin with the protons on the polymer chain. The situation has been found to be exactly the case by the direct measurement of the hyperfine coupling using ENDOR spectroscopy, as below [13, 14].

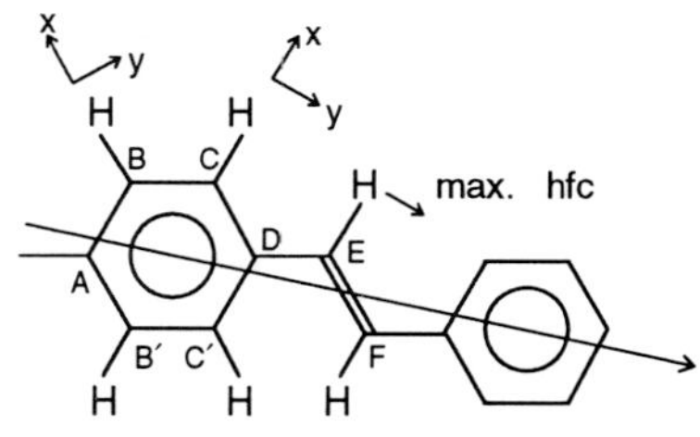

Fig. 2 Chemical structure of poly(paraphenylene vinylene), PPV. Principal axes of proton hyperfine coupling of a π–electron are also shown for two inequivalent C-H bond orientations

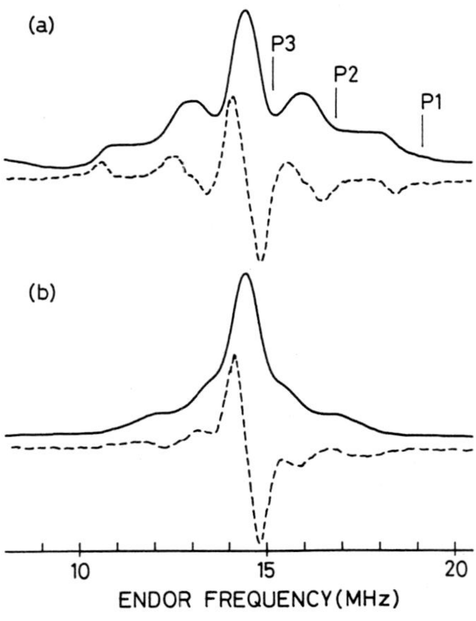

Fig. 3 Proton ENDOR spectra at 4K in a 10-times stretched undoped PPV films [13]. The external field is parallel to the stretch direction in (a) and perpendicular to it in (b). In each figure the thick-line curve represents the spectrum, which was obtained by integrating the observed frequency-derivative spectrum show by a dotted line. Three distinct spectral turning points are marked as P1, P2 and P3 in the spectrum of the stretch direction.

Figure 3 shows the anisotropy of the proton ENDOR spectra at 4K in a 10-times stretched undoped PPV film. The external magnetic field is parallel to the stretch direction for the curves in Fig. 3(a) and perpendicular to the stretch direction for those in Fig. 3(b). In each figure the thick-line curve represents the spectrum, which was obtained by integrating the observed frequency-derivative spectrum shown by a dotted line. The broad distribution of the spectral frequency up to 19.5 MHz directly shows the distribution of the spin density on the polymer chain from eq. (4). An important feature of the spectra in Fig. 3 is that the spectrum taken with the field parallel to the stretch direction shows the larger hyperfine coupling. This provides direct evidence for the π–electron nature of the spin, similar as in the case of polyacetylene mentioned in §2. As pointed above, the proton coupling of a π–electron shows the maximum value along the y-axis. Figure 2 shows that the y-axis is more inclined to the direction of the polymer axis for the four protons (B', C, E, F), while it points intermediate between the polymer axis and the direction perpendicular to it for the two protons (B, C'). Therefore, as a whole, larger proton coupling will be expected for the stretch direction, along which the polymer axis may be preferentially oriented.

Another important feature of Fig. 3 is that the spectrum along the stretch direction has three distinct spectral turning points, P1 to P3 as marked in the figure. These spectral turning points have been confirmed to be intrinsic structures associated with the single spin species observed by cw ESR, using ENDOR-induced ESR technique [13, 14]. The values of the spin density can be deduced from the ENDOR frequencies at these turning points using eq. (4) by assuming a McConnell's constant of 70 MHz (25 gauss). The largest spin density, determined from the largest coupling at P1 reaches 0.09. This quantity, defined as ρ_{max}, is nearly inversely proportional to the full width at the half maximum of the spin density distribution, defined as $\ell_{h.w.}$ [21]. In the case of the soliton in Shirakawa polyacetylene, ρ_{max} reaches to 0.17 (with A=70 MHz) corresponding to $\ell_{h.w.}$ of about 18 carbon sites. Of course, in the case of polyacetylene, the value of ρ_{max} is enhanced due to electron correlation effect and the bare value of ρ_{max} in the limit of no electron correlation will be slightly smaller. Thus, if we assume that the electron correlation in PPV is of comparable as that in polyacetylene, $\ell_{h.w}$ of the present species becomes close to 40 carbon sites, corresponding to several PPV monomer units.

The appearance of the structures at P2 and P3 shows that there are three distinct groups in the magnitude of spin density, the largest one with the peak value given by ρ_{max}, the intermediate one with the peak value of 0.05, nearly half of ρ_{max}, obtained from P2, and the group with small spin densities with the peak density of 0.01, obtained from P3. The existence of these resolved structures shows that the variation of the spin density is not monotonic.

As for the origin of the observed π–electron spins, it may be natural to relate them to certain defect states of the polymer, since there is no intentional generation of spins by doping, photo-excitation, etc. in the present case. If we consider that PPV has a nondegenerate ground state structure, one candidate of the observed π–electron spin is the polaron trapped in the dark state of the polymer. It is interesting to note that the above obtained extension of the spin density, i.e. the maximum value of the hyperfine coupling, is comparable to but slightly smaller than that observed for the anion radical (negative polaron) generated in doped oligomers of phenylene vinylene with the longest case of 7 monomer units, studied by ENDOR [23, 24]. Although the detailed shape of the spin distribution in the reported oligomer naturally does not coincide with the present results obtained in the polymer, the approximate coincidence of the spatial extension further suggests the possibility of the polaron. Moreover, the observed ENDOR spectra are actually well reproduced by the theoretically calculated spin distribution of polarons in the case of finite electron correlation, as to be explained below. This strongly suggests that the observed paramagnetic species are polarons trapped in the dark state of the polymer.

4 Polaron spin distribution

Figure 4 shows the theoretically calculated spin density distribution of a polaron using PPP(Pariser-Parr-Pople) hamiltonian that includes Coulomb interaction terms. The spin distribution well reproduces the observed ENDOR spectra as shown in Fig. 5. In this figure, the calculated spectra are obtained by using the theoretical distribution with an ENDOR spectrum simulation method for a partially oriented system [7, 9], by taking the existence of the two inequivalent sites of C-H bond orientation into account.

Important features of the ENDOR spectra, described in the previous section are all reproduced by calculation. The half width of distribution amounts to about 4 phenyl rings, as discussed above. As for the parameter values of PPP hamiltonian that give best results as in Fig. 5 are on-site and nearest neighbor Coulomb interactions of U=2.5t and V=1.3t with t being transfer integral, respectively, and more long-range interactions given by Ohno formula and

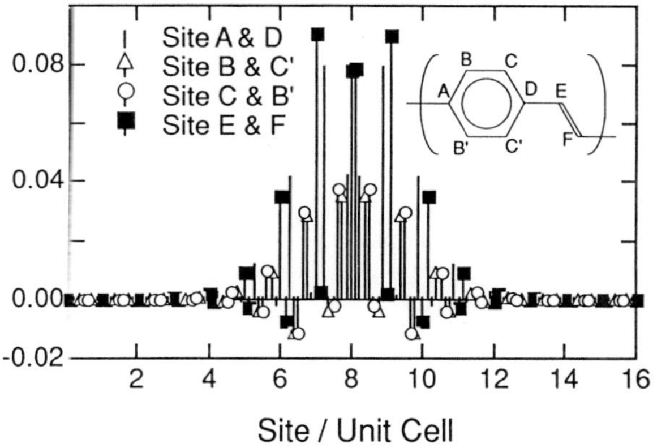

Fig. 4 Spin density distribution of a polaron in a PPV chain calculated by the PPP model with λ=0.16, U=2.5t, and V=1.3t. See text for the definition of parameters. Vinylene sites (sites E and F in Fig. 2) are marked by the solid squares.

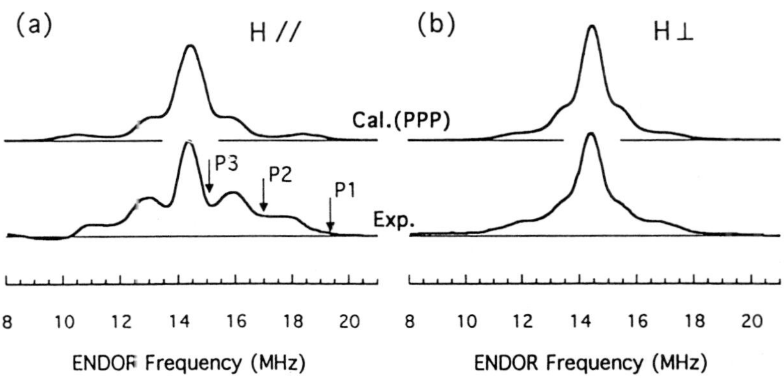

Fig. 5 Comparison between the calculated ENDOR spectra using the polaron spin distribution shown in Fig. 4 and the observed spectra that are similar as those of integrated ones in Fig. 3. The external field is parallel to the stretch direction for the curves in (a) and perpendicular to it in (b).

dimensionless electron-phonon coupling constant of λ=0.16. These parameter values are fairly close to those in the case of the soliton in polyacetylene obtained from the ENDOR analyses and the values of U and V show that the π−systems in polyacetylene and PPV fall in the intermediate coupling regime. The present information of electron correlation is consistent with the analyses of linear and nonlinear optical properties of PPV and other conjugated polymers [25]. In this situation the nearly twice as large extension of the polaron in PPV as compared with the soliton in polyacetylene would be reasonably understood if we consider that polaron in conjugated polymers can be viewed as the bound soliton and anti-soliton pair, one neutral and the other charged, resulting in the larger extension of the polaron than the soliton.

The values of the largest and the second largest densities are 0.09 and 0.08, respectively, both reside on the carbons of vinylene sites. The relatively close values of these largest densities, that give rise to the wing portions of the spectra, can not be resolved and make the line shape of the wing portion of the spectra around 18-19.5 MHz and 9.5-11 MHz of the parallel direction varying slowly. The observed distinct anisotropy of the wing portion is consistent with the theoretical assignment of the largest densities to those on vinylene sites, that should show larger anisotropy than those of phenyl rings, as discussed in the previous section. The theoretical distribution certainly shows the three different groups of the magnitude of the spin density that is consistent with the three resolved structures at P1 to P3 in the observed spectrum of the parallel direction.

Thus the coincidence between the observed and calculated spectra strongly suggests that the observed paramagnetic species existing in undoped PPV samples are polarons trapped in the dark state of the polymer. At present, the generation mechanism of polarons such as thermal generation, accidental doping by remaining oxygen etc. is not clear. The trapped sites of the polarons would be the defects such as chain ends or remaining sp^3 defects in the polymer [26]. Another supporting evidence for the trapped polarons is the resemblance between the dark and light-induced ESR spectra, since it is highly possible that the latter signal arises from positive and negative polarons generated by photo-induced charge transfer between polymer chains. The dark ESR spectra of trapped polarons, giving rise to ENDOR signals, are selectively observed by ENDOR-induced ESR and they are close to photoinduced ESR spectra [13, 14]. In fact, our recent studies of photoinduced ESR by employing variable-energy light source has revealed the characteristic excitation energy dependence, which supports the photoinduced polarons [16]. The results are described in the next section.

5 Photo-generated polarons - Light-induced ESR

We first describe our earlier light-induced ESR results using non-monochromatic visible light. Figure 6 shows the ESR signals recorded at 4K with and without illumination of visible light in a stretched PPV sample. The light intensity was less than 2mW/cm^2. The external field is parallel to the stretch direction for the upper spectra and perpendicular to it for the lower ones. In each case, thick-line curve represents the dark ESR signal and thin-line curve represents the signal under illumination. The increment of the ESR signal under illumination is evident. This increment is actually transient in nature, as expected for the photo-induced ESR signal.

Figure 7 shows the temperature dependence of the time response of the photo-induced ESR signal, observed at the lower-peak position of the first-derivative ESR signal for the field perpendicular to the stretch direction. Three curves corresponds to the temperatures of 4K, 100K and 140K, from the top curve to the bottom. At 4K, the signal rises in a few minutes. After turning off the light, about 30% of the signal decays with the time constant of several minutes and the rest shows a much slower decay. More than 60% of the signal remains after an hour. At higher temperatures, the rise and decay time constants become shorter and the signal intensity becomes weaker. At the same time, the fraction of the slowly decaying component becomes also smaller, negligible at 140K. These behaviors are qualitatively consistent with those expected for the photo-

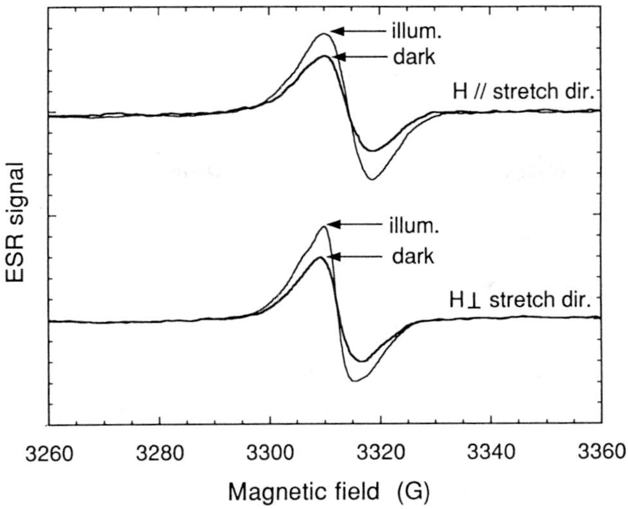

Fig. 6 Orientation dependence of the ESR spectra at 4K with and without illumination of visible light in a 10-times stretched undoped PPV film [14]. The external field is parallel to the stretch direction for the upper curve and perpendicular to it for the lower curve.

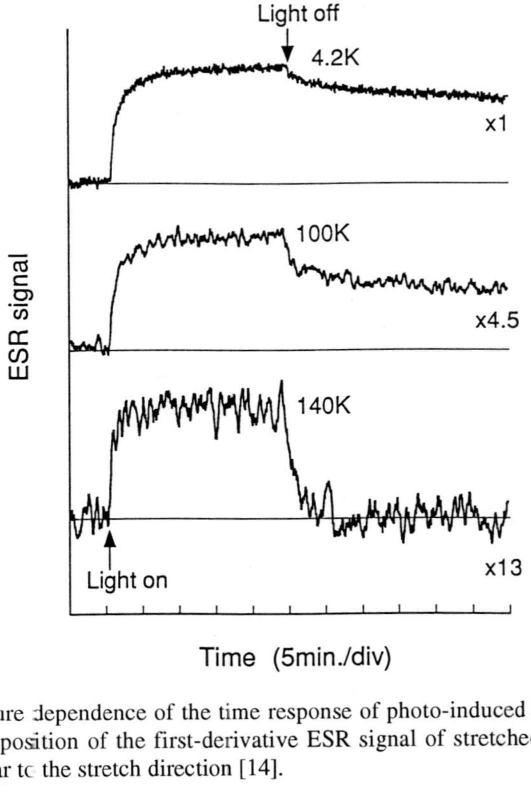

Fig. 7 Temperature dependence of the time response of photo-induced ESR signal, taken at the lower-field peak position of the first-derivative ESR signal of stretched PPV with the external field perpendicular to the stretch direction [14].

induced ESR signals [27]. That is, the initial rise results from the charge separation and subsequent charge-trapping and the decay after the light-off corresponds to the recombination of charges. In such case the faster the time constant, the smaller the signal intensity, consistent with observation. In fact, the light-induced signal became undetectable for the temperatures higher than 150K.

Thus the observed ESR signal under illumination is reasonably ascribed to the photo-induced ESR signal. Then it is interesting to see whether the photoinduced signal is actually identical to the dark ESR signal. In Fig. 6, it is seen that the observed photoinduced signal shows similar spectral span as that of the dark ESR signal. As mentioned above, it has been shown that the ESR spectra of trapped polarons responsible for ENDOR signals, observed by ENDOR-induced ESR, are close to photo-induced ESR spectra. It should be also pointed that the spectral line shape of these dark and photo-induced ESR spectra are close to the reported g=2 ODESR spectra [28], which have been assigned to recombination of 'distant' polaron pairs. These observations may further support the assignment of the dark spins to trapped polarons.

In this context, it would be interesting to get information on the dependence of the light-induced ESR intensity on the excitation energy to see whether the light-induced ESR spectra show similar action spectra of charge carriers in PPV. Figure 8 shows the variation of light-induced ESR intensity on the excitation energy of light together with the excitation spectra of photo-carriers as well as optical absorption spectrum reported in a PPV derivative, MEH-PPV. The optical absorption of both compounds are close to each other, although the energy gap of PPV is slightly larger than that of MEH-PPV, exciton peak at 2.5 eV of the former vs. 2.4 eV of the latter. It is seen that the light-induced ESR intensity increases significantly with increasing the photon energy above about 3.2 eV despite of lower optical absorption coefficient. This threshold energy in PPV has been assigned to the Hartree-Fock gap by the recent PPP calculation [15, 29]

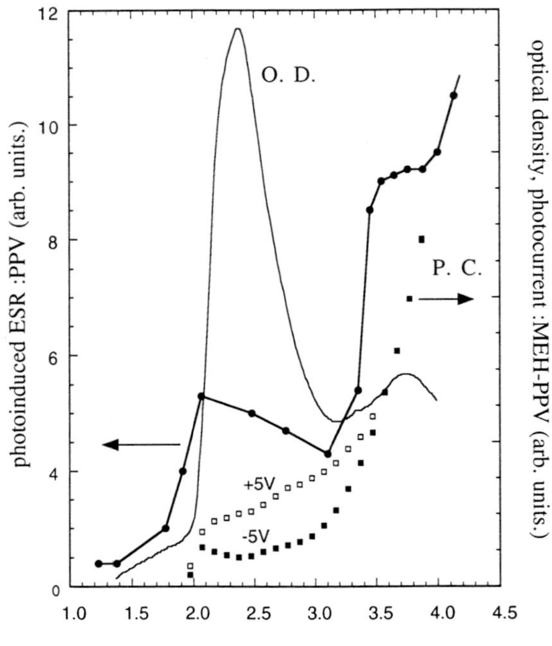

Fig. 8 Photon energy dependence of the photo-induced ESR signal at 4K in undoped PPV together with the action spectra of photo current reported for a PPV derivative, MEH-PPV.

and then it is reasonably expected that electron-hole pairs are efficiently created above this energy and may be relaxed into polarons. We note that in the action spectra of photoinduced ESR, a structure is resolved at 3.7 eV, where a structure is also observed in the optical absorption of PPV. Similar behavior is also reported in the action spectra of photo-current in MEH-PPV where prominent increase at higher photon energy with the threshold around 3.2 eV is observed [17] as also shown in Fig. 8. These results suggest that the photoinduced spins are charged species. For the excitation energy lower than the threshold, lower but non-negligible light-induced ESR is seen with some increase around 2eV that is below the exciton peak of optical absorption at 2.5eV in PPV. These signals may be related to some defect related generation mechanism of polarons. More detailed discussion is beyond the scope of this paper and will be described elsewhere. Thus the observed similarity of the action spectra of light-induced ESR and photo-currents strongly suggests that the light-induced ESR originates from photo-generated and subsequently trapped polarons because they are paramagnetic and ESR active. This assignment can be further examined if we notice that the observed light-induced ESR spectra should be reproduced by calculating the proton hyperfine structures expected from polaron spin distribution in this case.

Solid lines in Fig. 9 show the orientation dependence of light-induced ESR signal at 60K in a 10-times stretched undoped PPV film under the illumination of 300 nm monochromatic light. The spectra were obtained by the subtraction between the dark signals and those under illumination. At this temperature, the contribution from the second component with the larger line width than that of the primary component [14] becomes smallest in magnitude because of different temperature dependencies of each component and thus the spectra are suited for the analysis. The external field is parallel to the stretch direction for the upper curve and perpendicular to it for the lower curve. Clear anisotropy of the linewidth, the larger for the stretch direction, is observed, which is consistent with the proton hyperfine coupling of π–electron of the conjugated chain oriented along the stretch direction, according the discussion given above. In other words, the observed distinct anisotropy shows that the light-induced species are π–electrons generated on the oriented conjugated chains. Dotted lines in Fig. 9 show the calculated spectra using the theoretical spin distribution of polarons shown in Fig. 4, by assuming the preferential orientation of chain axes with the ESR simulation method developed previously [7].

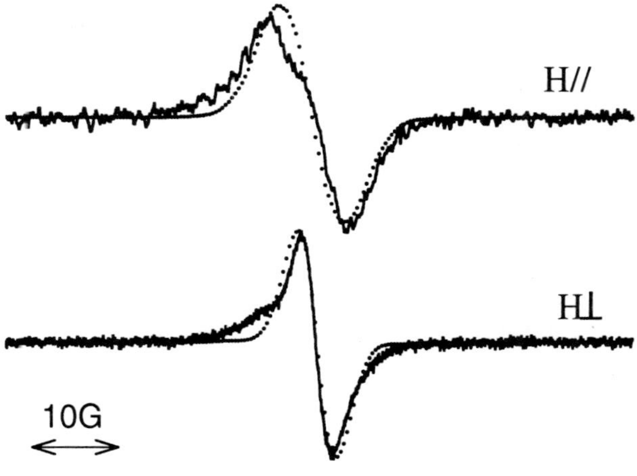

Fig. 9 Orientation dependence of photo-induced ESR spectra at 60K (solid curves) obtained by subtracting the dark spectrum from that under 300 nm monochromatic light illumination. The external field is parallel to the stretch direction for the upper curve and perpendicular to it for the lower curve. Dotted curves show the calculated ESR spectra by using the spin distribution of a polaron shown in Fig. 4.

The agreement between the observed and calculated curves is fairly well, providing strong evidence that the photo-generated spins are polarons. The remaining discrepancy, in particular for the lower field side results from the slight asymmetry of the observed curve as well as the overlapping shoulders. The former can be ascribed to the existence of the misoriented fraction of chains and the latter to so-called spin-flip transitions in ESR signals. The details may be described elsewhere. From the results described in this section concerning the action spectra and ESR line shapes, it is therefore concluded that the light-induced ESR signals in undoped PPV arise from photo-generated polarons.

6 Concluding remarks

The present studies show that the π–electron spin species existing in the dark and photo-generated in undoped PPV are polarons trapped in the system and therefore our ENDOR spectra of undoped PPV provide the first example of the observation of polaron extension in conjugated polymers. The results also show the existence of Coulomb interactions in the intermediate coupling regime in PPV. The existence of various derivatives of PPV's, developed for the electroluminescence devices, as well as other conjugated polymers with nondegenerate ground-state structures may provide further opportunities to study the polaron states by using ESR and ENDOR spectroscopies.

References

[1] A.J. Heeger, S. Kivelson, J.R. Schrieffer and W.P. Su, *Rev. Mod. Phys.* **60** (1988) 781.

[2] J.H. Burroughes, D.D.C. Bradley, A.R. Brown, R.N. Marks, K. Mackay, R.H. Friend, P.L. Burns and A.B. Holmes, *Nature* **347** (1990) 539.

[3] D.R. Baigent, R.H. Friend, A.B. Holmes and S.C. Moratti, in these proceedings and references there in.

[4] T. Ohnishi, S. Doi, Y. Tsuchida and T. Noguchi, in these proceedings and references there in.

[5] L. Yu, *Solitons and Polarons in Conducting Polymers* (World Scientific, Singapore, 1988).

[6] S. Kuroda and H. Shirakawa, *Solid State Commun.* **43** (1982) 591.

[7] S. Kuroda, *Int. J. Mod. Phys.* **B9** (1995) 291.

[8] S. Kuroda, M. Tokumoto, N. Kinoshita and H. Shirakawa, *J. Phys. Soc. Jpn.* **51** (1982) 693.

[9] S. Kuroda, H. Bando and H. Shirakawa, *J. Phys. Soc. Jpn.* **54** (1985) 3956

[10] S. Kuroda and H. Shirakawa, *Phys. Rev.* **B35** (1987) 9380.

[11] S. Kuroda and H. Shirakawa, *J. Phys. Soc. Jpn.* **61** (1992) 2930.

[12] M. Mehring, A. Grupp, P. Höfer and H. Käss, *Synth. Met.* **28** (1989) D399.

[13] S. Kuroda, T. Noguchi and T. Ohnishi, *Phys. Rev. Lett.* **72** (1994) 286.

[14] S. Kuroda, K. Murata, T. Ohnishi and T. Noguchi, *J. Phys. Soc. Jpn.* **64** (1995) 1369.

[15] Y. Shimoi, S. Abe, S. Kuroda and K. Murata, *Solid State Commun.* **95** (1995) 137.

[16] K. Murata, S. Kuroda, Y. Shimoi, S. Abe, T. Ohnishi and T. Noguchi, Paper to be presented at 51st Annual Meeting of the Physical Society of Japan, Kanazawa, 1996.

[17] M. Chandross, S. Mazumdar, S. Jeglinski, X. Wei, Z.V. Vardeny, E.W. Kwock and T.M. Miller, *Phys. Rev.* **B50** (1994) 14702.

[18] A. Carrington and A.D. McLachlan, *Introduction to Magnetic Resonance* (Harper and Row, New York, 1967).

[19] L. Kevan and L.D. Kispert, *Electron Spin Double Resonance Spectroscopy* (John Wiley & Sons, New York, 1976).

[20] J. R. Morton, *Chem. Rev.* **64** (1964) 453.
[21] K. Yonemitsu, Y. Ono and Y. Wada, *J. Phys. Soc. Jpn.* **57** (1988) 3875.
[22] S. Kuroda, I. Murase, T. Ohnishi and T. Noguchi, *Synth. Metals* **17** (1987) 663.
[23] P. Brendel, A. Grupp, M. Mehring, R. Schenk, K. Müllen and W. Huber, *Synth. Metals* **45** (1991) 49.
[24] M. Baumgarten, L. Gherghel and S. Karabunarliev, *Synth. Metals* **69** (1995) 633.
[25] S. Abe, in these proceedings.
[26] H.A. Mizes and E.M. Conwell, *Phys. Rev. Lett.* **70** (1993) 1505.
[27] C. Kittel, *Introduction to Solid State Physics* (Wiley, New York, 1966) 3rd ed., Chap. 17.
[28] L.S. Swanson, J. Shinar, A.R. Brown, D.D.C. Bradley, R.H. Friend, P.L. Burn, A. Kraft and A.B. Holmes, *Phys. Rev.* **B46** (1992) 15072.
[29] Y. Shimoi and S. Abe, *Synth. Metals.,* in press.

Index of Contributors

Springer Proceedings in Physics

Managing Editor: H. K. V. Lotsch

1 *Fluctuations and Sensitivity in Nonequilibrium Systems*
Editors: W. Horsthemke and D. K. Kondepudi
2 *EXAFS and Near Edge Structure III*
Editors: K. O. Hodgson, B. Hedman, and J. E. Penner-Hahn
3 *Nonlinear Phenomena in Physics*
Editor: F. Claro
4 *Time-Resolved Vibrational Spectroscopy*
Editors: A. Laubereau and M. Stockburger
5 *Physics of Finely Divided Matter*
Editors: N. Boccara and M. Daoud
6 *Aerogels* Editor: J. Fricke
7 *Nonlinear Optics: Materials and Devices*
Editors: C. Flytzanis and J. L. Oudar
8 *Optical Bistability III*
Editors: H. M. Gibbs, P. Mandel, N. Peyghambarian, and S. D. Smith
9 *Ion Formation from Organic Solids (IFOS III)*
Editor: A. Benninghoven
10 *Atomic Transport and Defects in Metals by Neutron Scattering*
Editors: C. Janot, W. Petry, D. Richter, and T. Springer
11 *Biophysical Effects of Steady Magnetic Fields*
Editors: G. Maret, I. Kiepenheuer, and N. Boccara
12 *Quantum Optics IV*
Editors: J. D. Harvey and D. F. Walls
13 *The Physics and Fabrication of Microstructures and Microdevices*
Editors: M. J. Kelly and C. Weisbuch
14 *Magnetic Properties of Low-Dimensional Systems*
Editors: L. M. Falicov and J. L. Morán-López
15 *Gas Flow and Chemical Lasers*
Editor: S. Rosenwaks
16 *Photons and Continuum States of Atoms and Molecules*
Editors: N. K. Rahman, C. Guidotti, and M. Allegrini
17 *Quantum Aspects of Molecular Motions in Solids*
Editors: A. Heidemann, A. Magerl, M. Prager, D. Richter, and T. Springer
18 *Electro-optic and Photorefractive Materials*
Editor: P. Günter
19 *Lasers and Synergetics*
Editors: R. Graham and A. Wunderlin
20 *Primary Processes in Photobiology*
Editor: T. Kobayashi
21 *Physics of Amphiphilic Layers*
Editors: J. Meunier, D. Langevin, and N. Boccara
22 *Semiconductor Interfaces: Formation and Properties*
Editors: G. Le Lay, J. Derrien, and N. Boccara
23 *Magnetic Excitations and Fluctuations II*
Editors: U. Balucani, S. W. Lovesey, M. G. Rasetti, and V. Tognetti
24 *Recent Topics in Theoretical Physics*
Editor: H. Takayama
25 *Excitons in Confined Systems*
Editors: R. Del Sole, A. D'Andrea, and A. Lapiccirella
26 *The Elementary Structure of Matter*
Editors: J.-M. Richard, E. Aslanides, and N. Boccara
27 *Competing Interactions and Microstructures: Statics and Dynamics*
Editors: R. LeSar, A. Bishop, and R. Heffner
28 *Anderson Localization*
Editors: T. Ando and H. Fukuyama
29 *Polymer Motion in Dense Systems*
Editors: D. Richter and T. Springer
30 *Short-Wavelength Lasers and Their Applications*
Editor: C. Yamanaka
31 *Quantum String Theory*
Editors: N. Kawamoto and T. Kugo
32 *Universalities in Condensed-Matter*
Editors: R. Jullien, L. Peliti, R. Rammal, and N. Boccara
33 *Computer Simulation Studies in Condensed-Matter Physics: Recent Developments*
Editors: D. P. Landau, K. K. Mon, and H.-B. Schüttler
34 *Amorphous and Crystalline Silicon Carbide and Related Materials*
Editors: G. L. Harris and C. Y.-W. Yang
35 *Polycrystalline Semiconductors: Grain Boundaries and Interfaces*
Editors: H. J. Möller, H. P. Strunk, and J. H. Werner
36 *Nonlinear Optics of Organics and Semiconductors*
Editor: T. Kobayashi
37 *Dynamics of Disordered Materials*
Editors: D. Richter, A. J. Dianoux, W. Petry, and J. Teixeira
38 *Electroluminescence*
Editors: S. Shionoya and H. Kobayashi
39 *Disorder and Nonlinearity*
Editors: A. R. Bishop, D. K. Campbell, and S. Pnevmatikos
40 *Static and Dynamic Properties of Liquids*
Editors: M. Davidović and A. K. Soper
41 *Quantum Optics V*
Editors: J. D. Harvey and D. F. Walls
42 *Molecular Basis of Polymer Networks*
Editors: A. Baumgärtner and C. E. Picot
43 *Amorphous and Crystalline Silicon Carbide II: Recent Developments*
Editors: M. M. Rahman, C. Y.-W. Yang, and G. L. Harris
44 *Optical Fiber Sensors*
Editors: H. J. Arditty, J. P. Dakin, and R. Th. Kersten
45 *Computer Simulation Studies in Condensed-Matter Physics II: New Directions*
Editors: D. P. Landau, K. K. Mon, and H.-B. Schüttler